To Sarah Alicia "Cosi" Carrizales
Never was a child more wanted

and

To Juan Javier "Jay" Carrizales
For your love, support, and editing

Global Population AND Reproductive Health

Edited by

Deborah R. McFarlane, DrPH, MPA

Professor and Regents Lecturer
University of New Mexico
Albuquerque, New Mexico

JONES & BARTLETT
LEARNING

World Headquarters
Jones & Bartlett Learning
5 Wall Street
Burlington, MA 01803
978-443-5000
info@jblearning.com
www.jblearning.com

Jones & Bartlett Learning books and products are available through most bookstores and online booksellers. To contact Jones & Bartlett Learning directly, call 800-832-0034, fax 978-443-8000, or visit our website, www.jblearning.com.

Production Credits
Executive Publisher: William Brottmiller
Publisher: Michael Brown
Associate Editor: Chloe Falivene
Production Editor: Jill Morton
Senior Marketing Manager: Sophie Fleck Teague
Manufacturing and Inventory Control Supervisor: Amy Bacus
Composition: diacriTech
Cover Design: Theresa Manley
Manager of Photo Research, Rights & Permissions: Amy Rathburn
Cover and Title Page Images: Background, © Gregor Buir/ShutterStock, Inc.; World map, ©Toria/ShutterStock, Inc.; Upper left, © Muellek Josef/ShutterStock, Inc.; Upper right, © Andy Dean Photography/ShutterStock, Inc.; Lower left, © iStockphoto/Thinkstock; Lower right, ©iStockphoto/Thinkstock
Printing and Binding: Edwards Brothers Malloy
Cover Printing: Edwards Brothers Malloy

Library of Congress Cataloging-in-Publication Data
Global population and reproductive health / edited by Deborah R. McFarlane.
    p. ; cm.
Includes bibliographical references and index.
ISBN 978-1-4496-8520-1 (pbk. : alk. paper)
I. McFarlane, Deborah R., 1951- editor.
[DNLM: 1. Reproductive Health. 2. Environment. 3. Reproductive Health Services. WQ 200.1]
RA408.W65
362.198—dc23
                               2014007999
6048
Printed in the United States of America
18 17 16 15 14   10 9 8 7 6 5 4 3 2 1

Contents

Preface

This text addresses an important and timely public health topic—global population and reproductive health. World population passed the 7 billion mark in 2011, yet millions of women and couples still lack access to reproductive health services. In 2010, an estimated 222 million sexually active women who wanted to use modern contraception were unable to access family planning services. As this book goes to press, the global estimate for unmet family planning need has been raised to 233 million.[1]

These facts have profound implications for maternal and child health, the status of women, environmental quality, climate change, and food security, each of which is explored here. This text is innovative because it integrates population and reproductive health—areas that are inextricably linked, yet often discussed separately as if they have no relationship to each other. It also elucidates the connections between population and reproductive health and environmental issues. Overall, we strive to show that the relationships among all of these issues demand integrated policies and international cooperation.

The *Population and Reproductive Health* chapter introduces the connections between population and reproductive health. The components of reproductive health services are presented in an expanded version of the Davis Blake model of human reproduction. This chapter also outlines linkages between population, reproductive health, women's status, and sustainability as well as other environmental issues. It includes a brief glossary of some of the terms that are used throughout the text. The chapters that follow cover the multifaceted relationship between population and reproductive health in far greater detail.

The *History and Future of World Population* chapter presents the history of human population growth and discusses projections for the future. Growth rates vary widely across the globe, so individual continents and countries are examined over time. This chapter also explains the world's demographic divide (e.g., high fertility in developing countries and low fertility in industrialized countries) as well as the importance of both immigration and urbanization.

Understanding the dynamics of human populations globally and within specific regions or nations requires familiarity with data and measures. The *Measuring Populations: Mortality, Fertility, and Migration* chapter provides a thorough introduction to the measurement of populations. Included here are the three components of demographic change—fertility, mortality, and migration—as well as data sources and measures to describe the size, dynamics, and structure of populations.

The *Measuring Reproductive Health* chapter discusses key indicators of reproductive health, their concomitant measures, and their importance. This chapter includes measures of and data sources for sexual behavior, contraceptive use, pregnancy, abortion, and births. Further recognizing the breadth of reproductive health, this chapter presents measures of the status of women as well as for sexually transmitted infections.

The *Population Theories and Dynamics* chapter presents major theoretical paradigms that have guided thinking about human population since the Enlightenment. Included here are the Malthusian theory of population, the Marxist view of population, demographic transition theory, the neo-Malthusian perspective, and the theory of demographic change and response. This chapter also covers the age and sex structure of populations, explaining the importance of youth bulges and population aging.

The *Contraceptive History and Practice* chapter introduces the history of contraceptive practice and policies, describes currently used contraceptive methods, and discusses contraceptive effectiveness, safety, and prevalence.

The *Abortion and Reproductive Health* chapter focuses on induced abortion. It outlines the history of induced abortion, explains its demography and epidemiology and discusses the legal and policy aspects of abortion globally as well as in the United States.

The *Benefits of Family Planning* chapter explains the maternal and child health benefits of family planning. It also discusses the synergy between the demand for birth limitation and the global contraceptive revolution of the past half-century as well as the relationship between contraceptive availability and induced abortion. The contributions of contraceptive availability and girls' education as components of economic development are also considered.

The *Women's Status and Reproductive Rights* chapter addresses gender inequities around the world, examining how they are related to sexual and reproductive rights. Topics covered include girls' education, child marriage, female genital mutilation/cutting, obstetric fistula, and sex-selective abortion. The chapter also examines the historic movement toward sexual and reproductive rights, articulated at the 1994 ICPD meeting in Cairo, from the perspectives of both proponents and critics. This chapter is dedicated to the late Barbara Pillsbury, who agreed to write it before her unexpected death in 2012.

The *Sustainability, Population, and Environmental Degradation* chapter expounds the role of population growth and increased consumption in environmental degradation, presents methods for measuring environmental impact, and describes some of the natural resources at risk. Global efforts to address environmental issues are discussed, including sustainable practices and green growth as well as the importance of women's leadership in sustainable development.

The *Climate Change, Population, and Reproductive Health* chapter explains how population is related to climate change. First, population serves as a potential driver of carbon emissions. Second, population growth translates into increasing numbers of people exposed to the impact of climate change. Finally, population dynamics influence human adaptation to changes in climate. This chapter also examines how reproductive health is linked to climate change mitigation and adaptation.

Within the context of the 21st century, the *Food Security, Population, and Reproductive Health* chapter addresses an old Malthusian postulate: the relationship between food supply and population. The indicators of and factors affecting food supply are discussed as well as global and regional projections of food supply. This chapter also delineates the relationships between food security and women's reproductive health. The chapter concludes with necessary steps for achieving food security.

The *Global Population and Reproductive Health Policies* chapter provides a history of, a current analysis of, and future prospects for international and domestic policies throughout the world. This chapter describes the first population and family planning programs and explains how these issues became global concerns. It discusses how and why family planning evolved into reproductive health as well as the ups and downs of international assistance for population and reproductive health.

Toward the Future is the text's conclusion. Discussing the synergies between population and reproductive health, this chapter delineates what must be done to address global population growth, promote reproductive health, and move toward sustainability. Addressing these issues in a timely, integrated, and humane manner is not a foregone conclusion. We hope this text makes a contribution toward developing effective global solutions.

As editor, I have been privileged to work with generous and knowledgeable scholars who care deeply about the issues discussed in this text. A special note of gratitude is due to John R. Weeks, who willingly wrote two chapters. Thank you to Barbara Crane from IPAS for suggesting contributors, reading sections, and always being willing to help. Thank you to Tom Hatfield,

Department of Environmental and Occupational Health at California State University at Northridge, who listened carefully and explained sustainability to me. Sherrill Redmon, recently retired Director of the Sophia Smith Collection at Smith College, merits special mention as well, for taking a chance on the Population and Reproductive Health Oral History Project—information that underlies the subject matter covered in this text.

Special thanks are due to students from the University of New Mexico who signed up for PS 377 year after year and test-drove these chapters and other materials. This text is largely a response to your interest and enthusiasm. The following students deserve special mention: Gabe Alarid, Josh Anolick, Adrian Avila, Erin Barringer-Sterner, Vanessa Bautista, Sunny Bergh-Holmes, Luis Carrasco, Matt Carter, Isabel Coello, Beth Fischer, Herbert Friedman, Anna Gonzales, Margaret Gonzales, Shelby Greaser, Lisa E. M. Hofheinz, Camille Hemstreet, C. Grace Isner, Glenda Kodaseet, Emily Lies, Jessie Loera, Javier Martinez, Heather Metcalf, Margot Wilson-Meyer, Bill Mutidjo, Lan Nguyen, Lindsey Platero, Luis Rocha, Anita Sager, Ashley Seibert, Shane Shariff, Josh Sheak, Bryant Shuey, Samantha Stevens, Helena Taflin, Willie Vidmar, and Julia Vlajic. Through his quiet example, Ed Baklini at Albuquerque Academy has reminded me of the joy and privilege of teaching young people.

I want to acknowledge my family: my husband, Jay Carrizales, a first-rate editor when he's not flying an Airbus; my daughter, Cosi Carrizales, who hopes I will cook and clean more now that the book is done; and my mother, Bette McFarlane, who eagerly reads anything I write.

The team at Jones & Bartlett Learning merits special recognition as well. Thank you to Mike Brown for supporting this project and understanding its importance, to Chloe Falivene for patiently answering my many (and often repetitious) questions, and to Jill Morton and Amy Rathburn for going many extra miles to produce this book.

REFERENCES

Alkema L, Kantorova V, Menozzi C, Biddlecom A. National, regional, and global rates and trends in contraceptive prevalence, and unmet need for family planning between 1990 and 2015: a systematic and comprehensive analysis, *Lancet.* 2013;381:1642–1652.

Contributors

John E. Becker, DPDS
Department of Environmental and Occupational Health
California State University, Northridge
Northridge, CA

E. Hazel Denton, PhD
Department of International Health
Georgetown University and
School of Advanced International Studies
Johns Hopkins University
Washington, DC

Richard Grossman, MD, MPH
Durango, CO

Erin R. Hamilton, PhD
Assistant Professor
Department of Sociology
University of California, Davis
Davis, CA

Karen Hardee, PhD
Senior Fellow
Futures Group
Washington, DC

Andrzej Kulczycki, PhD
Associate Professor
Department of Health Care Organization Policy
University of Alabama at Birmingham
Birmingham, AL

John F. May, PhD
Visiting Scholar
Population Reference Bureau
Washington, DC

John R. Weeks, PhD
Distinguished Professor of Geography
Director
International Population Center
San Diego State University
Clinical Professor of Global Public Health
San Diego School of Medicine
University of California
San Diego, CA

Richard E. White, PhD
Professor Emeritus
Smith College
Northampton, MA

Sara Yeatman, PhD
Assistant Professor
Department of Health and Behavioral Sciences
University of Colorado at Denver
Denver, CO

Part

I

OVERVIEW

Chapter 1

Population and Reproductive Health

Deborah R. McFarlane

INTRODUCTION

World population passed the 7 billion mark in 2011. Most of the earth's human inhabitants were unaware of this demographic landmark, so there was little fanfare upon its achievement. For those who did recognize the new population threshold, 7 billion people was not a cause for celebration. Instead, concerns were voiced about sustaining this population, the "demographic divide" between rich and poor countries, differences in women's health and well-being throughout the world, unequal access to contraceptive services and safe abortion, environmental degradation, climate change, and increasing food insecurity in many regions. With a population projection of 8 billion for 2024 depicted in Figure 1–1, these issues and our collective wherewithal to address them assume even more urgency.[1]

This text introduces the topics of global population and reproductive health, which are intertwined throughout the world. Countries with high birth rates tend to have rapidly growing and younger populations and, in many cases, a scarcity of reproductive health services. In Uganda, for example, the average woman has between six and seven births in her lifetime, and she begins childbearing early (see Figure 1–2). The median age of the population is just 15.5 years.[2] Only one fourth of reproductive-age women use modern methods of birth control, so fertility rates remain high. Most Ugandan women give birth alone, without the benefit of trained attendants or access

Figure 1–1 Population of Our World

© Tom and Kwikki/Shutterstock.com

Figure 1–2 Young African Mother with Child

© Dana Ward/Shutterstock.com

Figure 1–3 Physician-Attended Hospital Birth

© Wavebreak Media/Thinkstock

to hospital services if needed.[3] Even considering emigration, Uganda's population is projected to double in about two decades.[4]

In countries where most individuals are able to plan their births and be assured of their children's survival, family sizes tend to be smaller, thus contributing to lower population growth rates. In France, for example, the average woman has two births, and the population's median age is just over 40 years. Birth control methods are widely available, with three-fourths of all married women of reproductive age using highly effective, modern methods. Nearly all births occur in well-equipped hospitals, where women are attended by highly trained physicians as shown in Figure 1–3.[3] Even with immigration, France's population is not projected to double for nearly a century and a half.

HUMAN POPULATION GROWTH

Three demographic phenomena—fertility, mortality, and migration—determine human population size. Fertility refers to live births produced by a population. Mortality refers to deaths occurring in a population. Migration is the movement of people across a specified boundary for the purpose of establishing a permanent or semi-permanent residence. For most human societies throughout time, fertility has been the most significant component of population growth.[5]

Human beings have existed for at least 200,000 years, and probably much longer. For nearly all of that history, the pressing population issue has been the survival of the species. In early human life, high fertility rates were necessary to counteract high mortality rates, particularly among infants and children, just so populations could sustain themselves. For example, between 8000 BCE and 5000 BCE, just over 300 people were added to the world's population annually. By the time of Christ, the world's population had reached 200 million and was increasing by almost 300,000 people annually. Nevertheless, this rate of growth waxed and waned for more than a millennium and a half.[6]

Following the end of the Black Plague in the mid-17th century, the world population grew rapidly, although not evenly, throughout the globe. With the advent of the Industrial Revolution in the mid-1700s, mortality rates began to drop, first in Europe and then in North America. Economic development in these regions led to better housing, improved nutrition, and better sanitary practices. Consequently, there was lower exposure to disease and increased resistance to contagion. By 1804, the world population surpassed 1 billion. With the Industrial Revolution in full swing, the 2 billion mark was reached in 1927, only 123 years after the first billion was achieved.[6,7]

After World War II, much of the world's population benefited from public health programs, such as widespread immunization campaigns, as well as increased access to food and better sanitation. Death rates plummeted, especially among children. At this point, the combination of traditionally high birth rates and rapidly decreasing mortality rates propelled population growth on a scale heretofore unknown in human history.

This phenomenon has been called the population explosion,[8] and thus far, each additional billion mark has been reached at an accelerating pace (Figure 1–4). By 1960, world population stood at 3 billion, only 33 years following the achievement of the second billion. It took only 14 years to reach the fourth billion in 1974. Thirteen years later (in 1987), the world population surpassed 5 billion, and 12 years later (in 1999), 6 billion. By 2011, 7 billion humans resided the planet; the eighth billion is projected to be reached by 2025. The United Nations estimates that the world population will reach 9 billion before 2045, and "projections portend growth beyond that number."[6(p 36)]

While the rate of growth has slowed, the current and future size of the human population remains a critical issue. The current growth rate of 1.2% means that approximately 79 million persons are being added to the world population each year. This increase is occurring almost completely in countries with the least developed economies and the fewest resources to allocate to feeding and educating their youth. Their situation stands in marked contrast to that in Europe and Japan; both of these areas are expected to have fewer people in 2050 than they do now.[6,9]

In absolute numbers, the demographic contrast between developing and developed countries has been widening and continues to increase. In 1950, there was 1 person in the richer developed world for every 2 in the

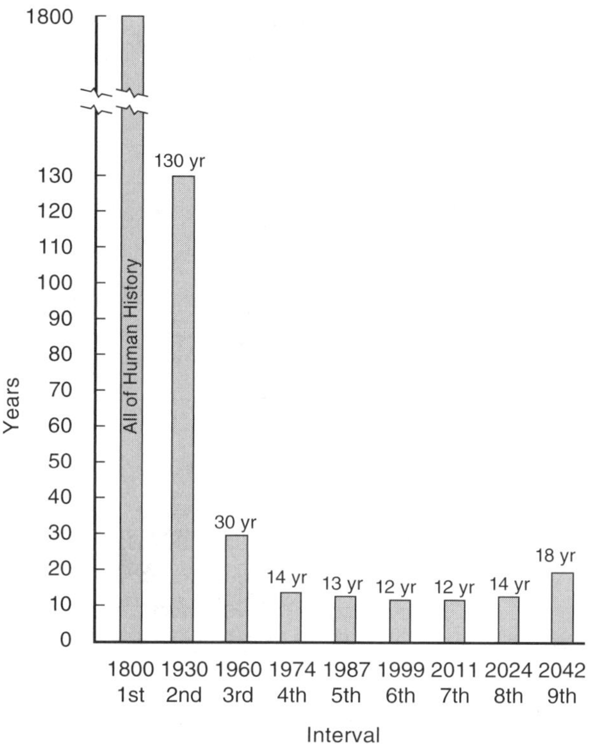

Figure 1–4 Years to Add Each Billion to World Population

Modified from the Population Reference Bureau. Years to add each billion to world population. 2013. http://www.prb.org/Publications/GraphicsBank/PopulationTrends.aspx.

developing world. By 2010, that ratio was 1 to 4.5. By 2050, it will be 1 to 6. Another way of looking at this demographic contrast is to point out that Africans and Asians currently account for 75% of the world's population, but this proportion is projected to rise to nearly 80% by 2050.[10] Despite being the epicenter of the global HIV/AIDS epidemic, the population of sub-Saharan Africa is growing faster than that in any other region. The total fertility rate (TFR)—that is, the number of children that women are having today—is 5.2 in sub-Saharan Africa, compared to the world TFR of 2.5.[3,11]

Population Momentum

Even if high fertility rates are curtailed in countries like Uganda, population growth would continue for decades because of the young age structure of the population, as shown in Figure 1–5. This phenomenon,

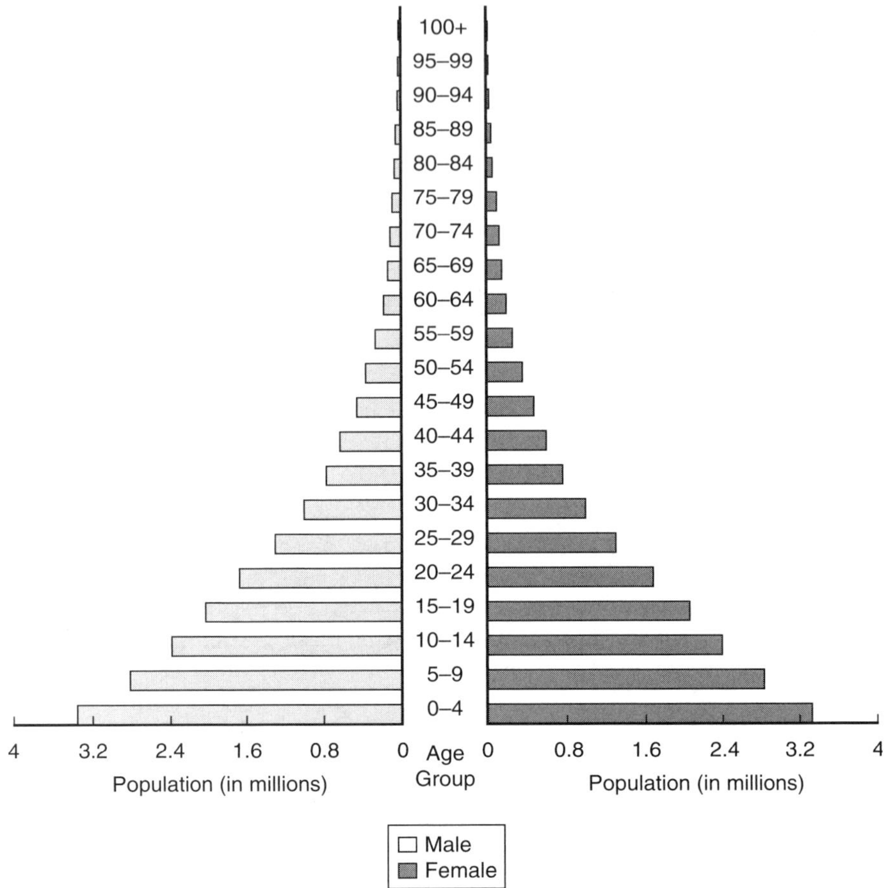

Figure 1-5 Population Pyramid for Uganda, 2013

Reproduced from the U.S. Census Bureau. International programs: international data base. December 2013. http://www.census.gov/population/international/data/idb/informationGateway.php. Accessed December 30, 2013.

known as *population momentum*, occurs because younger age groups or cohorts are so large relative to older cohorts. Figure 1–5 shows that there are now and will continue to be larger numbers of people moving into their reproductive years. Even when replacement-level fertility is achieved (roughly 2.1 children per woman), the total population will continue to grow due to the large numbers of women in their reproductive years.[10] In the case of Uganda, current fertility is much higher than replacement level.

Proximate Determinants of Fertility

The likelihood that these fertility rates will persist is related to the proximate determinants of fertility.[6,9] These factors include the proportion of the population that is married or in sexual unions, the use of contraceptives, the incidence of abortion, and involuntary fecundity, usually related to breastfeeding. The relative importance of each determinant varies by country and by culture. Nevertheless, contraceptive prevalence and abortion incidence explain most of the current differences among countries' fertility rates.

FERTILITY CONTROL

The demand for fertility control is apparently universal. In every known culture, literate or preliterate, individuals have attempted to control their family sizes.[12] What have varied are the efficacy, safety, and prevalence of preconception and postconception practices. History shows a wide range of fertility control practices, including infanticide ("known to have been practiced in much of Asia"[6(p 221)]), delayed marriage, celibacy, sexual taboos, contraception, and induced abortion. Short histories of the latter two practices follow.

Contraception

The oldest known recipes for contraceptives come from an Egyptian papyrus dating back to 1850 BCE. This document, known as the Kahun papyrus, contained several recipes, including one for a pessary or diaphragm and one for a vaginal spermicide.[12] Ancient Hebrews practiced coitus interruptus (withdrawal), and they used vaginal sponges. Ancient Greeks used herbs, such silphium (a giant fennel), to make birth control substances, and they passed their birth control knowledge to ancient Romans. Modern analyses suggest that many of the methods used by ancients were effective for reducing the risk of pregnancy.[2,13]

Scant evidence exists of controversy associated with early contraceptive practice. In AD 416, St. Augustine (AD 354–430) did denounce birth control practice, but it is unclear to whom this critique was directed.[14] In the Middle Ages, an estimated 200 fertility control methods were in common use in Europe, and at least a few were apparently effective. The average household in medieval Italy had only 2.44 children; that number was 2.36 in the German territories. Much of the region's knowledge about fertility control was transmitted by lay midwives, who served as healers and childbirth assistants.[2,15]

Before 1350, birth control was openly discussed in both canonical and secular documents. A classic medieval textbook, *Canons of Healing*, describes birth control methods. Eminent clerical scholar Albert the Great (1200–1280) explained the contraceptive effects of certain plants and herbs, with no indication that he was addressing a delicate subject.[15] Chaucer also mentions birth control, another sign of its widespread practice.[12]

By the 15th century, however, hardly any Western European documents refer to birth control—a void that has long perplexed scholars. Because midwives were their major target, the witch hunts that occurred between 1360 and 1700 offer a plausible explanation of the disappearance of herbal contraceptives in that region. Reliable estimates of the killings committed during the witch hunts range from 1 million women to multiples of that number. The pervasiveness of these massacres can be better appreciated when one considers that in 1600 the entire population of northwestern Europe, the center of witch hunting, was less than 50 million.[15]

Given the scale of these executions, one can reasonably assume that by 1700, when these killings ceased, birth control and abortion knowledge had been almost obliterated in Western Europe.[15] Consequently, the European population grew rapidly for the next 300 years, a trend described by Reverend Thomas Malthus in 1798. He warned about unchecked population growth and subsequent poverty. Considering birth control immoral, Malthus advocateddeferred marriage and sexual abstinence as means to curtail rapid growth.[16]

Despite Malthus's beliefs, birth control practice began to rebound. In 18th-century Europe, condoms came into widespread use. In 1822, Frances Place launched the English birth control movement, circulating explicit contraceptive pamphlets throughout English industrial districts.[17] Across the Atlantic, the American birth control movement would not take hold for nearly a century. Puritan disapproval of birth control as well as other social conventions conspired against open discussion of these practices. Gradually, these mores softened, and a thriving commercial market for contraceptive devices developed after the Civil War.[2,18]

In 1873, passage of the federal Comstock Law brought American birth control to a standstill. This legislation prohibited interstate trade in obscene literature and articles of immoral use, including contraceptives. Over the next two decades, this legislation spawned 22 "little Comstock" state laws, many of which went further in prohibiting birth control practice.[19] These laws stayed on the books for nearly a century, thwarting the dissemination of contraceptive information as well as chilling public and professional attitudes toward birth control. Fearing legal prosecution, early 20th-century medical textbooks did not mention contraception[20]; it was 1959 before an American television show dared to bring up birth control.[21] Comstockery underlay the context in which change occurred and portended current ideological battles.

In 1959, a presidential commission on foreign aid, chaired by General William Draper, recommended that the United States provide family planning assistance to countries requesting it. Catholic bishops denounced this policy advice, and President Dwight Eisenhower initially opposed it. Nevertheless, the Draper Commission served as a catalyst in making birth control methods more widely available: by 1967, U.S. international population assistance was enacted with bipartisan consensus.[22] Contraceptive technology also had advanced: the U.S. Food and Drug Administration approved the birth control pill in 1960, and plastic intrauterine devices (IUDs) came on the market in the early 1960s.[2,19]

American women rapidly accepted these new methods. With international aid, initially spearheaded by the United States, modern contraceptive methods were disseminated globally. In 1960, only 10% of the world's women of reproductive age used any method (including traditional) of contraception. By 2008, however, 57% had adopted a modern method. Asia's contraceptive prevalence was 61%, with 63% prevalence noted in Latin America and the Caribbean. Consequently, fertility rates dropped in the developing world (including China) from about six children per woman to fewer than three children per woman.[23]

Induced Abortion

Throughout history, induced abortion has been used to end unwanted pregnancy.[10] An Egyptian papyrus from 1550 BCE described several abortifacients. Abortion was also common in ancient Greek society,[12] and Greek philosophers, notably Aristotle, were greatly interested in the timing of ensoulment or animation. Aristotle postulated that animation occurred 40 days following conception in male fetuses and after 80 days in female fetuses. His theories influenced Christian thought, including the works of St. Augustine and St. Thomas Aquinas.[2,24]

The latter's view influenced Pope Innocent IV, who declared in 1257 that abortion before the infusion of the soul was not homicide. This dictum remained the Catholic Church's policy for more than three centuries. In 1586, Pope Sixtus V declared that abortion was premeditated murder; two years later, he recommended excommunication and execution for those who practiced it. Only three years later, Pope Gregory XIV withdrew these penalties, believing them too severe in light of the animation debate. His 1591 declaration stood until 1869, when Pope Pius IX restored the policy of Pope Sixtus V.[2,24] The current teaching of the Catholic Church is that abortion is not morally permissible.

Clerical debates had little impact on abortion practice or policy in early America. In 1800, abortion was not illegal in any state,[25] but the practice of abortion remained both widespread and risky. The medical profession began to advocate for restrictions, and by 1910 abortion was illegal in every state except Kentucky.[26] As the 20th century progressed, abortion remained prevalent[27] and its safety increased.[2]

Backed by organized medicine, state laws began to change in the late 1960s.[28] In 1973, the U.S. Supreme Court's decision in *Roe v. Wade* legalized abortion throughout the land; it also eventually unleashed a powerful anti-abortion lobby that began to target international population assistance. In 1973, Congress passed the Helms Amendment, forbidding U.S. international population assistance funds from being used to pay for abortion in any manner.[2,29,30]

Despite U.S. abortion politics, induced abortion remains globally prevalent. Induced abortion is likely to occur when an unintended pregnancy occurs, when contraception is not used or available, or when contraception fails. Abortion is safe when performed in sanitary conditions by trained medical staff. However, about half the world's abortions are performed illegally, so they have a high probability of being unsafe.[6,31] It has been known since the 1960s that countries undergoing a transition in fertility rates can expect initial increases in both contraception and abortion. As contraception becomes more widespread, abortion rates eventually decline.

REPRODUCTIVE HEALTH

This text juxtaposes population issues and reproductive health. While the reproductive behavior of individuals certainly drives world population growth, reproductive health is a vital human rights concept itself, independent of demographic concerns. In 1994, the International Conference on Population and Development (ICPD) in Cairo offered the following comprehensive definition: "Reproductive health is a state of complete physical, mental and social well-being and not merely the absence of disease or infirmity, in all matters relating to the reproductive system and to its functions and processes."[31] The World Health Organization (WHO)[32] and the United Nations Population Fund (UNFPA)[33,34] have embraced this definition and have used it as a basis to broaden their programs to include maternal health and sexually transmitted infections. In some cases, the definition has been used to extend programs to men or couples, but for the most part reproductive health services remain women-centered. Part of the reason for this, of course, is that modern contraceptive technology is largely focused on women.

In many parts of the world, however, individual women and married couples still lack access to modern contraception and other reproductive health services. An estimated 222 million sexually active women who report they would like to avoid or postpone another pregnancy lack access to modern contraceptive methods.[34] More than one-third of pregnancies worldwide are not intended. One consequence of unintended pregnancies is induced abortion, which is all too often, performed in unsafe conditions. Another is short intervals between births, which can have serious—even deadly—consequences for women and their children.[33]

The Status of Women, Fertility, and Reproductive Health

The opportunities available to women and girls affect fertility rates. Some of the most important factors are age at marriage, educational opportunities, and income. In societies where girls marry at very young ages (often involuntarily), birth rates are high. Very young women who enter a marriage or sexual union are likely to have partners who are much older (up to 15 years older in some countries) than they are. "This difference reduces the chance that the woman will be able to participate in decisions about childbearing or negotiate the use of contraceptives."[34] Similarly, when girls lack educational opportunities, they are likely to spend a greater proportion of their lives bearing children. Higher levels of education are linked to higher levels of contraceptive use, lower fertility levels, better child health, and higher aspirations for one's children.[10] Related to age at marriage and education is income. Throughout the world, low-income women have higher fertility than their wealthier counterparts.[9,34]

Each of these factors offers a lever to national governments and international organizations as they develop their reproductive health policies. Age at marriage can be raised, a change probably best implemented by increasing opportunities for girls' education. Educational attainment is highly associated with income, and both contribute to child survival, which in the long run lowers fertility aspirations. The millions of women who report that they would like to limit their family size but lack access to contraception information present an obvious policy challenge, as do those who want to use birth control methods but lack access to contraceptive services.

Reproductive Health Services

Reproductive health services address sexuality, contraception, induced abortion, and childbirth. Figure 1–6 depicts these services in relation to the three major steps in human reproduction. With rare exception, the first requirement for human reproduction is sexual intercourse. The second step is conception, and the third is gestation and parturition.[19,35]

The vast majority of reproductive health services are provided *after* women or couples become sexually active, but there are interventions before Step 1, aiming to reach young women and men before the initiation of sexual activity.[36] Between Steps 1 and 2 is contraception, which is designed to decrease the likelihood that sexual intercourse will lead to conception. Between Steps 2 and 3 is induced abortion, which terminates a pregnancy.

Most factors (such as gender, poverty, and mobility[37]) related to the age at which young women or men become sexually active are beyond the reach of reproductive health services. Throughout the world, sexual activity begins

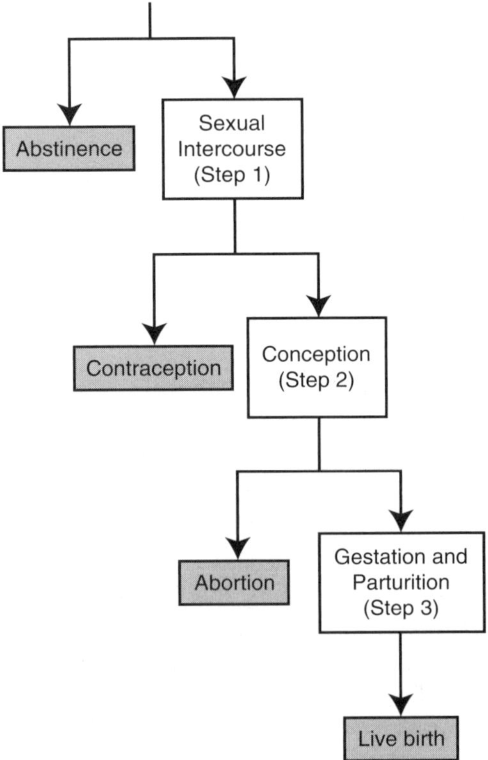

Figure 1–6 The Human Reproductive Process

Data from Davis K, Blake J. Social structure and fertility: an analytic framework. *Econ develop cultural change,* 1956;4:211–235.

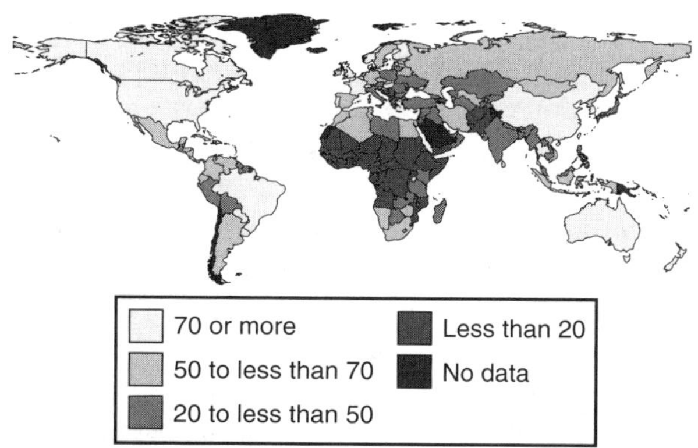

Figure 1–7 Percentage of Women Using Any Modern Method of Contraception Among Those Aged 15–49 Who Are Married or in a Union: Most Recent Data Available

Reproduced from World Contraceptive Use 2011. Population Division, Department of Economic and Social Affairs, United Nations. http://www.un.org/esa/population/publications /contraceptive2011/anymodern.pdf. Accessed December 30, 2013.

for most people in the later teenage years (ages 15–19 years), but there are substantial regional and gender differences.[36] Many health educators and health providers try to consult with adolescents before sexual activity is initiated, so as to help them reduce the risks of unwanted pregnancy and sexually transmitted infections. Considerable overlap exists with sexuality education, which usually occurs outside of a clinical setting. In recent years, there has been an increased emphasis on sexual health within clinical practice.[37]

Access to contraceptive or family planning services varies greatly throughout the world (Figure 1–7). In the United Kingdom, for example, 84% of women not wishing to become pregnant use modern and effective contraceptive methods. This proportion is 74.8% in France, but in Afghanistan, modern contraceptive prevalence is only 15%. The corresponding figure for Uganda is 17.9%.[38]

LINKING POPULATION AND REPRODUCTIVE HEALTH AND ENVIRONMENTAL SUSTAINABILITY

Some will regard this text's approach of linking population and reproductive health as controversial. For example, some feminist critics of national family planning programs have rightfully pointed out that when programs are too zealous in reaching demographic targets, then women's reproductive rights have been ignored or other reproductive health services are neglected.[39-41] The view taken in this text is that both population and reproductive health have vital public health implications, and that each needs to be discussed within the context of the other. For example, realistic program planning for reproductive health services, particularly at the national level, requires demographic data.[42] Moreover, in many situations, reproductive rights and population concerns are complementary.[42] This text examines both the controversies and the synergies afforded by population and reproductive health issues.

The size of a country's population as well as its consumption habits have important implications for the global environment. Reproductive health services fit into this discussion because they enable people to plan the number of children that they will have as well as raise the chances that these children will survive to adulthood. Providing access to reproductive health services alone, however, will not solve the problem of unsustainable consumption practices, particularly in many industrialized countries.

Throughout the world, carbon footprints vary widely. The richer countries with slowing or stable growth consume far more per capita than countries with rapidly growing populations. Figure 1–8 shows total carbon

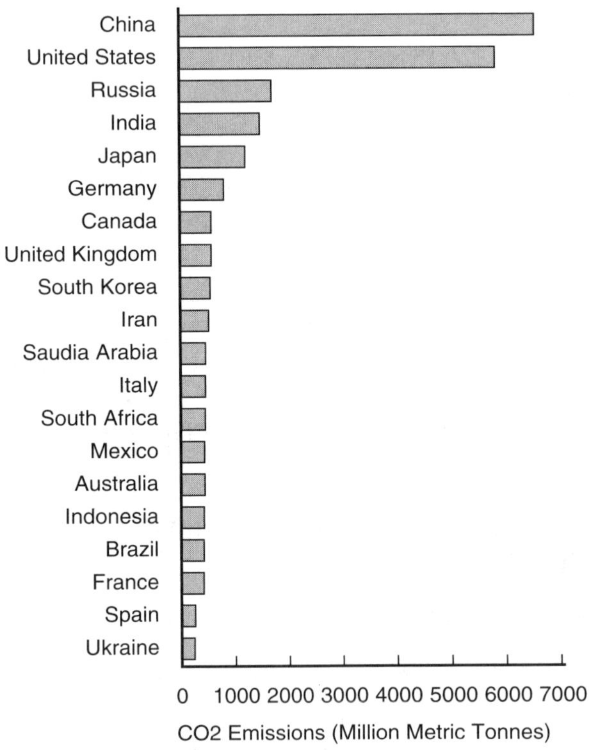

Figure 1–8A Total Carbon Dioxide Emission by Country

Reproduced from Union of Concerned Scientists. Each country's share of global emissions. 2012. http://www.ucsusa.org/global_warming/science_and_impacts/science/graph-showing-each -countrys.html. Accessed August 21, 2013.

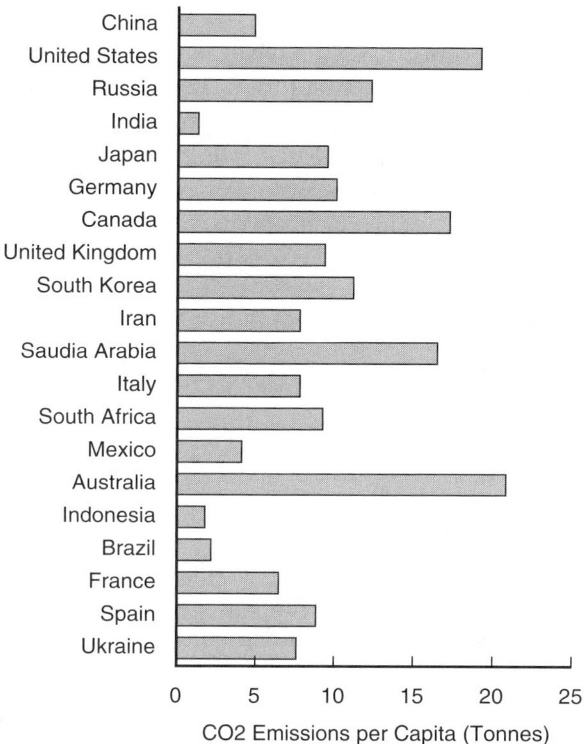

Figure 1–8B Carbon Dioxide Emissions per Capita by Country

Reproduced from Union of Concerned Scientists. Each country's share of global emissions. 2012. http://www.ucsusa.org/global_warming/science_and_impacts/science/each-countrys-share-of -co2.html. Accessed August 21, 2013.

dioxide emissions by country as well as per capita carbon dioxide emissions by country. As can be seen in the graphs, developed countries and major emerging economy nations (e.g., China) lead in total carbon dioxide emissions.[43] Developed nations typically have much higher carbon dioxide emissions per capita, while some developing countries lead in the growth rate of carbon dioxide emissions. These uneven contributions to the climate problem are at the core of the challenges the world community faces in finding effective and equitable solutions.

Population growth exacerbates these problems. The environmental impact of population growth on the environment coupled with the high consumption of wealthy countries is relentless.[44] The human footprint has never been greater than it is today, and it is even changing the earth's climate. Food insecurity for growing numbers of people is one of the results—yet discussions of environmental sustainability and population growth seldom occur in the same fora.

International population and reproductive health assistance has been in place for nearly a half-century, with wealthy countries providing assistance to poorer countries to help them address their burgeoning populations and vast reproductive health needs. This is an area fraught with political minefields, yet also one of the great public health accomplishments of the last century.[45] Since the inception of American population assistance, the world population has grown by more than 3 billion. Nevertheless, there is widespread agreement that this figure would have been far higher in the absence of this aid.

INTERNATIONAL POPULATION ASSISTANCE

The U.S. Congress first authorized funding for international population assistance in 1967 with the passage of Title X of the Foreign Assistance Act. After three years of Senate hearings (1965–1967) headed by Senator Ernest Gruening and the 1959 recommendation of the 1959 Draper Commission, legislators in both parties and President Lyndon Johnson were convinced that rapid human population growth was a critical problem.[22]

Since the mid-1980s, U.S. policy toward international population growth has been inconsistent, increasingly partisan, and intertwined with domestic abortion politics.[46] In the 1970s, the United States led the world in addressing this global problem. Preferring a multilateral approach, President Richard Nixon supported the newly created United Nations Fund for Population Activities (UNFPA) at the same time that U.S. bilateral population assistance was increasing.[22] In 1974, at the United Nations (UN) conference on population held in Bucharest, U.S. officials and others argued that investments in family planning programs would reap large benefits for developing economies. By 1984, at a similar UN forum held in Mexico City, the U.S. delegation surprised the world by declaring that population growth was a "neutral phenomenon" for economic development. The Americans also announced what has become known as the Mexico City policy or the Global Gag Rule: the United States would withdraw population assistance from any foreign nongovernmental organization that supported abortion, even if it did so with its own funds.[9,10]

Since 1984, the Mexico City policy and U.S. funding for UNFPA have become partisan battlegrounds. Within days of being sworn in, Democratic President Bill Clinton rescinded the Mexico City policy from the Republican Reagan–Bush years and supported reinstating UNFPA funding. In 1996, disagreements about population assistance between President Clinton and the Republican-dominated Congress literally shut down the federal government. Upon assuming office, Republican President George W. Bush reinstated the Mexico City policy and withdrew American support from UNFPA.[46]

Democratic President Obama has followed the partisan lead on both issues as well, by countermanding his predecessor's policy.

The United States no longer dominates international population assistance. Other wealthy nations have increased assistance for contraceptive services and even safe abortion care. When the United States curtailed funding for such programs in the early 21st century, the European Union stepped in to shore up UNFPA and other nongovernmental organizations, such as the International Planned Parenthood Federation, a perennial victim of the Mexico City policy.[47,48] Many poor countries support some level of family planning services, but they are unable meet the demand from their citizens for such services. Worldwide, support for voluntary population programs has not kept pace with global demographic growth.[49]

OTHER POPULATION GROWTH ISSUES

Population growth is linked with many other issues, including human rights, national identity, and environmental sustainability. Feminist groups brought the link between human rights and population to the forefront during the 1994 International Conference on Population and Development in Cairo.[10] At issue was whether world population programs had exploited women while they were trying to bring down birth rates.[6] Reproductive health replaced population as a paradigm, but largely due to the Catholic Church's influence, no consensus emerged on abortion. At the present time, the largest human rights issue is that at least 222 million women who want to use contraception still lack access to family planning services.[34]

Within a handful of countries, there have also been concerns about coercion and fertility control. China's enforcement of its "one child policy" is the largest and best-known example. India's vasectomy camps in the 1970s were overzealous at best. Forced sterilizations have been documented in Peru and elsewhere. Unfortunately, human rights concerns are often raised in the political arena by birth control opponents trying to denigrate voluntary family planning programs.[50]

Uneven population growth in the world's regions has precipitated international migration, usually from poor nations to wealthy countries. The latter increasingly need young workers as their own populations age. Immigrants are often ethnically and culturally different from the native populations in receiving countries, so their presence often raises social and political questions related to national identity.[9]

Strategies to address population growth and even reproductive health remain controversial within many religious and political circles, despite the fact that scientific knowledge and worldwide demand could alleviate, if not eliminate, the negative impact of population growth on the environment. Because population growth was stagnant for 99% of human history, it is no wonder that some religious and political institutions have had difficulty addressing the current facts of population growth.

A WORD ABOUT TERMS

Terminology in population and reproductive health is fraught with controversy. We should note that this is nothing new for topics concerning human fertility. For example, Margaret Sanger, founder of Planned Parenthood Federation, disliked both the terms *family planning* and *planned parenthood*, preferring *birth control*. The use of consistent terminology is a goal for this text, but the yield from this effort is imperfect. In other words, in certain usage and in some contexts, the same terms may have different meanings. Given these caveats, some concrete definitions are in order. These terms are loosely grouped here into population terms, reproductive health terms, and environmental sustainability terms.

Population Terms

Population refers to groups of and numbers of people.[11]

Population assistance is international assistance for (1) enhancing a country's demographic capacity or (2) supporting family planning or reproductive health services. Population assistance may be **bilateral aid** (from one country to another country) or **multilateral aid** (from more than one country to a global organization, such as UNFPA, which distributes funds, commodities, or technical assistance to individual countries).

Originally, the term **population control** was used to refer to policies "geared at reductions of fertility levels through family planning programs, as a means to improve socioeconomic outcomes in developing countries."[10(p 93)] The term has been little used since the 1994 International Conference on Population and Development. Today, the term is almost always used pejoratively by critics of population assistance, with the implication that coercive measures are being used.[50]

Population policies are "actions taken explicitly or implicitly by public authorities in order to prevent, delay, or address imbalances between demographic changes, on the one hand, and social, economic, and political goals on the other."[10(p 280)]

Overpopulation refers to "a situation in which the population has overshot a region's carrying capacity."[6(p 545)] Clearly this term is also pertinent to environmental **sustainability**.

Reproductive Health Terms

Abortion refers to either a miscarriage (spontaneous abortion) or to an induced abortion, an intentional termination of pregnancy through surgical or pharmaceutical means.

Unsafe abortion is "abortion being performed by an unskilled person, under unsanitary conditions, or both."[51(p 502)]

Contraception is the intentional prevention of pregnancy through the use of various devices, agents, drugs, sexual practices, or surgical procedures. For the most part, contraception is practiced by those who are sexually active.[52(p 409)]

Family planning is most often used to refer to voluntary programs that enable individuals to access birth control information and contraception. This term is often used synonymously with **birth control**, although family planning programs usually provide more health services than just contraception.

Fertility control refers to limiting births through abstaining from sexual activity, the use of contraception, or induced abortion.[53]

Reproductive rights refer to the "basic right of all couples and individuals to decide freely the number, spacing, and timing of their children and to have the information and means to do so. It also includes their right to make decisions concerning reproduction free of discrimination, coercion, and violence."[54(p 3)]

Reproductive health is the "state of complete physical, mental and social well-being and not merely the absence of disease or infirmity, in all matters relating to the reproductive system and its functions and processes."[31]

Reproductive health services may include sexuality education, contraceptive services, and maternity care. Whether they include safe abortion services has been an ongoing political battle for more than two decades.[46]

Environmental Sustainability Terms

Sustainability "creates and maintains the conditions under which humans and nature can exist in productive harmony, that permit fulfilling the social, economic and other requirements of present and future generations. Sustainability is important to making sure that we have and will continue to have the water, materials, and resources to protect human health and our environment."[55]

Sustainable development "meets the needs of the present without compromising the ability of future generations to meet their own needs."[55]

Carrying capacity refers to "the number of people the Earth can support."[44(p 4)]

Climate change refers to "any significant change in the measures of climate lasting for an extended period of time. In other words, climate change includes major changes in temperature, precipitation, or wind patterns, among others, that occur over several decades or longer."[56]

Climate change adaptation refers to initiatives and measures to reduce the vulnerability of natural and human systems against actual or expected climate change effects.[57]

Climate change mitigation is defined by the U.S. Environmental Protection Agency (EPA) as "a human intervention to reduce the human impact on the climate system; it includes strategies to reduce greenhouse gas sources and emissions and enhancing greenhouse gas sinks."[57]

Food security exists when all people, at all times, have physical and economic access to sufficient safe and nutritious food that meets their dietary needs and food preferences for an active and healthy life. The 1996 World Food Summit identified four dimensions of food security: "food availability, affordable access, household and physiological utilization, and temporal stability."[58(p 1)]

Food insecurity consists of *chronic food insecurity*, which is long term and persistent, and *transitory food insecurity*, which is short term and temporary.[58]

CONCLUSION

Population and reproductive health are inextricably linked. Global population continues to grow, projected to reach 8 billion by 2025. A vast demographic divide exists between poor countries with rapidly growing populations and wealthier nations with slower or no growth. Reproductive health services, including contraception, are far more widely available in the latter group of nations than in the former. Indeed, a recent UNFPA estimate shows that at least 222 million women and couples who would like to practice family planning lack access to modern contraceptive methods.

Birth control is not a new practice. History shows that people have attempted to limit their fertility for millennia. What has varied is the efficacy and safety of extant methods of fertility control, including induced abortion. Since 1960, contraceptive technology has become far more effective, and there have been significant international efforts to increase its availability.

Support for international population and reproductive health assistance have waxed and waned during the past half century. Initially, the United States dominated those efforts, but domestic partisan politics have diminished the American role. In recent decades, multilateral efforts have become far more important and the focus has broadened from family planning to reproductive health.

Throughout the world, carbon footprints vary greatly. The size of a country's population as well as its consumption patterns affect the natural environment. Richer countries with slow or stable growth consume far more per capita than poorer countries, and they contribute disproportionately to global warming. Reproductive health services fit into this discussion because they enable people to plan the number of children that they will have as well as raise the chances that these children will survive to adulthood. Providing access to reproductive health services, however, will not solve the problem of unsustainable consumption practices occurring in many industrialized countries.

DISCUSSION QUESTIONS

1. Discuss at least three ways in which population growth and reproductive health are related.
2. How do the lives of women in the developed world differ from those in the developing world?
3. How are human rights related to world population growth and reproductive health?
4. Discuss the relative impact of population growth and human consumption upon the environment.
5. Describe how population momentum is related to world population growth.

REFERENCES

1. Population Reference Bureau. Years to add each billion to world population. 2013. http://www.prb.org/Publications/GraphicsBank/PopulationTrends .aspx.
2. Central Intelligence Agency. Median age. In: *World Factbook*. 2013. https://www .cia.gov/library/publications/the-world-factbook/fields/2177.html.
3. Population Reference Bureau. Datafinder. 2013. http://prb.org.
4. U.S. Census Bureau. International programs: international data base. 2013. http://www.census.gov/population/international/data/idb /informationGateway.php.
5. Haub C, Gribble J. The world at seven billion. In: *Population Bulletin*, Vol. 66(2). Washington, DC: Population Reference Bureau; 2011. http://www.prb.org /Publications/Datasheets/2011/world-population-data-sheet.aspx.
6. Weeks JR. *Population: An Introduction to Concepts and Issues*. 11th ed. Belmont, CA: Wadsworth Cengage Learning; 2012.
7. McKeown T. *The Modern Rise of Population*. London: Edward Arnold; 1976.
8. Ehrlich PR, Ehrlich AH. *The Population Explosion*. New York: Simon and Schuster; 1990.

9. Population Reference Bureau. *Transitions in World Population*. Washington, DC: 2004. http://www.prb.org/.

10. May JF. *World Population Policies*. New York: Springer; 2012.

11. Haupt A, Kane TT. *Population Handbook*. 11th ed. Washington, DC: Population Reference Bureau; 2004. http://www/prb.org. Accessed April 26, 2014.

12. Himes N. *Medical History of Contraception*. New York: Schoken Books; 1970.

13. Riddle JM, Estes JW, Russell JC. Ever since Eve: birth control in the ancient world. *Archeology*. 1994;47:29–35.

14. Noonan JT. *Contraception: A History of Its Treatment by Catholic Theologians and Canonists*. Cambridge, MA: Harvard University Press; 1986.

15. Heinsohn G, Steiger O. The elimination of medieval birth control and the witch trials of modern times. *Int J Women's Stud*. 1982;5:193–214.

16. Thomlinson R. *Population Dynamics: Causes and Consequences of World Demographic Change*. New York: Random House; 1965.

17. Finch BE, Green H. *Contraception Through the Ages*. London: Peter Owen; 1963.

18. Brodie JF. *Contraception and Abortion in Nineteenth Century America*. Ithaca, NY: Cornell University Press; 1994.

19. McFarlane DR, Meier KJ. *The Politics of Fertility Control*. Washington, DC: Congressional Quarterly Press (Chatham House); 2001.

20. Dienes CT. *Law, Politics, and Birth Control*. Urbana: University of Illinois Press; 1972.

21. Jaffe FS. Public policy on fertility control. *Sci Am*. 1973;229:1,17–23.

22. Piotrow PT. *World Population Crisis: The United States Response*. New York: Praeger; 1974.

23. Sinding S. Introduction. In: Robinson W, Ross JA, eds. *The Global Family Planning Revolution: Three Decades of Population Policies and Programs*. Washington, DC: International Bank for Reconstruction and Development; 2007.

24. Sheeran PJ. *Women, Society, the State, and Abortion: A Structuralist Analysis*. New York: Praeger; 1987.

25. Mohr JC. *The Origins of and Evolution of National Policy, 1800–1900*. New York: Oxford University Press; 1978:vii.

26. Craig BH, O'Brien DM. *Abortion and American Politics*. Rev. ed. Chatham, NJ: Chatham House; 1993.

27. Luker K. *Abortion and the Politics of Motherhood*. Berkeley: University of California Press; 1984.

28. Tribe LH. *Abortion: The Clash of Absolutes*. New York: Norton; 1992.

29. U.S. Agency for International Development. USAID's family planning guiding principles and U.S. legislative and policy requirements. USAID–Family Planning. 2012. http://transition.usaid.gov/our_work/global_health/pop/restrictions.html. Accessed November 17, 2012.

30. Taylor J, Kumar AK. How existing US policy limits global health and the achievement of Millennium Development Goals to improve maternal health and promote gender equality. *Yale J Int Affairs*. Winter 2011:43–52.

31. Sieverding M. Gender and reproductive health. In: Smelser NJ, Baltes PB, eds. *International Encyclopedia of the Social and Behavioral Sciences*. Vol. 9. Oxford, UK: Elsevier; 2001:5969–5972.

32. World Health Organization. Sexual and reproductive health. 2013. http://www .who.int/reproductivehealth/hrp/en/.
33. United Nations Population Fund. Improving reproductive health. 2013. http://www.unfpa.org/rh/index.htm.
34. Greene M, Joshi S, Robles O. *By Choice, Not by Chance: Family Planning, Human Rights, and Development.* New York: United Nations Fund for Population; 2012.
35. Davis K, Blake J. Social structure and fertility: an analytic framework. *Econ Develop Cultural Change.* 1956;4:211–235.
36. Wellings K, Collumbien M, Slaymake E, et al. Sexual behaviour in context: a global perspective. *Lancet.* 2007;369(9558):274.
37. Higgins JA, Davis AR. Sexuality and contraception. In: *Contraceptive Technology.* 20th ed. Atlanta, GA: Ardent Media Inc.; 2011:1–28.
38. United Nations Department of Economic and Social Affairs, Population Division. *World contraceptive use, 2011.* New York, NY: United Nations; 2013. http://www.un.org/esa/population/publications/contraceptive2011 /contraceptive2011.htm.
39. DeJong J. The role and limitations of the Cairo International Conference on Population and Development. *Soc Sci Med.* 2000;51:941–953.
40. Kabir S. Oral history, interviewed by McFarlane D. Northampton, MA: Sophia Smith Collection, Smith College; 2004. http://www.smith.edu/libraries/libs /ssc/prh/transcripts/kabir-trans.pdf.
41. Germaine A. Oral history, interviewed by Sharpless R. Northampton, MA: Sophia Smith Collection, Smith College; 2003. http://www.smith.edu/libraries /libs/ssc/prh/transcripts/germain-trans.pdf.
42. Sai FT. Oral history, interviewed by McFarlane D. Northampton, MA: Sophia Smith Collection, Smith College; 2004. http://www.smith.edu/libraries/libs /ssc/prh/transcripts/sai-trans.pdf.
43. Union of Concerned Scientists. Each country's share of global emissions. 2012. http://www.ucsusa.org/global_warming/science_and_impacts/science/graph -showing-each-countrys.html. Accessed August 21, 2013.
44. Heinrichsen D, Robey B. *Population and the Environment: The Global Challenge.* Population Reports, Series M, No. 15. Baltimore, MD: Johns Hopkins School of Public Health, Population Information Program; Fall 2000.
45. Centers for Disease Control and Prevention. Ten great public health achievements in the twentieth century. http://www.cdc.gov/about/history/tengpha .htm. Accessed May 30, 2013.
46. McFarlane DR. Reproductive health policies in President Bush's second term: old battles and new fronts in the United States and internationally. *J Public Health Policy.* 2006;27:4:405–426.
47. European Parliamentary Forum on Population and Development. *Euromapping 2012: Mapping European Development Assistance and Population Assistance.* Brussels: European Parliamentary Forum on Population and Development; November 2012. http://www.euroresources.org/fileadmin/user_upload/Euromapping /Euromapping_2012/Euromapping_2012_LoRes.pdf.

48. McFarlane DR. Human population growth. In: Steele BS, ed. *Controversies in Science and Politics.* Washington, DC: Congressional Quarterly Press; 2014:450–455.

49. Population Action International. *Trends in US Population Assistance.* Washington, DC: Population Action International; October 4, 2011. http://populationaction .org/articles/trends-in-us-population-assistance/. Accessed August 21, 2013.

50. Connelly M. *Fatal Misconception: The Struggle to Control World Population.* Cambridge, MA: Belknap Press; 2008.

51. Murthy P, Smith CL. *Women's Global Health and Human Rights.* Sudbury, MA: Jones and Bartlett; 2010.

52. *The American Heritage Dictionary of the English Language.* 3rd ed. Boston, MA: Houghton Mifflin; 1992.

53. McFarlane DR, Meier KJ. *The Politics of Fertility Control: Family Planning and Abortion Policies in the American States.* Chatham, NJ: Chatham House Press (now Congressional Quarterly Press); 2001.

54. Pillsbury B, Maynard-Tucker G, Nyguen F. *Women's Empowerment and Reproductive Health: Links Throughout the Life Cycle.* New York, NY: United Nations Population Fund and Pacific Institute for Women's Health; 2000.

55. United States Environmental Protection Agency. What is sustainability? http:// www.epa.gov/sustainability/basicinfo.htm. Accessed August 21, 2013.

56. United States Environmental Protection Agency. Glossary of climate change terms. http://www.epa.gov/climatechange/glossary.html. Accessed August 21, 2013.

57. International Panel on Climate Change. *Glossary of Terms Used in the IPCC Fourth Assessment Report.* 2007.

58. Food and Agriculture Organization of the United Nations. *An Introduction to the Basic Concepts of Food Security.* Rome: FAO Food Security Programme; 2008. http://www.fao.org/docrep/013/al936e/al936e00.pdf.

Chapter 2

History and Future of World Population

John R. Weeks

INTRODUCTION

Human beings have been around for a long time, perhaps a million or more years,[1] but for most of that time the imprint of humans on the planet was scarcely noticeable. Humans were hunter-gatherers living a primitive existence marked by high fertility, high mortality, and only very slow population growth. Given the large amount of space required by a hunting-gathering society, it seems unlikely that the earth could support more than several million people living like that,[2-4] so it comes as no surprise that the world population on the eve of the Agricultural Revolution (sometimes called the Neolithic Agrarian Revolution) about 10,000 years ago (8000 BCE) is estimated at about 4 million.

POPULATION GROWTH OVER THE PAST 10,000 YEARS

Since hunting and gathering use resources *extensively* rather than *intensively*, it was natural that over tens of thousands of years humans would move into the remote corners of the earth in search of sustenance. Eventually, people in most of those corners began to use the environment more intensively, leading to the more sedentary, agricultural way of life that has characterized most of human society for the past 10,000 years, starting first in what is now the Middle East, and then in what is now the eastern part of China, and apparently popping up at roughly the same time in what are now Central and South America.[5]

The population began to grow more noticeably after the Agricultural Revolution, although it is hard to detect this trend in Figure 2-1 because the population was still tiny by comparison to today's numbers. Between 8000 BCE and 5000 BCE, about 333 people on average were added to the world's total population each year, but by 500 BCE, as major civilizations were being established in China, India, and Greece, the world was adding 100,000 people each year to the total. By the time of Christ (the Roman Period, AD 1), there may well have been more than 200 million people on the planet, with their number increasing by nearly 300,000 each year. There was some backsliding in the third through fifth centuries AD, when increases

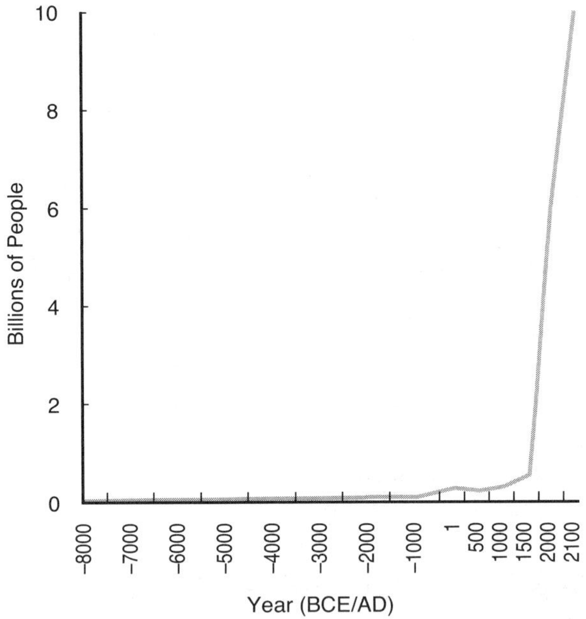

Figure 2-1 World Population Size from 8000 BCE to the Present and Projected to AD 2100

Data from the following sources: The population data from 8000 BCE through AD 1900 are the average of the three different sets of estimates: Evedy C, Jones R. *Atlas of World Population History.* New York: Penguin Books; 1978; Biraben J-N. Essai sur l'évolution du nombre des hommes. *Population.* 1979;34(1):13–24; Durand JD. The modern expansion of world population. *Proc Am Philosoph Soc.* 1967;3:137–140. Numbers for 1910 through 1940 are from the United Nations Population Division. The world at six billion. 1999. http://www.un.org/esa/population/publications/sixbillion/sixbillion.htm. Population figures for 1950 through 2100 are medium variant projections from the United Nations Population Division. *World Population Prospects in World Population Policies 2009.* New York: United Nations; 2010.

in mortality, probably due to the plague, led to declining population size in the Mediterranean area as the Roman Empire collapsed, and in China as the Han empire collapsed from a combination of flood, famine, and rebellion.[3] Population growth recovered its momentum only to be set back by yet another plague, the Black Death, that arrived in Europe in the middle of the 14th century and did not leave until the middle of the 17th century.[6] After that, during the period from about 1650 to 1850, Europe as a whole experienced population growth as a result of the disappearance of the plague, the introduction of the potato from the Americas (which added calories and improved resistance to death), and evolutionary (although not revolutionary) changes in agricultural practice—probably a response to the receding of the Little Ice Age.[7] On the eve of the Industrial Revolution (about 1750), the population of the world was approaching 1 billion people and was increasing by more than 2 million every year. This was unprecedented in human history.

Figure 2–1 shows rather dramatically that the beginnings of world population growth were associated with the emergence of the Enlightenment in Europe.[8-10] Indeed, this initial population growth was occurring primarily in Europe and it is quite likely that the Enlightenment, and the scientific and industrial revolutions that took place as part of that huge change in Europe, occurred in part because of this population growth. An early proponent of the idea that population growth could be the trigger of economic development was the Danish economist Ester Boserup. In a set of extremely influential writings,[11-13] she advanced the idea that, in the long run, a growing population is more likely than either a nongrowing population or a declining population to lead to economic development. The history of Europe shows that the Industrial Revolution and the increase in agricultural production were accompanied almost universally by population growth. Boserup's argument is based on the thesis that population growth is the motivating force that brings about the clearing of uncultivated land, the draining of swamps, and the development of new crops, fertilizers, and irrigation techniques, all of which are linked to revolutions in agriculture.

Europe of 300 or 400 years ago was reaching the carrying capacity of its agricultural society. Europeans first spread out looking for more room, which led to cultivation of crops outside of Europe for Europeans to use, and then they began to invent more intensive uses of their resources to meet the needs of a growing population.[12,14] The major resource was energy, which, with the discovery of fossil fuels (first coal, then oil, and more recently natural gas), helped to fire up the modern world.

The early seeds of global population growth were also importantly sown by the Enlightenment, with its emphasis on science, which, among many other things, started the process of bringing death under control. At the same time, World War II was a global turning point: this seemingly catastrophic world event led to the spread of death control around the globe and triggered widespread population increase (Figure 2-2).

For tens of thousands of years the population of the world grew slowly—and then, within less than 300 years, the number of people mushroomed to more than 7 billion, with the majority of that growth occurring after World War II. There can be little question why the term *population explosion* was coined to describe these historically recent demographic events. The world's population did not reach 1 billion until after the American Revolution—the United Nations fixes the year at 1804[9]—but since then we have been adding each additional billion people at an accelerating pace. As noted in the *Population and Reproductive Health* chapter, the 2 billion mark was hit in 1927, just before the Great Depression and 123 years after the first billion;

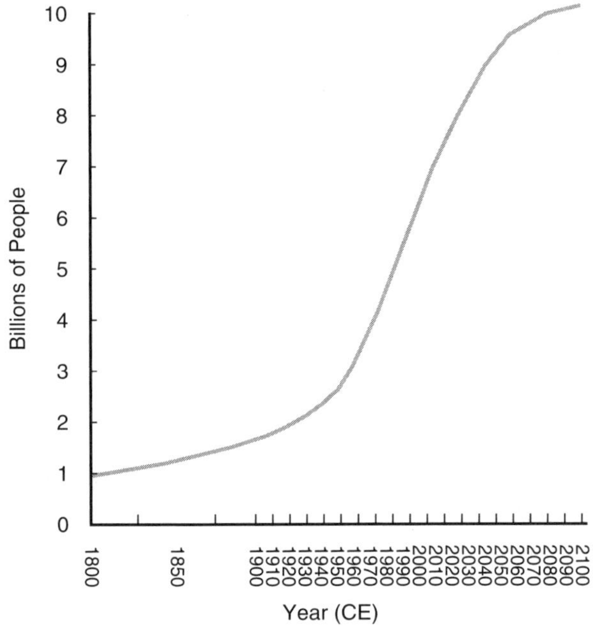

Figure 2–2 Detail of World Population Growth from 1800 to 2100

Data from the following sources: The population data from 1800 through 1900 are the average of the three different sets of estimates: Evedy C, Jones R. *Atlas of World Population History*. New York: Penguin Books; 1978; Biraben J-N. Essai sur l'évolution du nombre des hommes. *Population*. 1979;34(1):13–24; Durand JD. The modern expansion of world population. *Proc Am Philosoph Soc.* 1967;3:137–140. Numbers for 1910 through 1940 are from the United Nations Population Division. The world at six billion. 1999. http://www.un.org/esa/population/publications/sixbillion/sixbillion.htm. Population figures for 1950 through 2100 are medium variant projections from the United Nations Population Division. *World Population Prospects in World Population Policies 2009*. New York: United Nations; 2010.

3 billion in 1960 (33 years); 4 billion in 1974 (14 years); 5 billion in 1987 (13 years); 6 billion in 1999 (12 years); and 7 billion in 2011 (12 years). The United Nations Population Division projects that the world population will reach 8 billion by 2025, 9 billion before 2045, and 10 billion by 2085.

It is obviously difficult to say what will happen a half-century or more from now, because even small differences in the number of children born to women or in the death rate can create huge differences in long-range projections. Nevertheless, nearly everyone agrees that global population growth is likely to come to an end sometime late in this century. Furthermore, as discussed later in the chapter, almost all of the population increase between now and the end of this century will occur in the developing nations.

POPULATION GROWTH: RATES VERSUS NUMBERS

There is no question that the rapid rate of growth over the past 200 years has been explosive. The revolutionary consequence of that explosion is that the numbers of people are destined to stay vastly higher than they were 200 years ago, creating huge problems that have to be dealt with. If we look back to the year 1550—250 years before the world population reached 1 billion—we find that the population was about half what it was in 1800; we were clearly in the early stages of the population explosion. But if we look ahead 250 years from that point in 1800, we see there will be nine times as many people in 2050 as there were in 1800. Dealing with this dramatic rise in numbers has driven changes taking place everywhere in the world.

Regardless of the *rate* of growth (which is the explosive part), the *numbers* are what we actually cope with. Of course, the combination of rates and numbers means that a big increase in population in a short period of time is more challenging to deal with than a big increase over a longer span of time. The rate of population growth for the world peaked around 1970 and has been declining since then. This ought to be good news, but as we build on an ever larger base of human beings, the lower rates of growth are still producing very large absolute increases in the human population. When you build on a base of more than 7 billion (the current population), the seemingly slow rate of growth of about 1.1% per year for the world still translates into the annual addition of more than 78 million people, whereas "only" 71 million were added annually when the rate of growth peaked in 1970. There were scarcely more than 20 million people being added each year when the world population was last growing at about 1% per year, back in the early 1950s. Put another way, during the next 12 months, approximately 134 million babies will be born in the world, while 56 million people of all ages will die, resulting in a net addition of about 78 million people. Just in the two seconds that it took you to read that sentence, 9 babies were born while 4 people died, so the world's population increased by 5.

WHY WAS POPULATION GROWTH SO LOW FOR MOST OF HUMAN HISTORY?

The reason the population grew so slowly during the first 99% of human history was that death rates were very high, yet very few populations tried to maximize the number of children born.[15,16] During the hunting-gathering phase of human history (hundreds of thousands of years), it is likely that life expectancy at birth averaged about 20 years.[16,17] At this level of mortality, more than half of all children born will die before age 5, and the average woman who survives through the reproductive years will have to bear nearly 7 children to assure that 2 will survive to adulthood. This is a lot of children, to be sure, but only about half of what might be thought of as the biological maximum for a group of humans. Of course, given the high mortality rates, an average of 7 means that many women will have died giving birth to their first or second child, balanced at the other end by those women surviving to have 12 or 13 children.

Research in the 20th century among the last of the hunting-gathering populations in sub-Saharan Africa suggests that a premodern woman might have deliberately limited the number of children born by spacing them a few years apart to make it easier to nurse and carry her youngest child and to permit her to do her work.[18] She may have accomplished this by abstinence, abortion, or possibly even infanticide.[19,20] Hern's work among the Shipibo in the Peruvian Amazon region illustrates the desire of women in premodern societies to limit fertility, but with almost no success. Indeed, it is likely that the only truly effective means of family size limitation prior to the modern era was infanticide, which is "child control" (occurring as it does after a live birth), but not actually "fertility control."[21] The withdrawal method is mentioned in the Old Testament, but the standard joke (among demographers, at least) is that people who use this strategy, like any "natural" fertility control method, are called "parents."

It seems logical to think that the Agricultural Revolution increased growth rates among human populations as a result of people settling down in stable farming communities, where death rates were lowered. Sedentary life was assumed to have improved living conditions because of the more reliable supply of food. The idea would be that birth rates remained high but death rates declined slightly, with the end result that the population grew. However, archaeological evidence combined with studies of extant hunter-gatherer groups suggests another explanation for growth during this period of human history.[22] It is likely that fertility rates rose as new diets improved the ability of women to conceive and bear children. Frisch[23] was among the first to suggest that a certain amount of fat must be stored as energy before menstruation and ovulation can occur on a regular basis. Thus, if a woman's level of nutrition is not sufficient to permit fat accumulation, she may experience a temporary absence or suppression of menstruation and/or anovulatory cycles, in which no egg is released. For younger women, the onset of puberty may be delayed until an undernourished girl reaches a certain critical weight.[24]

A related feature of sedentary life is that it became easier to wean children from the breast earlier because of the greater availability of soft foods, which are easily eaten by babies. This would have shortened the birth intervals, and the birth rate could have risen on that account alone, and to a level higher than the death rate, thus promoting population growth. At the same time, the sedentary life and the higher-density living associated with farming probably raised death rates, rather than lowering them, by creating sanitation problems and heightening exposure to communicable diseases, at the same time that the higher birth rate would have put pressure on what were surely limited resources. Nonetheless, growth rates probably went up even in the face of higher mortality as the constraints of hunter-gatherer life were reduced and fertility rates rose to a level slightly higher than the death rate.

It should be kept in mind, of course, that only a small difference between birth and death rates is required to account for the slow growth achieved between 8000 BCE and AD 1750, when the world was adding an average of only 67,000 people each year to the population. Currently, that many people are being added every 7.5 hours.

WHAT EXPLAINS THE EXPLOSION IN NUMBERS SINCE THE 18TH CENTURY?

The acceleration in population growth after 1750 was due almost entirely to the declines in the death rate that accompanied the scientific revolution that was a significant part of the Enlightenment. First in Europe and North America, and more recently in the rest of the world, death rates have decreased sooner and much more rapidly than have fertility rates. The result has been that many fewer people die than are born each year. In the more-developed countries, declines in mortality at first were due to the effects of economic development and a rising standard of living—people were eating better, wearing warmer clothes, living in better houses, bathing more often, drinking cleaner water, and so on.[25] These improvements in the human condition helped to lower exposure to disease and build up resistance to illness.

Clean water, toilets, bathing facilities, systems of sewerage, and buildings secure from rodents and other disease-carrying animals are all public ingredients for better health. We now accept the importance of washing our hands as common sense, but the important work of Semmelweis in Vienna, Lister in Glasgow, and Pasteur in Paris in validating the germ theory actually took place only in the mid-19th century—just a heartbeat away from us in the overall timeline of human history. Public health is largely a matter of preventing the spread of disease, and these kinds of measures have been critical in the worldwide decline in mortality (see Box 2–1). The medical model of curing disease gets much more attention in the modern world, but its usefulness is predicated on the underlying foundation of good public health.[26] Cutler and Miller[27] point to the particularly important role played by the introduction of clean water technology (chlorination and filtration) in cities of the United States in the late 19th and early 20th centuries, the time period when life expectancy made its single biggest jump in U.S. history. This was, of course, a direct application of the germ theory.

As already noted, declines in death rates first occurred in those regions of the world experiencing economic development in the 19th and early 20th centuries, primarily Europe and North America. Fertility also began to decline in these parts of the world within at least one or two generations after their death rates began its drop. However, since World War II, medical and public health technology has been available to virtually all

---

Box 2–1   Key Elements in Postponing Death Among Humans

19th Century

- Improved nutrition (occurred first in Western Europe)
- Clean water (Snow in London)
- Sewerage in cities (sanitation studies in Liverpool)
- Smallpox vaccinations (Jenner in England)
- Validation of germ theory (Semmelweis in Vienna, Lister in Glasgow, Pasteur in Paris)

20th Century

- Health as a social movement
- Antibiotics
- More vaccinations
- Oral rehydration therapy for infants
- Advanced diagnoses, drugs, and other treatments for degenerative diseases to keep older people alive longer

countries of the world regardless of their level of economic development. In the less-developed countries, although the risk of death has been lowered dramatically, birth rates have gone down less quickly, and the result is continuing population growth. For at least 200 years, from 1750 to 1950, the rich countries were the places where population was growing the fastest. Since then, the pattern has reversed and it is now the poorer countries whose populations are growing most quickly.

MIGRATION AND POPULATION CHANGE

As populations have grown disproportionately in different areas of the world, the pressures or desires to migrate have also grown. Migration streams generally flow from areas where there are too few jobs to areas where there is a greater availability of jobs—that is, from poorer to richer economies. Thus, especially since the end of World War II, we have seen migration from Latin America and Asia to the United States; from Asia and Latin America to Canada; from Africa, Latin America, and Asia to Europe; and within Europe, from the east to the west.

In earlier decades, a shortage of jobs generally occurred when the population grew dense in a particular region, and people then felt pressured to migrate to some other less populated area. This pattern of migration characterized the expansion of European populations into other parts of the world, especially North America, as European farmers sought land in less densely settled areas. This phenomenon of European expansion is, of course, critically important because as Europeans moved around the world, either as settlers or conquerors, they altered patterns of life, including their own, wherever they went.

European Expansion

Beginning in the 14th century, migration out of Europe started gaining momentum, and this movement virtually revolutionized the entire human population. With their gun-laden sailboats, Europeans began to stake out the less-developed areas of the world in the 15th and 16th centuries, with the English and French settling North America and the Spanish and Portuguese exploiting (more than settling) Central and South America. Migration of Europeans to other parts of the world on a massive scale took hold in the 19th century, when the European nations began to industrialize and swell in numbers due to the decline in mortality. At the same time (and for the same reasons), the invention of steam-powered ships made ocean travel faster and safer, thus facilitating long-distance migration.

Before the great expansion of European people and culture, Europeans represented about 18% of the world's population, with almost 90% of these people living in Europe itself. By the 1930s, at the peak of European dominance in the world, people of European origin in Europe, North America, and Oceania accounted for 35% of the world's population.[8] By the beginning of the 21st century, this share had declined to 16%, and it is projected to drop to 13% by the middle of this century. However, even that may be a bit of an exaggeration, since the rate of growth in North American and European countries is increasingly influenced by immigrants and births to immigrants from developing nations.

South-to-North Migration

Since the 1930s, the outward expansion of Europeans has ceased. Until that time, European populations had been growing more rapidly than the populations in Africa, Asia, and Latin America, but since World War II that trend has been reversed. The less-developed areas now have by far the most rapidly growing populations. It has been said that "population growth used to be a reward for doing well; now it's a scourge for doing badly."[28] This change in the pattern of population growth has resulted in a shift in the direction of migration. For the past half-century, there has been far more migration from less-developed countries (the "South") to developed areas (the "North") than the reverse. Furthermore, since migrants from less-developed areas generally have higher levels of fertility than natives of the developed regions, their migration makes a disproportionate contribution over time to the overall population increase in the developed areas to which they have migrated. As a result, the proportion of the population whose origin is one of the modern world's less-developed nations tends to be on the rise in nearly every developed country. Within the United States, for example, non-Latino whites (the European-origin population) are no longer the majority in the state of California, and it is likely that Latinos (largely of Mexican ancestry) will represent the majority of Californians by the middle of this century given that the majority of all births in California (as in all southwestern states) are now to Latina mothers.

When Europeans migrated, they generally filled up territory that had very few people, because they tended to move to land used by hunter-gatherers who, as noted earlier, use land extensively rather than intensively. Those seemingly empty lands or frontiers have essentially disappeared today, and as a consequence, migration into a country now results in more noticeable increases in population density. Moreover, just as the migration of Europeans was typically greeted with violence from the indigenous populations upon whose lands they were encroaching, migrants today routinely meet with prejudice, discrimination, and violence in the places to which they have moved, even as their labor is valued by the host community. An important difference is that today's migrants tend to settle in cities and suburbs, rather than the countryside.

The Urban Revolution

Until very recently in world history, almost everyone lived in basically rural areas. Large cities were few and far between. For example, Rome's population of 650,000 in AD 100 was probably the largest in the ancient world.[29] It is estimated that as recently as 1800, less than 1% of the world's population lived in cities of 100,000 or more. Nearly half of all humans now live in cities of that size.

The redistribution of people from rural to urban areas occurred earliest and most markedly in the now developed nations, and this pattern was closely linked to industrialization. In 1800 about 10% of the English population lived in urban areas, primarily London. Two hundred years later, 90% of the British lived in cities. Similar patterns of urbanization have been experienced in other European countries, the United States, Canada, and Japan as they have industrialized. In the less-developed areas of the world, urbanization was initially associated with a commercial response to industrialization in Europe, America, and Japan. In other words, in many areas where industrialization was not occurring, Europeans had established colonies or trade relationships. The principal economic activities in these areas were not industrial but commercial in nature, associated with buying and selling. The wealth acquired by people engaged in these activities naturally attracted attention, and urban centers sprang up all over the world as Europeans sought populations to whom they could sell their goods, often in exchange for raw materials to keep industrialization going.

During the second half of the 20th century, when the world began to urbanize in earnest, the underlying cause was the rapid growth of the rural population. The rural population in every less-developed nation has outstripped the ability of the agricultural economy to absorb it. Paradoxically, to grow enough food for an increasing population, people have been replaced by machines in agriculture—a trend has sent the now jobless rural youth off to the cities in search of work.

DEMOGRAPHIC DIVIDE

Ever since the Enlightenment and its aftermath, the West has been different demographically and economically than the rest of the world.[30] These differences are now narrowing in many respects, but we are still coping with a world in which the haves and the have-nots—that is, developed and developing countries—are divided not only by income and rates of population growth, but also by distinctly different population structures. The legacy of several decades of low fertility and low mortality in the developed countries is an increasingly older population, whereas a much later start to the fertility decline, combined with ever lower mortality, keeps the populations of developing areas (countries "in transition," as they are sometimes called) with significantly younger population age structures. As these younger people grow up and look around for jobs, the response of the world to their situation will tell the tale of what the future will be.

In the less-developed nations, the population continues to grow quickly, especially in absolute terms. In sub-Saharan Africa, this growth is happening even in the face of the HIV/AIDS pandemic. In contrast, in the more-developed countries, population growth has slowed, stopped, or in some places even started to decline. As we look around the world, we see that the more rapidly growing countries tend to have high proportions of people who are young, poor, prone to disease, and susceptible to political instability. The countries that are growing slowly or not at all tend to have populations that are older, richer, and healthier; in addition, these nations are politically more stable and are calling the shots in the world right now. However, keep in mind that there is almost certainly something to the idea that "demography is destiny": a country cannot readily escape the demographic changes put into motion by the universally sought-after decline in mortality. Each country has to learn how to read its own demographic situation, and cope as well as it can with the inevitable changes that will take place as it evolves through all phases of the demographic transition. Too many people look to the past and wish they could revive those earlier demographic structures, rather than realizing that the "past" was actually a period of transition that would inevitably lead to a different future.

HOW IS THE POPULATION CURRENTLY DISTRIBUTED IN THE WORLD?

The 20 largest countries in the world currently account for nearly three-fourths (71% as of the year 2010) of the world's population, but only 38% of the world's land surface. The top 5 countries in terms of total population include

China, India, the United States, Indonesia, and Brazil (Table 2–1; Figure 2–3). Rounding out the top 10 are Pakistan, Bangladesh, Nigeria, Russia, and Japan. Mexico heads up the next 10, followed by the Philippines, Vietnam, Ethiopia, Germany, Egypt, Turkey, Congo (Kinshasa), Iran, and Thailand. The remaining 29% of the world's population is spread out among more than 200 other countries that account for the remaining 62% of the earth's terrain.

China and India (or more technically the Indian subcontinent, including the modern nations of India, Pakistan, and Bangladesh) were already the

Table 2–1 Population and Land Surface of the 20 Most Populous Nations, 2010

Country	Population 2010	Land Surface (km^2)
China	1,330,141,295	9,425,846
India	1,173,108,018	3,178,010
United States	310,232,863	9,489,567
Indonesia	242,968,342	1,908,515
Brazil	201,103,330	8,520,377
Pakistan	177,276,594	803,188
Bangladesh	158,065,841	141,217
Nigeria	152,217,341	910,614
Russian Federation	139,390,205	16,631,081
Japan	126,804,433	380,746
Mexico	112,468,855	1,952,898
Philippines	99,900,177	308,612
Vietnam	89,571,130	328,026
Ethiopia	88,013,491	1,259,921
Germany	82,282,988	362,301
Egypt	80,471,869	989,725
Turkey	77,804,122	778,859
Congo (Kinshasa)	70,916,439	2,344,649
Iran	67,037,517	1,625,528
Thailand	66,404,688	516,378
Top 20	4,846,179,538	61,856,058
Total as of 2010	6,851,673,664	164,662,261
Top 20 as Percentage of Total	71	38

Data from the United Nations Population Division. *World population prospects 2010.* 2011. http://esa.un.org/unpd/wpp/index.htm. Accessed April 28, 2014.

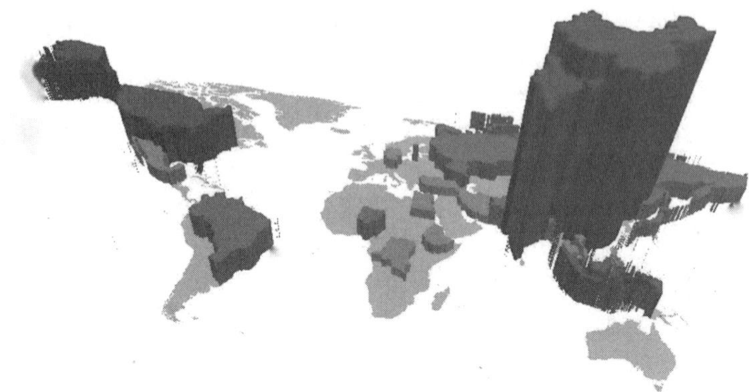

Figure 2–3 Top Twenty Countries in 2010 Mapped in Comparative Population Size

Data from the United Nations Population Division. World population prospects 2010. 2011.
http://esa.un.org/unpd/wpp/index.htm.

most populous places on earth in the year 1500, with Asia accounting for 53% of the world's 461 million people in that year. Five centuries later, the population in Asian countries accounts for 61% of all the people on earth, although this share is projected to shrink to 58% by the year 2050, as China's growth halts while Africa surges ahead.

Sub-Saharan Africa had about as many people as did Europe in 1500. However, contact with Europeans tended to be deadly for Africans (as it was for the native peoples in North and South America) because of disease, violence, and slavery. In the 20th century, sub-Saharan Africa rebounded in population size, to the point that it account for 10% of the total world population in 2000. In recent decades, high mortality from HIV/AIDS has slowed the rate of population growth in sub-Saharan Africa, but the United Nations projects that because of the continued above-replacement-level fertility levels in the region, sub-Saharan Africa could account for more than 18% of the world's population in 2050—even beyond where it had been in percentage terms in the year 1500, and with more than twice as many people as are projected to reside in Europe in 2050.

Before the Great Depression of the 1930s, the populations of Europe and, especially, North America were the most rapidly growing in the world. During the decade of the 1930s, growth rates declined in those two areas until they approximated the rates found in most of the rest of the world. Since the end of World War II, the situation has changed again: now Europe is on the verge of depopulation, while rapid growth in the less-developed countries of Africa, Asia, and (to lesser extent) Latin America is now responsible for almost all of the world's population increase.

PATTERNS OF POPULATION GROWTH

The world's population is currently growing at a rate of 1.1% annually—78 million people per year, as noted earlier, albeit with a lot of variability underlying those global numbers. We expect Europe, as a region, and Japan, as a nation, to have fewer people in 2050 than they have now. Populations in all other areas of the world will continue to grow in size or stop growing, but are not expected to start declining.

Are We Headed for a Population "Implosion"?

An implosion is something that collapses into itself—the opposite of an explosion. As the rate of population growth has slowed down over the past three decades, there has been talk of a population implosion,[31,32] implying that "the world is in for some rapid downsizing."[33(p22)] Although the world's population is in no danger of imploding anytime soon, the same cannot be said for the populations in much of Europe and East Asia. Several countries in these areas are either already declining in population or on the verge of doing so. The populations in Europe and East Asia all have birth rates that are below replacement level and have been that way for some time now, leading to a declining number of people in the younger age cohorts. It appears that the low fertility in countries like Russia is not just a temporary phenomenon. Rather, the evidence suggests that the motivation to have large families has disappeared and has been replaced by a propensity to try to improve the family's standard of living by limiting the number of children.[34,35] The situation in Russia is further complicated by the loss of a few years of life expectancy over the past couple of decades, which has further accelerated its population implosion. Although Russia's decline in population is moderated by consistent immigration from Asia (especially the Central Asian republics that were part of the former Soviet Union), most other Eastern European nations add to low fertility the demographic complication that people are leaving to go elsewhere—primarily to Western Europe, but also to North America.

According to data from the United Nations Population Division, 16 countries in 2010 had fewer people than they did in 2000. All 16 of these countries were in Eastern Europe; they were led by Russia and several former members of the Soviet Union, including Ukraine, Belarus, Georgia, Kazakhstan, the Republic of Moldova, Lithuania, Latvia, and Estonia. It is probably safe to say that the former Soviet Union has imploded.

The more controversial issue has been the aging of the populations in Europe and East Asia as a consequence of the low birth rate. A number of countries are expected to have at least 20% of their populations be age 65 or older in 2025 while at the same time having less than 20% of their populations be younger than age 15. These include some of the biggest and most dynamic economies in the world, but that status is threatened by the fact that the proportion of the population that is 65 and older is growing rapidly, whereas the younger population is shrinking.

One reaction to this situation is to suggest that it is a good thing for the planet as a whole, if not necessarily for Europeans. Residents of these countries are among the highest per-person consumers of the earth's resources; if these populations eventually decline in size, their impact on the environment will be lower, thereby lessening the chance of global environmental and economic collapse.[36] Within most of these countries, however, there is a concern about the economic impact of what many people call a "silver tsunami." Who will earn the money that is to be paid to retirees as pensions? Who will pay for the healthcare and social needs of the elderly? Who will keep the economy going so that the standard of living does not drop even as the expenses associated with population aging increase?

Among the proposed solutions are (1) trying to raise the birth rate, (2) trying to increase labor force participation among older persons, and (3) replacing the "missing" population with immigrants. With respect to the birth rate solution, those countries with the lowest fertility rates turn out to be those in which the least accommodation has been made to permit women to simultaneously have a job and a family. The availability of daycare programs, maternity leave, and family leave, combined with societal pressure for men to help with childrearing and housework, tend to increase the ability of women to participate in the labor force and still have children. Men have obviously always had that ability, but many countries, especially in Southern and Eastern Europe and East Asia, have opened up the labor market to women without making it easy to combine a woman's participation in the labor force with a family. The result has been birth rates depressed below what they might otherwise be. Researchers have also noted that the effect of a low birth rate would be a little less severe if women simply had children at a younger age, even if they had the same number of offspring as they are currently having.[37] This would shorten the time between generations and would actually increase the growth rate by a slight amount.

The impact of an aging population on a nation's economy is exacerbated by the preference for retirement at an early age. For most of human history, people simply worked until they were physically no longer able to do so. Retirement has been widely available in the rich countries of the world only for the past half-century or so, but ever since that option was offered, people have been grabbing it—people prefer retirement to work (no surprise!).

Thus, we have witnessed the situation in which even as life expectancy has increased, people have been choosing to retire earlier. This trend would not be a problem if all of these people had actually saved up enough money to live comfortably during a protracted retirement, but this is largely not the case. For the most part, people have been promised a retirement pension that is based on the transfer of money from people currently in the labor force (through taxation) to people who are retired (the "pay-as-you-go" or PAYGO scheme). As long as the population was growing and the economy was improving, these promises were easy to keep (almost like a Ponzi scheme); when these very same people who now want to collect a pension have not had enough children to supply the needs of the labor force, however, there is a problem. One solution being promoted is to raise the age at retirement,[38,39] perhaps to as high as 75.[40] Encouraging older people to work longer means that they will continue to pay taxes to fund the pensions of those who are retired. At the same time, these older workers would not be burdening the system with their own pension demands. Vaupel and Loichinger[39] have even suggested that if older people stayed in the work force longer, the number of hours worked by younger people could be reduced somewhat, which might in turn encourage a rise in the birth rate.

The short-term solution to labor shortages in the world has always been to import labor. This pattern explains the history of slavery in the Americas, and then the history of waves of immigrants to the United States from England, Germany, Italy, Mexico, and elsewhere. It also describes the history of England, Germany, France, and several other European countries that needed labor to rebuild their economies after World War II. Between 1945 and the early 1970s, European nations allowed migration from former colonies, and they instituted guest worker programs, in which people contract to work for a few years and then go home again. The rub is that many workers choose not to go home. They stay, build families, and become part of the fabric of their adopted society. If workers came for a while, worked, and then left as they got older and were replaced by younger people, immigration would not be a particularly thorny issue. The Gulf States in the Middle East have managed to accomplish this outcome largely by prohibiting workers from having families with them, and by forcing the deportation of workers who overstay their contracts.[41]

Europeans have rarely been willing to take those extreme measures, so guest workers are likely to stay past the end of their contract to become undocumented immigrants. The reality, then, is that replacement migration in Europe means the immigration of not just workers but also their families. Within a generation or two, the children of these immigrants can become a major force in the demographic makeup of the receiving countries. In the meantime, many of these immigrants are not in the labor force, so they are not working and paying taxes—thus they are not exactly "replacing" the older population.[42]

France and the United Kingdom have both taken in significant numbers of permanent immigrants from former colonies and, as a result, neither country is projected to experience a decline in population over the next several decades. But the fact that an estimated 10% of France's population is now Muslim has created a variety of political and social dilemmas for that country. Studies conducted by the Pew Research Center have shown that immigration is strongly opposed throughout Europe, but paradoxically (and perhaps more positively, in terms of future policies) there is more support for it among the young than among the old.[43]

Japan, like other Asian countries, has an extremely restrictive immigration policy because of an explicit desire to preserve the country's ethnic homogeneity. Although Japan does tolerate a small number of immigrants, it is unlikely that the country will soon allow an invasion of workers to prevent its impending population implosion, despite concerns that the declining native population may significantly harm the nation's economy.[44] Japan's low birth rate is slightly counterbalanced by the extreme longevity of the Japanese—they have the second highest life expectancy in the world (only recently overtaken by Hong Kong)—but that translates into one on four Japanese being age 65 or older—the highest percentage in the world.

WHAT DOES THE FUTURE HOLD DEMOGRAPHICALLY?

Now, and for the foreseeable future, almost all of the growth of the world's population is originating in less-developed nations. The term *originating* is an apt one because some of that growth then spills into the more-developed countries (especially the United States and Canada) through migration. Even more significantly, almost all of that growth is expected to show up in cities of developing nations.

Figure 2–4 maps the countries of the world based on the projected increase in total population *size* between 2010 and 2050, according to projections from the United Nations Population Division, with the underlying data shown in Table 2–2. You can see clearly that India is projected to be the single most important source of population increase between now and the middle of the 21st century. It takes the prize for being on a trajectory to add nearly half a billion people to its current population. India's death rate has been steadily declining at a faster rate than its birth rate has been declining. The average woman in India is

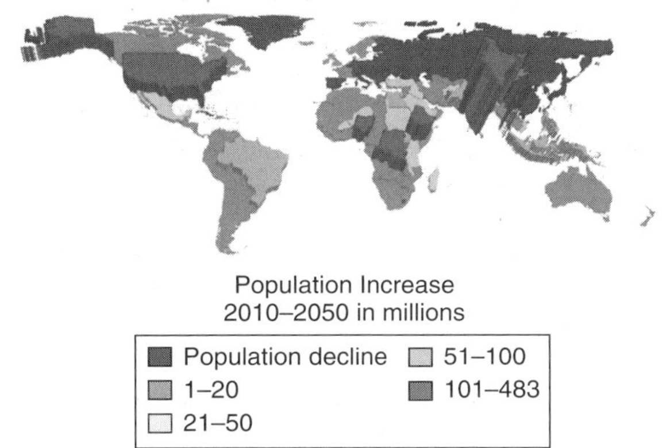

Figure 2–4 Countries According to Population Growth

Data from the United Nations Population Division. World population prospects 2010. 2011. http://esa.un.org/unpd/wpp/index.htm.

Table 2–2 Top 20 Countries of the World in Terms of Projected Population Growth Between 2010 and 2050

Country	Projected Population Increase, 2010–2050
India	483,445,614
Ethiopia	190,269,646
United States	128,777,390
Congo (Kinshasa)	118,394,410
Nigeria	112,045,064
Pakistan	99,152,164
Uganda	94,608,832
Bangladesh	75,521,438
Philippines	72,064,010
Indonesia	70,052,505
Brazil	59,589,163
Egypt	57,400,653
Sudan	46,247,579
Niger	39,426,178
Mexico	35,438,795
Madagascar	35,231,983
Burkina Faso	31,187,698

(Continues)

Table 2–2 Top 20 Countries of the World in Terms of Projected Population Growth Between 2010 and 2050 (*Continued*)

Country	Projected Population Increase, 2010–2050
Iraq	26,644,724
Kenya	25,129,298
Tanzania	24,950,417
Total for Top 20	1,825,577,561
Percentage of World Increase	76.7

Data from the United Nations Population Division. World population prospects 2010. 2011. http://esa.un.org/unpd/wpp/index.htm. Accessed April 28, 2014.

having 2.5 children—far fewer than just a few decades ago, but now most of those children are surviving to adulthood. Given that World Bank data[45] show that more than two out of every three residents of India currently live on less than $2 per day, you can appreciate the struggle that will be part of India's future. It will share that struggle with its subcontinental neighbors of Pakistan and Bangladesh, whose own populations will collectively grow by about 175 million people through 2050. The southeast Asian nations of Indonesia and the Philippines are both expected to add tens of millions more people to the world population. Sub-Saharan Africa is projected to contribute another half billion people in just four countries alone, led by Ethiopia, Congo (Kinshasa), Nigeria, and Uganda. The countries in western Asia—regionally between sub-Saharan Africa and South Asia—are also projected to grow by tens of millions of people.

The United States is in the top 10 list of countries in terms of expected growth between now and 2050, with its total population growing from 310 million in 2010 to almost 400 million by 2050. In 2009, the U.S. Census Bureau made projections assuming zero net international migration; it resulted in a very different scenario characterized by an increase to only 322 million by 2050, if there were no more immigrants after 2010. Thus, if the U.S. population really does approach 400 million by 2050, that growth will have been fueled almost entirely by immigrants and their children. On these grounds alone, it is understandable that immigration is both a hot topic and an issue that cannot easily be resolved. Brazil and Mexico are expected to be other high-demographic-growth areas in the Americas, even when taking into account the outmigration from those countries to the United States and elsewhere.

Conspicuously absent from future growth is China. This country experienced a huge increase in its population (750 million people) after the end of World War II, but it is expected to soon reach its goal of halting population growth. Finally, if we consider rates of growth instead of sheer population size, sub-Saharan Africa, the Middle East, and Western Asia jump out as the places that are anticipated to be growing most quickly in the world—places where local communities will certainly be under the greatest pressure to cope with their burgeoning populations. And, of course, it is inconceivable that rapid growth in these places will not forcefully impact the richer, slower (or non-) growing nations as well.

CONCLUSION

For most of human history, the world population grew very slowly (indeed, sometimes not at all, and sometimes it even declined) as a result of universally high death rates. To maintain themselves, humans created a variety of social institutions to ensure the continuation of the group, including early and universal marriage for girls, and pressure on women to conceive and bear children. These pronatalist views are reified in the texts of all major religions of the world. In the face of historically high death rates, the average woman had to bear 5 to 7 live children just to have 2 who would survive to adulthood. Since getting pregnant is one of the most life-threatening things a young woman can do, especially anytime prior to the past few decades, you can appreciate why some strong motivation might be required to repeatedly bear children.

That situation changed with the Enlightenment and the scientific revolution, which formed a core part of thinking about the world that emerged in Europe and North America scarcely more than 200 years ago. Our new ability to control disease and postpone death to ever older ages was the most transformative thing that ever happened to humans. It helped to change every major aspect of human existence, leading eventually to lower fertility and the resulting freedom of women to be more than reproductive machines, while creating massive changes in the size and age structure of populations, and leading to migration and especially urbanization, which is associated with new ways of organizing everyday life around these new demographic realities.

Although we are on a trajectory to eventually level off in terms of global population size, we still have a long way to go, given our current expectations about future population growth. Between now and the end of the 21st century, the world population is likely to grow by an additional 3 billion people. At the beginning of the 22nd century, the world's population will be 10 times greater than it was just 300 years earlier. Furthermore, the majority of people will live in cities of developing countries, rather than in the countryside. The future will indeed be a foreign country.

DISCUSSION QUESTIONS

1. Why did the human population grow so slowly over most of its history?
2. Which is more important: the rate of growth or the increasing numbers of people? Explain your answer.
3. Is population implosion a positive development? Compare its effects to those of population explosion.
4. Discuss the importance of urbanization for the future of world population.
5. What is the "demographic divide" and why is it significant?

REFERENCES

1. Cann RL, Wilson AC. The recent African genesis of humans. *Sci Am Spec Ed.* 2003;13(2):54–61.
2. Coale A. The history of the human population. *Sci Am.* 1974;231(3):41–51.
3. Evedy C, Jones R. *Atlas of World Population History.* New York: Penguin Books; 1978.
4. Biraben J-N. Essai sur l'évolution du nombre des hommes. *Population.* 1979;34(1):13–24.
5. Mann CC. *1491: New Revelations of the Americas Before Columbus.* 2nd ed. New York: Vintage Books; 2011.
6. Cantor NF. *In the Wake of the Plague: The Black Death and the World It Made.* New York: Free Press; 2001.
7. Fagan B. *The Little Ice Age: How Climate Made History 1300–1850.* New York: Basic Books; 2000.
8. Durand JD. The modern expansion of world population. *Proc Am Philosoph Soc.* 1967;3:137–140.
9. United Nations Population Division. The world at six billion. 1999. http://www.un.org/esa/population/publications/sixbillion/sixbillion.htm.
10. United Nations Population Division. *World Population Policies 2009.* New York: United Nations; 2010.
11. Boserup E. *The Conditions of Agricultural Growth.* Chicago: Aldine; 1965.
12. Boserup E. *Population and Technological Change: A Study of Long-Term Trends.* Chicago: University of Chicago Press; 1981.
13. Boserup E. *Woman's Role in Economic Development.* New York: St. Martin's Press; 1970.
14. Harrison P. *The Third Revolution: Population, Environment and a Sustainable World.* London: Penguin Books; 1993.
15. Abernethy V. *Population Pressure and Cultural Adjustment.* New York: Human Sciences Press; 1979.
16. Livi-Bacci M. *A Concise History of World Population.* 3rd ed. Malden, MA: Blackwell; 2001.

17. Petersen W. A demographer's view of prehistoric demography. *Curr Anthrop.* 1975;16:227–246.

18. Dumond D. The limitation of human population: a natural history. *Science.* 1975;232:713–720.

19. Howell N. *Demography of the Dobe Kung.* New York: Academic Press; 1979.

20. Lee RB. Population growth and the beginnings of sedentary life among the Kung Bushmen. In: Spooner B, ed. *Population Growth: Anthropological Implications.* Cambridge, MA: MIT Press; 1972:329–342.

21. Skinner GW. Family systems and demographic processes. In: Kertzer D, Fricke T, eds. *Anthropological Demography: Toward a New Synthesis.* Chicago: University of Chicago Press; 1997:53–95.

22. Spooner B, ed. *Population Growth: Anthropological Implications.* Cambridge, MA: MIT Press; 1972.

23. Frisch RE. *Female Fertility and the Body Fat Connection.* Chicago: University of Chicago Press; 2002.

24. Komlos J. The age at menarche in Vienna: the relationship between nutrition and fertility. *Hist Meth.* 1989;22:158–163.

25. McKeown T. *The Modern Rise of Population.* London: Edward Arnold; 1976.

26. Meade MS, Earickson RJ. *Medical Geography.* 2nd ed. New York: Guilford Press; 2000.

27. Cutler D, Miller G. The role of public health improvements in health advances: the twentieth-century United States. *Demography.* 2005;42(1):1–22.

28. Blake J. Personal communication, 1979.

29. Chandler T, Fox G. *3000 Years of Urban Growth.* New York: Academic Press; 1974.

30. Ferguson N. *Civilization: The West and the Rest.* New York: Penguin Books; 2011.

31. Eberstadt N. The population implosion. *Foreign Policy.* March–April 2001, 42–53.

32. Wattenberg BJ. *Fewer: How the New Demography of Depopulation Will Shape Our Future.* Chicago: Ivan R. Dee; 2004.

33. Singer M. The population surprise. *Atlantic Mthly.* August 1999:22–25.

34. Avdeev A, Monnier A. A survey of modern Russian fertility. *Population: An English Selection.* 1995;7:1–38.

35. Kholer H-P, Billari FC, Ortega JA. Low fertility in Europe: causes, implications, and policy options. In: Harris FR, ed. *The Baby Bust: Who Will Do the Work? Who Will Pay the Taxes?* Lanham, MD: Rowman & Littlefield; 2006.

36. Diamond J. *Collapse: How Societies Choose to Fail or Succeed.* New York: Viking; 2005.

37. Lutz W, O'Neill BC, Scherbov S. Europe's population at a turning point. *Science.* 2003;299:1991–1992.

38. Moffett S. Fast-aging Japan keeps its elders on the job longer. *Wall Street Journal.* June 15, 2005: A9.

39. Vaupel JW, Loichinger E. Redistributing work in aging Europe. *Science.* 2006;312(5782):1911–1913.

40. United Nations Population Division. *Replacement Migration: Is It a Solution ot Declining and Ageing Populations?* New York: United Nations; 2000.

41. Castles S, Miller MJ. *The Age of Migration.* 4th ed. New York: Guilford Press; 2009.

42. Bijak J, Kupiszewska D, Kupiszewska M. Replacement migration revisited: simulatons of the effects of selected population and labor market strategies for the aging Europe, 2002–2052. *Pop Res Policy Rev.* 2008;27:321–342.

43. Pew Research Center. A global generation gap. 2004. www.pewglobal.org/2004/02/24/a-global-generation-gap/, 2010.

44. Moffett S. Abe has remedy to spur Japan: open up the economy. *Wall Street Journal.* November 2, 2006:A4.

45. World Bank. World development indicators. 2013. http://data.worldbank.org/data-catalog/world-development-indicators.

Part

II

MEASURES AND THEORIES

Chapter 3

Measuring Populations: Mortality, Fertility, and Migration

Erin R. Hamilton

INTRODUCTION

Measurement is the basic and essential task of population science, or *demography*. Demography is defined by its *population perspective*, or the description of characteristics of and relationships in the population at large.[1] *Populations* are aggregates, either the sum total of people living in a specific place at a specific point in time or, more commonly, the collectivity of people that persists over time in a place even though its specific members are changing. Description of the size, change, and structure of populations provides basic facts about society that are needed for a variety of audiences—for public health practitioners and social workers, social scientists and policy makers, politicians and social commentators, and the public at large.

This chapter provides a brief introduction to the measurement of populations. It covers the basic tools of demographic study, including the data that demographers use and the measures that demographers construct to describe the size, change, and structure of populations, as well as the three components of demographic change: fertility, mortality, and migration. In reading this chapter, you will become familiar with how basic measures of demographic processes are constructed, thereby enabling you to use and interpret these measures in your own work and to understand the empirical bases for a wide range of analyses of the social world.

POPULATION DATA

Demographic measurement begins with data, and for demographers to describe populations they require *population data*. The defining characteristic of population data is that the data include or represent an entire population. Population data come from sources that enumerate or count events for the entire population, such as censuses, registration systems, and administrative data, or sources that are representative of the population, such as samples drawn through probability sampling. Community or clinic samples, convenience samples, and theoretical samples—all techniques that are common to social science data collection and useful for certain purposes—are not considered population data because they do not represent a larger population. Demographers tend not to use these sources of data, at least not for the purpose of demographic measurement, because estimates constructed from them cannot be assumed to represent the entire population.

Most demographers do not collect population data themselves—it is too costly. Instead, they generally use data collected by government agencies or well-funded research institutions, a practice called *secondary data analysis*. Because of the high cost of collecting population data, and the fact that much funding for the collection of population data comes from public sources (e.g., the federal government), most sources of population data are available for public use online within a few years of the data collection. To comply with rules established to protect human subjects in research, key identifying information about individuals is typically omitted from publicly released data sets to protect the confidentiality of research participants. In other words, information like names, Social Security numbers, addresses, and other sensitive details are omitted from publicly available data sets. In some cases, permission is required for the use of public data, and to obtain permission users must comply with a series of security measures, such as storing the data in encrypted files on nonnetworked computers. Box 3–1 lists some of the major outlets for obtaining population data online, and the rest of this section describes the major sources of demographic data—censuses, registration and administrative data, and survey data—in more detail.

Census Data

Population censuses are complete enumerations, or counts, of all individuals living in a certain place at a certain time. They are usually administered at regular intervals, with every 10 years being standard. Census data are typically collected by governments for statutory purposes. In the United States, the census was mandated by the Constitution for the purpose of

---

Box 3-1 Online Sources of Population Data

U.S. Census data and data from the *American Communities Survey* are available directly from the U.S. Census Bureau at http://www.census.gov. The Census Bureau also has an interactive tool to create tables and maps using census data at http://www.factfinder2.census.gov.

U.S. and international census data, as well as data from the *American Communities Survey* and the *Current Population Survey*, are available from the Integrated Public Use Microdata Series run by the University of Minnesota at http://www.ipums.org.

U.S. vital statistics data as well as a variety of *health surveys*, including the *National Survey of Family Growth*, are available from the National Center for Health Statistics at http://www.cdc.gov/nchs/.

Demographic and Health Surveys, which substitute for census and vital registration data in many underdeveloped countries, are available at http://www.measuredhs.com/.

A vast database of *survey data sets* are available from the Inter-university Consortium for Political and Social Research at the University of Michigan at http://www.icpsr.umich.edu.

International demographic data in tabulated form are available from a variety of sources, including the U.S. Census Bureau at http://www.census.gov/population/international/data/, the Population Reference Bureau at http://www.prb.org/DataFinder.aspx, and the World Bank at http://data.worldbank.org/.

---

apportioning taxes and seats in the House of Representatives among the states. However, census data serve an additional purpose, which is to provide information about the number, distribution, and characteristics of a population. They also provide essential information necessary for the construction of demographic measures—that is, counts of the population at risk of mortality, fertility, and migration.

In the United States, census forms are sent by mail to every household, and census enumerators follow up in person with households that do not return their census form as well as with populations who do not have regular households, such as the homeless. Figure 3-1 shows a 1940 poster encouraging people to participate in the census that year. In theory, censuses count everyone who is part of a population at certain point in time, but in practice censuses fall short of achieving this difficult goal. To start, census agencies must decide who to count. "De jure" enumeration counts everyone who is legally resident in a place on census day, whereas "de facto" enumeration

Figure 3–1 A Poster Encourages People to Participate in the 1940 Census

Courtesy of the Library of Congress, LC-USZC4-1801

counts everyone who is actually present, regardless of whether they are legally resident in the place. In the United States, as in many other countries, the practice falls in between these two standards, counting people who are "usually resident" on census day, which in the United States is April 1. This counting standard excludes tourists and business visitors, but includes undocumented immigrants as well as military and diplomatic personnel serving overseas.

Although in theory people such as undocumented immigrants are counted by the usual residence standard, in practice they may not be. *Under-enumeration* occurs when people refuse to participate in the census or when the census misses people by accident. Some populations, such as undocumented immigrants who may be reluctant to fill out government forms, are at especially high risk of not being counted. By contrast, *over-enumeration* occurs when people are counted more than once. For example, college students may fill out a census form at their university dorm, and their parents may also report them at their home; in their case, confusion about where they are usually resident leads to over-enumeration. The U.S. Census Bureau has developed a variety of techniques to estimate the degree of census enumeration error, both over- and under-enumeration, and to adjust its population estimates accordingly. The broader point is to recognize that while census data are typically the best (and only) enumeration of the entire population, they nevertheless contain errors that can affect the validity or generalizability of estimates based on them. The same is true of all data sources. Secondary data users should investigate and be cognizant of the methods of data collection and the errors introduced by those methods to assess potential bias or other problems presented to the analysis by the data itself.

In the United States, the decennial census form that is sent to all households collects only a small amount of information. In 2010, it asked about the type of residence, and the age, sex, race/ethnicity, and relationships of all people living in the household. The vast majority of data collected by the U.S. Census Bureau is, in fact, collected through surveys; previously, a sample of about one-sixth of all U.S. households received the census "long-form" questionnaire in addition to the basic census form. After 2005, the Census Bureau implemented the American Communities Survey (ACS) as a replacement to the decennial census long-form. The ACS is a survey administered on an ongoing basis to a sample of the population, and it asks about a broad range of demographic, economic, and social characteristics.

Measurement error is another form of error that is common to all forms of social data collection. Measurement error refers to whether questions on a census form or survey reliably and validly measure the characteristic or behavior they intend to. In terms of the U.S. census, there has historically been a great deal of measurement error when it comes to measuring race and ethnicity,[2] and most observers agree that the racial/ethnic categories on the 2010 census form still fall far short of adequately capturing reality. This in part reflects the complexity of lived race and the inherent challenge of simplifying that complexity into categories on a questionnaire; in addition, there are political as well as seemingly arbitrary decisions underlying the racial/ethnic categories on the census form.[2] Perhaps most notably, the questions separate Hispanic ethnicity from race, even though other ethnicities—such as Asian ethnicities—are included as racial categories. There is no category for Arab Americans or Middle Easterners, who may choose Asian, white, or "some other race." Unlike Hispanics and Asians, white and black respondents are not given the option of reporting an ethnicity (such as Jewish or Jamaican). The limits to census measurement of race/ethnicity in turn limit demographers' ability to accurately describe the racial/ethnic composition of the U.S. population. They also have important political and social consequences for the construction of racial and ethnic identities and meanings. These kinds of measurement errors will affect census and survey data collection across a broad range of topics and contexts.

As noted earlier, censuses are typically conducted every 10 years, although in some countries they are conducted more or less frequently. In between census years, population size must be estimated. A variety of techniques are used to estimate population size in inter-census years. The most common technique uses the demographic equation, which is described in further detail in the next section. Because populations change only through fertility, mortality, and migration, this technique involves adding births and in-migrants to and subtracting deaths and out-migrants from the population counted in a census year. To achieve this, counts of births, deaths, and migrants are necessary for shorter time periods than decades. These counts come from registration systems.

Registration Systems

Like census data, *registration systems* collect data about key life events—births, deaths, migrations, marriages, and divorces—for all people in a population, or at least for all people experiencing those events. Figure 3–2 shows an early 20th-century poster encouraging birth registration from the United States. In the United States, the coverage of vital registration systems is considered nearly complete for birth and death registration through the birth and death certificates programs run by state governments. As mentioned in Box 3–1, these data are available through the National Center for Health Statistics (NCHS) for download, and the NCHS also provides detailed reports of the data. Birth and death certificates collect a small amount of information about the infant and his or her parents, and about the deceased, respectively, and the data are considered the best available for demographic measurement of fertility and mortality. For the study of infant mortality, the NCHS also provides linked birth–death files that connect birth and death certificates to identify infants who die within the first year of life. Some European countries use combined registration systems that identify the vital events of each individual—birth, migration, marriage, divorce, and death—over the life course in one central record.

Most countries do not have registration systems for migration. Instead, migration data are collected by agencies that administer and enforce immigration law. In the United States, this is the Department of Homeland Security (DHS). The DHS provides reports of basic counts and characteristics of migrants who obtain various forms of legal immigration status in a given year. Although these are the only counts of migrants in the United States on an annual basis, the data are limited in several important ways. First, they do not provide direct counts of immigrant flows, or immigrants entering or leaving the country in a given period, as many migrants included

Figure 3-2 Parents Are Encouraged to Register Their Babies' Births

© American Medical Association 1940. All rights reserved/Courtesy AMA Archives

in the data have been in the United States for years and are recorded in the data because they adjusted their legal status from a temporary to a permanent visa. The data also do not include counts of people who leave the country, which makes direct estimates of net migration impossible. Moreover, the data are available only in aggregate form—as counts, not as individual records—which makes analyzing relationships between characteristics of migrants challenging, and the types of information about migrants that are recorded and reported by the DHS are very limited. Finally, the DHS reports only information about apprehensions of undocumented immigrants—that is, those with whom the Immigration and Custom Enforcement (ICE) agency comes into contact. These counts include people who were apprehended more than once in a given year, so they represent events rather than people. As a measure of undocumented migration flow, they are also flawed because they are sensitive to variations in ICE efforts to apprehend undocumented migrants. Estimating the actual number of undocumented immigrants in the United States is a very difficult task as a result.

Because of the high costs of registration and administrative data systems, in many countries they are unavailable or very incomplete in coverage and, therefore, are considered unreliable as a basic source of population data.

In these countries, survey data are substituted for the purpose of describing basic demographic trends such as fertility and mortality rates. In the United States and many other countries, survey data are also used for estimating migration.

Survey Data

Unlike census data and registration systems, *survey data* are not collected for all members of a population, but rather for just a sample of the population. For survey data to be useful for demographic measurement, they need to be representative of the population from which they are drawn. This is achieved through probability, or random, sampling—that is, drawing a sample of the population where every individual in the population has a known and equal probability of being selected. Given probability sampling, sampling theory can be used to estimate the degree of error that is present using sample data to estimate population characteristics. This error is often called the margin of error or the confidence interval for a given estimate.

A vast multitude of survey data exists related to population, public health, and many other topics, and a great number of these data sets are publicly available (see Box 3–1 for a sample). For the purpose of measuring populations and their characteristics, survey data are essential when registration and/or census data are not available. Survey data are also essential for studying the characteristics of populations that are not recorded in censuses or registration systems, or are limited in the ways they are recorded in those systems.

Survey data are not perfect substitutes for vital registrations, and they present certain challenges for analysis. For the measurement of demographic events such as births, deaths, and migrations, survey data are generally retrospective, unless the survey follows individuals over time and records events in between waves of data collection; such *longitudinal data collection* is costly, however, and therefore uncommon. For analyzing mortality, retrospective data are rarely useful, as deaths have to be reported by family members; an exception to this is infant and child mortality, which is more reliably reported by parents.

In the United States, the NCHS has linked individuals in survey data, such as the National Health Interview Survey, to death records in the vital registration system. Using these linked files, analysts can examine the risk of death over a period following a survey, determining which characteristics of individuals measured at the time of the survey put them at risk of death during the follow-up period and generating mortality rates according to those characteristics. An excellent example of this type of analysis is an article by Robert Hummer and his colleagues, in which the researchers show that individuals who attend religious services weekly are significantly less likely to die in the follow-up period than individuals who never attend services—a difference that adds up to an average of 7 additional years of life for regular attenders of religious services.[3]

Another problem with survey data is that relatively small samples may be too small for analysis of important subgroups in the population. If a subgroup—for example, Asians—makes up only 4% of the entire population, and a survey collects nationally representative data on 1000 people, the sample will include about 40 (4% of 1000) Asian participants. Although the entire sample of 1000 is sufficiently large to estimate population characteristics, a sample of 40 may be too small to estimate characteristics for the Asian population with statistical confidence. Analysts using these data certainly cannot break the Asian population down further into subgroups based on nationality or ethnicity. These data power issues often affect analysts' decisions about the construction of race and ethnic groups, as well as other subgroups, in demographic analysis of secondary data.

To avoid this problem, data collectors can intentionally collect larger samples of minority groups, but this complicates probability sampling because not everyone in the population will have an equal probability of being selected. To adjust for this and other sampling procedures, mathematical weights are estimated that account for differential probabilities of being selected into the sample. Applying these weights means that survey data with over-sampled populations can yield nationally representative, or otherwise generalizable, estimates. When using secondary survey data, it is crucial to pay attention to the sampling method and properly use the weights provided.

MEASURING POPULATION SIZE AND GROWTH

Understanding the size and growth of populations is a basic task of demography, and this task has become ever more important as the demographic transition has ushered in an era of unprecedented and phenomenal global population growth over the last 50 years. We are living in truly unique moment in demographic history: those of us who were born by the 1960s witnessed the fastest ever global population growth rates, and, if demographic projections are correct, those of us alive through the middle of the 21st century will be part of the largest global population ever. Understanding these trends requires measuring population size and growth.

Population size is estimated through counting, and this is done primarily with censuses. As noted earlier, census data are typically collected at spaced intervals—every 10 years. In between censuses, counts of population size are estimated indirectly, through the basic logic of the *demographic equation*. Population growth occurs through only three components of population change: fertility (births), mortality (deaths), and migration (both in and out). The number of people added or subtracted to a population over some period of time reflects the combined influence of these forces: while people are added through fertility and in-migration, people are subtracted through mortality and out-migration. The demographic equation estimates population size at a later point in time (time 2) by beginning with a count of the population (at time 1) and then adding and subtracting births, deaths, and migrations that occur between time 1 and time 2:

$$\text{Population}_{t2} = \text{population}_{t1} + \text{births}_{t1\text{-}t2} - \text{deaths}_{t1\text{-}t2}$$
$$+ \text{in-migrants}_{t1\text{-}t2} - \text{out-migrants}_{t1\text{-}t2}$$

The demographic equation is thus a means for understanding population growth. *Population growth* is simply the number of people being added or subtracted to a population over some period of time. Thus population growth is the net difference between births and in-migrants, on the one hand, and deaths and out-migrants, on the other hand. Given two counts of population size at different times, the difference is equal to the number of people added or subtracted:

$$\text{Population growth}_{t1\text{-}t2} = \text{population}_{t2} - \text{population}_{t1}$$

Although *growth* commonly refers to positive change, demographers also use the term *population growth* when populations are declining in size.

At the global level, *net migration* (the difference between in-migration and out-migration) is irrelevant to population growth. Ignoring migration, population growth through births and deaths is called *natural increase*, and it is estimated by simply subtracting deaths from births over some period of time.

The number of people being added or subtracted to a population will vary according to population size (see Figure 3–3). In big populations such as that of China, more people may be added than in small populations such as that of Mongolia, but that does not necessarily mean that the Chinese population is growing more rapidly than the Mongolian population. In 2011, the Population Reference Bureau (PRB) estimated that China added 6,751,900 people, whereas Mongolia added less than 1% of that number—only 51,700 people.[4] However, Mongolia's population was growing at a rate more than three times faster than China's (1.8% versus 0.5%).[4] This example helps introduce the logic of *rates*, a basic demographic tool. Because the counts of population growth and the components of growth—births, deaths, and migrations—are affected by population size, to compare these demographic processes across populations, rates must be adjusted for differential populations "at risk" of the event. In the case of population growth, the entire population is at risk of changing; thus *population growth rates* adjust the number of people being added or subtracted by the population size at the beginning of the interval over which change is occurring:

$$\text{Growth rate}_{t1\text{-}t2} = \frac{\text{population growth}_{t1\text{-}t2}}{\text{population}_{t1}} \times 100$$

In 2011, annual population growth rates ranged globally between −2.7% in Lithuania and 3.7% in Burundi.[4] Expressed more formally, these rates mean that in 2011 Lithuania lost 2.7 people per 100 people and Burundi gained 3.7 people per 100 people. The multiplier 100 is a *constant*. Demographers regularly use constants to express rates in whole numbers; in this case, the constant creates a percentage (per 100). In 2011, the world's population was growing at a *rate of natural increase* of 1.2%.[4]

The growth rate can be transformed into a useful indicator called doubling time, which tells you how long it would take a population to double in size (or decline by half) if it continued to grow (or shrink) at the current rate with no change over time. Doubling time can be used to interpret a growth rate in any kind of scenario—for example, doubling time is often used to interpret how quickly investments will grow given a current interest rate, and it is a key concept for understanding compound interest. Like compound interest, populations grow exponentially because (positive) growth rates are acting on an ever-increasing base. Thus, the number of people being added over time increases, even if the growth rate remains the same. This explains why the shape of the curve showing population growth throughout human history takes the form of an exponential curve ($y = e^x$) rather than a linear curve (e.g., $y = 2x$). Doubling time is estimated with a simple equation that is derived from the fact that growth rates generate exponential growth:

$$\text{Doubling time} = \frac{70}{\text{growth rate expressed as a percentage}}$$

The numerator 70 comes from the natural logarithm of 2. If the growth rate is negative, then halving time is estimated by dividing −70 by the negative growth rate to yield a positive number of years over which the population will halve.

Table 3-1 Growth Rates and Doubling Times

Population Growth Rate	Years to Double
20%	3.5
10%	7
5%	14
2%	35
1%	70
0.1%	700
0.01%	7000
0.001%	70,000
0%	Never!

Note: The highest ever global population growth rate was estimated to be 2.2% in the mid-1960s.

Table 3-1 shows the doubling time for a handful of growth rates and reveals the exponential nature of differences across rates. Given the growth rates in Lithuania and Burundi, Lithuania's population would be halved in 25 years (–70/–2.7) and Burundi's population would double in 18 years (70/3.7). The world's population would double in 57.5 years (70/1.2). Although this seems like a relatively short time—particularly for the world's population to increase from 7 billion to 14 billion—such a doubling time scenario is unlikely to play itself out, as the global population growth rate has been steadily declining since it peaked around 2.2% in the 1960s. Doubling time assumes an unchanging growth rate; thus is it not a prediction but rather a way to interpret current growth. Given a declining growth rate over time, the world is adding proportionally fewer people in each subsequent year, but the world population will continue to grow so long as the growth rate is positive. Moreover, given a growing base, adding proportionally fewer people is still adding more people than have ever been added before.

POPULATION STRUCTURE

Population structure generally refers to the age and sex composition of a population, although other forms of structure such as the distribution of household and family types, and racial and ethnic composition, are also studied by demographers. Age and sex structure are essential demographic characteristics to understand because they are so highly related to mortality, fertility, and migration; these components of population change are demographically regimented by age and sex. In addition, age and sex structure are important to understand because of their own social and economic consequences. Age structure, in particular, is one of many factors that shape and constrain our individual lives. In a recent *New York Times* column, the prime minister of Indonesia, Najib Razak, argued that a young age structure, or "youth bulge," is one of two major problems facing youth in Muslim countries today.[5]

Age structure is measured in several ways. Most simply, the *median age* is the age at which half of the population is older and half of the population is younger. The *proportion of the population that is in certain age groups* is also used. For example, the proportion of the population that is between the ages of 10 and 24 gives a sense of the youth bulge referred to by Razak[5]: in Indonesia, 23% of the population is between these ages, compared to 17% in the Netherlands.[4] The proportion of the population younger than age 15 measures how young the population is. While only 17% of the Dutch population is younger than age 15, 27% of Indonesia's population is younger than age 15 (and, at the global extreme, 52% of Niger's population is younger than age 15).[4] The proportion of the population older than age 64, by contrast, measures population aging. Only 6% of Indonesia's population is older than age 64, while 16% of the Netherlands' population is. These age cut-offs are typically used to represent dependent, or nonworking, members of the population, and the *dependency ratio* is the ratio of the population that is younger than age 15 and older than age 64 to the population that is between these ages:

$$\text{Dependency ratio} = \frac{\text{population}_{<15} + \text{population}_{>64}}{\text{population}_{15\text{-}64}}$$

Ratios are another standard demographic tool. Unlike *proportions*, which express the relationship between a part and a whole, ratios express the relationship between two parts. Data for dependency ratios, age proportions, and median age typically come from population censuses.

Population pyramids are a visual representation of the age structure of a population. They use stacked bars to represent the relative size of each age group within a population, and typically men and women are separated on the left and right sides of the pyramid. Figure 3–3 shows current population pyramids for Niger, Indonesia, and the Netherlands—populations with very different age structures.[6] Niger's population is very young and growing rapidly, with the longest bars for the youngest age groups and an overall pyramid shape, meaning that the size of each incoming age group is larger than its older predecessor. The Nigerien age structure is the result of very fast population growth due to high fertility rates and declining mortality rates in the second stage of the demographic transition. Indonesia's age structure, by contrast, has more of a barrel shape for persons younger than age 35, with a visible bulge between ages 5 and 19 (the youth bulge Najib Razak worried

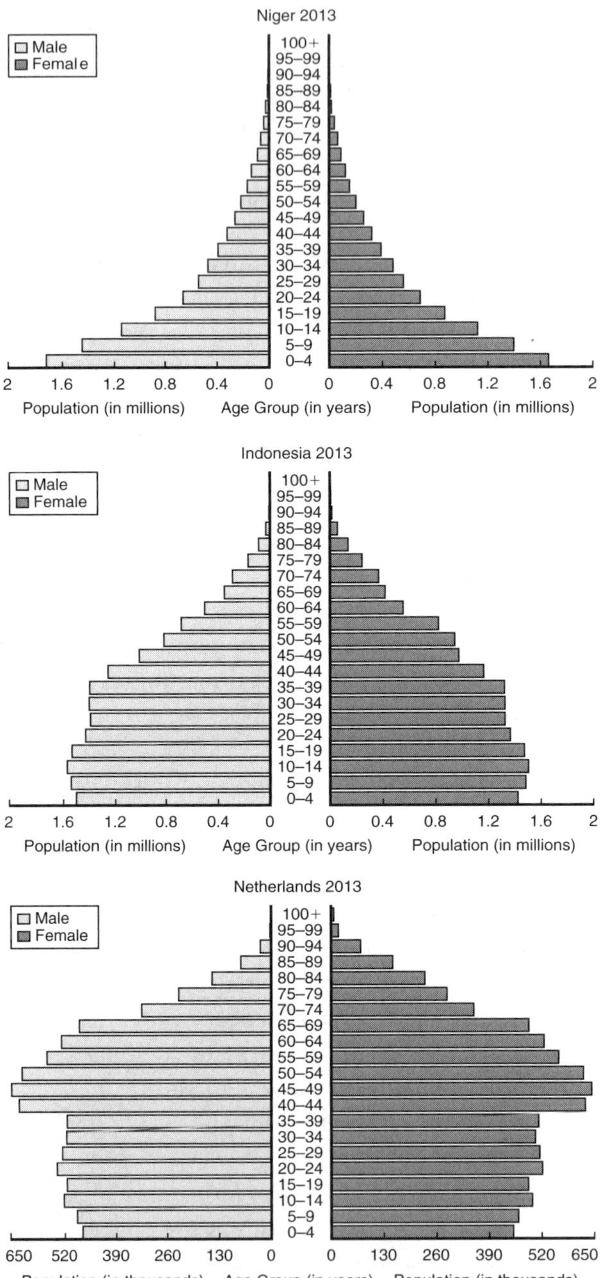

Figure 3–3 Population Pyramids for Niger, Indonesia, and the Netherlands, 2013

Data from U.S. Census Bureau International Programs. International data base. http://www.census.gov/population/international/data/idb/informationGateway.php. Accessed December 30, 2013.

about in his column).[5] The pyramid shape characterizing the entire Nigerien age structure is visible in Indonesia's population older than age 35, reflecting slowing population growth through declining fertility in the last 35 years, or the third stage of the demographic transition. Finally, the Dutch population pyramid shows an aging population with a bulge between the ages of 40 and 59, the result of the Dutch Baby Boom. The Dutch population is no longer growing through natural increase, as each older age group is larger than its younger counterpart, the opposite of the Nigerien pattern and the result of low fertility in the fourth stage of the demographic transition.

Age structure matters for population growth, particularly in countries that experience rapid population growth followed by rapid fertility decline, such as in Indonesia. This produces large youth cohorts that move through the population pyramid with drag due to their own fertility. This process, which is called *population momentum*, can be thought of as a slowing train that takes a while to stop. When rapid population growth results in a young age structure, large cohorts of reproductive-age people are reproducing, which tends to result in more births than deaths and, therefore, positive rates of natural increase, even when fertility rates are low. This is the effect of population momentum on population growth.

Population pyramids give some sense of the sex distribution of a population as well as the age structure. For example, the feminization of old age as a result of lower mortality among women is apparent in the Dutch population pyramid, where the bars on the right for women are longer in the oldest ages than the corresponding bars on the left for men. A more concise and common way to describe the sex distribution of the population is with the *sex ratio*, or the ratio of men to women:

$$\text{Sex ratio} = \frac{\text{men}}{\text{women}} \times 100$$

The sex ratio is expressed as the number of men per 100 women; given equal numbers of men and women, the sex ratio will be equal to 100.

In the United Arab Emirates, which relies heavily on male migration from places like Bangladesh, the sex ratio in 2012 was very high—219 men per 100 women.[6] In Bangladesh, at least in part due to this male emigration, the 2012 sex ratio of 94.5 men per 100 women means there were more women in that country than men.[6] Sex ratios less than 100 are common in places with aging populations because of the feminization of old age; the U.S. sex ratio in 2012 was 97.2 men per 100 women for the entire population, but for the 80–84 age group, the sex ratio was 69.1 men per 100 women.[6] The sex ratio at birth hovers above 100 because more male infants are born than female infants.

In the United States, the sex ratio for the 0–4 age group was 104.5 boys per 100 girls in 2012.[6] In China, the sex ratio for this age group in 2012, 113.8 boys per 100 girls,[6] reflects social practices in addition to natural differences in the rate of male and female births—in particular, a strong son preference that results in sex-selective abortion, female infanticide, and under-reporting of female infants.

MEASURING COMPONENTS OF CHANGE: MORTALITY, FERTILITY, AND MIGRATION

Although life begins at birth and ends at death, this chapter's discussion of measurement of the components of population change begins with mortality and proceeds to fertility. This is because mortality is easier to measure than fertility—death happens to everyone and it happens only once. Births, by contrast, occur to only a subset of the population (women of reproductive age) and, among that subset, some women will have no births while others will have many. Migration is far less demographically regimented than mortality or fertility—anyone at any age can theoretically (if not legally or practically) migrate, and it is the most difficult of the three components to define and measure.

Measures of mortality, fertility, and migration describe the frequency of these events in a population—how many deaths, births, or migrations are occurring within a given period of time in a given place. The basic elements of these measures are counts of the events, typically from vital registration systems (in the case of births and deaths) or administrative or survey data (in the case of migration), and counts of populations, typically from census data. Additional information—particularly the age and sex of those giving birth, dying, or migrating, and the age and sex of the population—is needed to construct more refined measures. Counts are then turned into rates by dividing the number of events (counts of births, deaths, or migrations) by the population at risk of that event (entire population, or subsets of the population most at risk, such as women of reproductive age).

There are two simple reasons to measure mortality, fertility, and migration. One is to understand the effect of these components on population change. For example, an analysis of components of change reveals the demographic determinants of the dramatic population growth that occurred after 1800: because mortality rates declined while fertility rates remained high and unchanged in the second stage of the demographic transition, populations grew through fewer people leaving, rather than from more people entering. Measuring fertility and mortality was essential for understanding this process.

A second reason why it is important to measure mortality, fertility, and migration is to understand these demographic processes in and of themselves. After demographers determined the demographic components of population growth in the demographic transition—that declining mortality rates preceded declining fertility rates—they turned their attention to the less proximate determinants of this transition. What caused mortality, and then fertility, to decline? To understand which causes and which effects these social forces have, one must begin with accurate description of their levels and trends over time and across space. Thus measures of mortality, fertility, and migration serve this purpose in addition to the purpose of describing, understanding, and predicting population change. Some measures are better suited to assessing population change, whereas others are better for understanding the component of change itself.

Measures of Mortality

The most basic measure of mortality is the *crude death rate* (CDR). The CDR is the number of deaths in a place at a given point of time, usually a year, typically expressed per 1000 people:

$$\text{CDR} = \frac{\text{deaths}}{\text{population}} \times 1000$$

In 2012, the CDR ranged globally between 1 death per 1000 people in Qatar and 17 deaths per 1000 people in the Democratic Republic of the Congo.[4] The CDR is a useful measure because it is easy to construct, requiring only two pieces of information, and because it gives an indication of how many people are exiting a population through mortality—that is, the effect of mortality on population growth. In 2012, far more people were exiting the population through mortality in the Congo than in Qatar.

While the CDR is useful for describing the effect of mortality on population growth, it is not a great measure of the force of mortality, or the actual risk of dying in a given year in a population, because it is biased by the age structure of a population. Because death is more common among the elderly, there will be more deaths in an older population than in a younger population even given equal risks of dying at every age. CDRs are thus biased upward (i.e., they are higher than the actual force of mortality) by older populations and biased downward (i.e., they are lower than the actual force of mortality) by younger populations. Many low-mortality, aging populations have higher crude death rates than high-mortality, young populations. For example, Japan's CDR in 2012 was 10 deaths per 1000 people, whereas Thailand's CDR was 7 deaths per 1000 people.[4] Similarly, the United Kingdom's CDR in 2012 was 8 and India's CDR was 7.[4] These comparisons yield misleading conclusions about the force of mortality across these populations: Japan and the United Kingdom do not have higher mortality than Thailand or India, but rather have older populations (which is, after all, partly a result of low mortality) and, therefore, expressed per the entire population, relatively more people are dying in Japan and the United Kingdom.

A better measure of the force of mortality should take into account and adjust for the age structure of a population. The relative force of mortality across populations can then be assessed net of the bias of age structure; instead of comparing a population of elderly people to a young population, you compare the old to the old and the young to the young. *Age-specific death rates* (ASDRs) do this by estimating a death rate at each age. The ASDR is the number of deaths to people of a certain age divided by the number of people in that age group expressed per 1000 people in that age group:

$$\text{ASDR}_{x-x+n} = \frac{\text{deaths}_{x-x+n}}{\text{population}_{x-x+n}} \times 1000$$

In this equation, x represents the age at the beginning of the corresponding age interval and n represents the length of the interval, which is typically 5 years. Thus the ASDR for 5- to 10-year-olds divides the number of deaths to 5- to 10-year-olds by the population of 5- to 10-year-olds and is interpreted as the number of 5- to 10-year-olds who died in a period of time per 1000 5- to 10-year-olds. Comparing ASDRs across two populations gives a more accurate picture of the force of mortality than the CDR. Figure 3–4 shows the ASDRs for Japan, Thailand, the United Kingdom, and India.[7] The figure confirms the bias in the CDR for these populations: the risk of death is, in fact, higher at every age in Thailand and India than it is in Japan or the United Kingdom.

Because there are so many ASDRs, it is more useful to summarize the force of mortality with a single indicator. The two most commonly used single indicators of mortality are the infant mortality rate and life expectancy. The *infant mortality rate* (IMR) is the ASDR that is most frequently used on its own. It is not equivalent to the ASDR for 0- to 1-year-olds because rather than taking the population at this age as the denominator, the count of live births is used. The IMR is formally the number of deaths to infants (up to age 1) divided by the number of live births in a population, typically expressed per 1000 live births:

$$\text{IMR} = \frac{\text{deaths}_{0-1}}{\text{births}} \times 1000$$

In 2011, the IMR ranged between 2 infant deaths per 1000 live births in Hong Kong and 131 infant deaths per 1000 live births in Afghanistan.[4] The IMR is a standard public health and development measure because it reflects

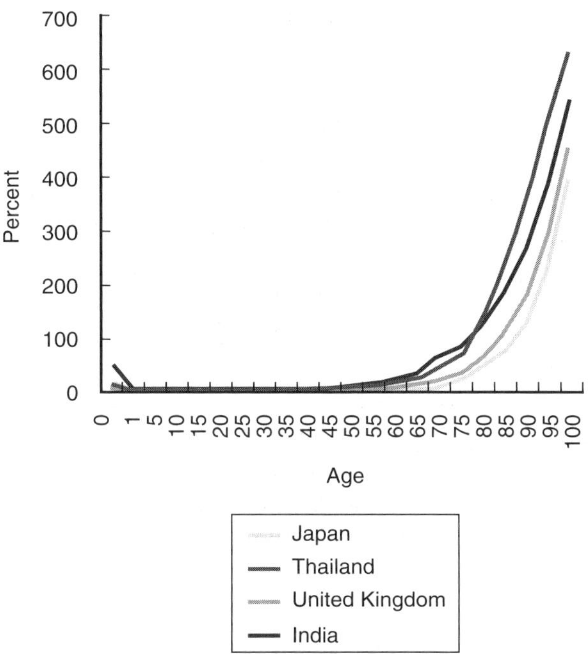

Figure 3–4 Age-Specific Death Rates of Japan, Thailand, the United Kingdom, and India, 2009

Data from World Health Organization. WHO mortality data and statistics. http://www.who.int/healthinfo/statistics/mortality/en/index.html. Accessed December 30, 2013.

how well the most vulnerable and dependent members of society (infants) are faring.

The most common single measure of mortality is *life expectancy at birth*, or the average length of life a person would live given current age-specific death rates. Life expectancy at birth uses *period* data—or data that come from a specific period, typically a year—to estimate a cohort-like measure of mortality. *Cohort* measures reflect the life-course experiences of a group of people born at the same time. To truly understand the mortality experiences of those born in 2012, for example, those individuals would have to be followed until the last member of the 2012 birth cohort died (potentially well into the 22nd century). Life expectancy at birth appears to be a cohort measure because it estimates an average length of life, but is actually based on period data and reflects current mortality risk. Most likely, individuals born in 2012 will live longer than life expectancy at birth in 2012 because over their lives the conditions affecting mortality will change (presumably for the better).

Demographers use a tool called the *life table* to estimate life expectancy. The basic way that the life table works is by mathematically subjecting an imaginary cohort of people—typically a cohort of 100,000—to the full set of age-specific death rates converted into probabilities. The age-specific probability of death is the ASDR adjusted for the differential risk of death across a year—in most age brackets, deaths are assumed to occur evenly across a year of life, but in early childhood and late life, deaths tend to concentrate at the beginning of the year. At the very beginning of the life table, at age 0, the cohort of 100,000 is subjected to the probability of death for the first year of life. Table 3-2 shows the first 10 years and last 10 years of the Belarusian life table from 2010.[8] Given a probability of dying between age 0 and age 1 in Belarus in 2010 of 0.00393, 393 members of the initial 100,000 cohort are estimated to die before age 1, and 99,607 (100,000 − 393) begin age 1 still alive. This number of 1-year-old survivors is then exposed to the probability of death for 1-year-olds to estimate the number of deaths to 1-year-olds; the number of deaths to 1-year-olds is subtracted from the number who began life at age 1 to estimate the number surviving to age 2; the number surviving to age 2 is then exposed to the probability of death for 2-year-olds; and so on. To estimate life expectancy, the life table sums the total number of years lived by the full cohort of 100,000 across all ages. For example, those who died in the 6th year of life contributed, on average, 6.5 years each (assuming that deaths were evenly distributed across the year), whereas those who died in the 70th year contributed 70.5 years each. Summing the total number of years lived across all people in the cohort (a number that demographers call *person-years*), and dividing by the initial cohort of 100,000, yields the average length of life.

A few characteristics of life expectancy and the life table are important to note. First, as already described, life expectancy is not a prediction of how long people who are born in a given year will live; instead, it is a summary measure of current death rates across all ages. Thus it reflects current conditions affecting life and death in a population across all ages. As those conditions change, so, too, will life expectancy. Over the past 100 years, life expectancy has steadily grown in most populations, meaning that people born in a given year typically live longer than the life expectancy estimated for that year. Two notable exceptions have been in some sub-Saharan African populations devastated by AIDS and in Russia, where social conditions in the post-Soviet era have contributed to increasing mortality, especially among men.

A second important point about life expectancy is that it is an unbiased measure of the force of mortality and, therefore, is highly useful for comparing levels of mortality across populations at a given point in time or within a specific population across different times. Life expectancy ranged globally between 44 years in Afghanistan and 83 years in Japan in 2011.[4] These estimates are unaffected by age structure; indeed, consistent with the higher ASDRs in Thailand and India than in Japan and the United Kingdom, in 2011 life expectancy at birth in was 74 years in Thailand, 64 years in India,

Table 3–2 The First and Last 10 Years of the Belarusian Life Table, 2010

Age	Probability of Dying Between Ages x and $x + n$	Number Surviving to Age x	Number Dying Between Ages x and $x + n$	Person-Years Lived Between Ages x and $x + n$	Total Number of Person-Years Lived Past Age x	Expectation of Life at Age x
x	$_nq_x$	l_x	$_nd_x$	$_nL_x$	T_x	e_x
0	0.00393	100,000	393	99,630	7,052,339	70.5
1	0.0005	99,607	50	99,582	6,952,709	69.8
2	0.00031	99,557	31	99,541	6,853,128	68.8
3	0.00032	99,526	32	99,510	6,753,586	67.9
4	0.00037	99,494	37	99,476	6,654,076	66.9
5	0.00024	99,458	24	99,446	6,554,600	65.9
6	0.00021	99,434	21	99,424	6,455,154	64.9
7	0.00023	99,413	23	99,402	6,355,730	63.9
8	0.00032	99,390	32	99,375	6,256,328	62.9
9	0.00015	99,358	15	99,352	6,156,953	62.0
…						
101	0.40543	163	66	130	301	1.84
102	0.42369	97	41	77	171	1.75
103	0.44142	56	25	44	94	1.68
104	0.45852	31	14	24	50	1.61
105	0.47492	17	8	13	26	1.54
106	0.49055	9	4	7	13	1.49
107	0.50536	5	2	3	7	1.44
108	0.51931	2	1	2	3	1.39
109	0.5324	1	1	1	1	1.36
110+	1	1	1	1	1	1.34

A note about notation: Each item in the life table is annotated by x on the right and some are additionally annotated by n on the left. x refers to age, and n refers to the length of the age interval, which in this case is 1 year. Items annotated by both x and n occur over the age interval x to $x + n$, whereas those annotated only by x represent the beginning of the age interval.

Life table calculations:
$_nq_x$ = the probability of dying between ages x to $x + n$ is equal to the ASDR adjusted by the uneven risk of dying across a year, or, stated differently, by the average length of life lived in that year by those dying in that year. In most years, this is assumed to be 0.5, but it is shorter in early childhood and longer in later life. The formula for calculating $_nq_x$ from an ASDR is $_nq_x = (n)$ (ASDR)/1 + $(_na_x)(n)$(ASDR), where n is equal to the length of the interval (in this case, 1 year) and $_na_x$ is equal to the average length of life lived in that interval by those dying in that interval.

(Continues)

Table 3–2 The First and Last 10 Years of the Belarusian Life Table, 2010 (*Continued*)

l_x = the number surviving to age x. At the beginning of the life table, l_0 is the size of the hypothetical cohort subjected to the life table, also called the *radix*. The radix is typically equal to 100,000, although it can be equal to any number. At each subsequent age, the number surviving is equal to the number beginning the age interval minus the number dying during that age interval, or $l_{x+n} = l_x - {_n}d_x$.

${_n}d_x$ = the number dying between ages x and x + n is equal to the number beginning life at that age times the probability of death, or ${_n}d_x = (l_x)({_n}q_x)$.

${_n}L_x$ = the person-years lived between ages x and x + n is the total number of years lived by the cohort surviving to age x during the interval x to x + n. Those who survive the age interval contribute 1 year of life each; those who die contribute the length of life they lived prior to death. Thus, ${_n}L_x$ is equal to the number beginning the cohort minus the number dying during the interval multiplied by their average length of life lived during that interval, or ${_n}L_x = l_x - ({_n}a_x)(d_x)$.

T_x = the total number of person-years lived above age x, or the sum of ${_n}L_x$ at age x and all ages greater than x, or $T_x = T_{x+n} + {_n}L_x$.

e_x = the expectation of life at age x, or life expectancy at age x, is equal to the average number of person-years lived in all remaining ages by each member of the cohort surviving to age x, or $e_x = T_x/l_x$.

Data from the Human Mortality Database, University of California at Berkeley. Belarus. http://www.mortality.org/cgi-bin/hmd/country.php?cntr=BLR&level=1. Accessed December 30, 2013.

83 years in Japan, and 80 years in the United Kingdom. Although life expectancy is useful for cross-population comparisons, it does not give any direct or clear indication of the effects of mortality on population growth. The CDR is the better measure for that purpose.

Third, life expectancy can be estimated at any age in the life table, although life expectancy at birth is most commonly used. An important pattern to note is that total life expectancy increases with age; for example, life expectancy at birth in the 2010 Belarusian life table in Table 3–2 is 70.5 years. Life expectancy at age 9, however, is 62 years—these 9-year-olds will live 0.5 year longer, on average, than the average infant (62 + 9 = 71 years versus 70.5 years). This result occurs because the 9-year-olds have survived some of the force of mortality—that between infancy and age 9.

Knowing life expectancy at a given age can be useful for analysis, as it can suggest mechanisms affecting mortality. For example, in a *New York Times* column, Nicholas Kristof argued on behalf of publicly funded health insurance with an analysis of life expectancy in the United States at birth versus life expectancy at age 65.[9] Whereas the United States ranks poorly compared to other industrialized nations in terms of life expectancy at birth, it ranks much better in terms of life expectancy at age 65. Kristof attributed this difference to Medicare, the U.S. public health insurance available to those persons older than age 65.

Finally, the life table can be used to estimate the life span of any number of conditions, such as marriage or unemployment. Because the life table was initially developed to assess mortality, analyses of length of time are called survival analyses.

Measures of Fertility

Like the CDR for mortality, the most basic measure of fertility is the *crude birth rate* (CBR). The CBR is simply the number of births in a population in a given period time (typically a year) divided by the population, and it is standard to express the rate as the number of births per 1000 people:

$$CBR = \frac{births}{population} \times 1000$$

In 2012, the CBR ranged globally from 6 births per 1000 people in Monaco to 46 births per 1000 people in Mali.[4] Again, like the CDR, the CBR is useful because it requires only two pieces of information to construct and it is a direct measure of how many people are entering a population via fertility. In its construction and interpretation, it mirrors the crude death rate perfectly; thus it can be used in the balancing equation (where the population growth rate is equal to the CBR minus the CDR plus the rate of net migration). If the CBR and the CDR are equal, the natural increase is zero and the population can change only through migration.

Whereas birth rates refer to measures of fertility adjusted for the entire population, fertility rates refer to measures of fertility adjusted by the population of women of childbearing age, a better approximation of the population at risk of fertility. The *general fertility rate* (GFR) does exactly this—the numerator is the number of children born in a particular period of time and place, and the denominator is the number of women in that same time and place, typically between the ages of 15 and 44:

$$GFR = \frac{births}{women_{15-44}} \times 1000$$

While the GFR is a conceptual improvement over the birth rate, it is not a substantive improvement because the proportion of the population that consists of women in their childbearing years does not vary dramatically across populations.

Although the proportion of women who are of childbearing age in a population does not vary dramatically across populations, the proportion who are in their key childbearing years does. Thus *age-specific fertility rates* (ASFRs), typically constructed for 5-year age groups, better adjust for age structure than the GFR. ASFRs take the same form as the GFR except that the numerator is specific to births to women in each 5-year age group and the denominator is the population of women in each 5-year age group; thus they mirror the ASDR.

$$ASFR_{x-x+5} = \frac{births_{x-x+5}}{women_{x-x+5}} \times 1000$$

As with ASDRs, the large number of ASFRs makes them unwieldy for describing and comparing fertility across populations. Instead, they are used as the building blocks of the most common and most easily interpreted measure of fertility, the total fertility rate.

The *total fertility rate* (TFR) is basically the sum of ASFRs: it is the number of births a woman would have if she bore children according to current age-specific fertility rates. The construction of the TFR can be understood conceptually as walking a woman through each ASFR and adding up the total number of births she would bear across all years of her reproductive life if she bore children according to those rates. If ASFRs were constructed for each age, the TFR would be the sum of all ASFRs divided by 1000. Because ASFRs are typically constructed for 5-year age groups, the TFR is the sum of each ASFR times 5:

$$\text{TFR} = \sum \frac{\text{ASFR}}{1000} \times 5$$

In 2011, the TFR ranged between 1.1 births per woman in Taiwan to 7.1 births per woman in Niger.[4]

The TFR reads as though it were a cohort measure—a completed fertility experience—but like life expectancy it is based on period data. Cohort fertility rates, or *completed fertility rates*, measure the total fertility of a cohort of women who have completed their childbearing years. Completed fertility rates are not useful for expressing current fertility patterns in a population because they rely on retrospective information for a subset of women—those who have completed their childbearing years and whose prime childbearing years were decades prior.

A woman's fertility experience across her own life course reflects both the number of children (*quantum*) she has and the timing of childbearing (*tempo*). Because the TFR is a period measure, it reflects year-to-year changes in not just the quantum of fertility but also the timing of fertility, although the TFR is typically interpreted as purely a reflection of quantum. This can present problems for analysis if a large number of women in a population are accelerating or postponing childbearing. If women are accelerating childbearing, the TFR in a given year will be higher than completed fertility; if they are delaying childbearing, the TFR in a given year will be lower than completed fertility for those cohorts. Many TFRs around the world are currently lower than completed fertility because of the *tempo distortion* introduced by postponement.

The TFR may be refined to measure only daughters born because only female babies will eventually bear children. This measure is called the *gross reproduction rate* (GRR). The GRR can be understood as the number of female children that a girl who was just born is expected to bear during her lifetime, with two assumptions. The first assumption is that all baby girls will survive to reach their reproductive years. The second assumption is that age-specific birth rates will remain the same as they are now. If the GRR is 1, then women will replace themselves. If it is greater than 1, then the next generation of child bearers will be greater than the current generation; conversely, if the GRR is less than 1, then there will be fewer women.[10]

The GRR assumes that all women will survive their childbearing years. In fact, some women will die before reaching the oldest age in which they might bear children. The *net reproductive rate* (NRR) takes the risk of dying into account. The NRR is always less than the GRR, since some women will die before the end of their childbearing years. Throughout the world, there is tremendous variation in the difference between the GRR and the NRR, and this difference is related to dissimilar death rates for women.[11]

From 2005 to 2010, the world NRR was 1.082 surviving daughters per woman, albeit with wide variations seen across the globe. The average number of surviving daughters per woman in Eastern Africa was 2.0, while the NRR in Eastern Europe was 0.664.[12]

The TFR, the GRR, and the NRR are most commonly used as measures of fertility itself, but they can also be used to understand population change. Given low mortality, a TFR of 2.1 is called *replacement fertility*, which is the level of fertility at which each cohort exactly replaces itself. Given 2.1 births, one birth replaces the mother, one replaces the father, and 0.1 compensates for the risk of mortality to children between birth and their own reproductive years. Replacement TFRs are higher in high-mortality populations. For example, if 1 out of every 3 children dies before reaching reproductive age, then women would need to bear, on average, 3 children each for 2 surviving children to reach reproductive age and replace the mother and father. Replacement NRRs are always equal to 1 because NRRs account for mortality.

Although replacement fertility suggests that populations are not growing—cohorts are simply replacing themselves with no additional members—many populations with replacement-level fertility today are still growing through natural increase as a result of population momentum. For example, in 2012 Brazil's TFR was 1.9 births per woman, below replacement level, but Brazil's population was still growing at a rate of 1% through natural increase (the difference between the crude birth rate and the crude death rate).[4] Because rapid population growth in Brazil resulted in a young age structure, large cohorts of reproductive-age people were having births in 2012. Even though those cohorts were simply reproducing at the replacement level, they were contributing equally large cohorts, yielding more births than deaths in a given year. This is the result of *population momentum*.

Measures of Migration

Migration is more difficult to measure than fertility or mortality because there is no straightforward definition of what migration is—how far one has to move, for how long, and with what purpose. Most scholars define *migration* as a change in usual place of residence that involves crossing an administrative boundary, such as a municipality, county, province, state, or international boundary. Thus, by this definition, tourists who cross international boundaries are not migrants because they do not change their usual place of residence, and people who make a residential move within a county are not migrants because their change of residence does not cross an administrative boundary. Cross-county and inter-state migrants are *domestic migrants*, whereas those moving between countries are *international migrants*.[13] Scholars often refer to domestic migrants leaving a place as *out-migrants*, and domestic migrants arriving to a place as *in-migrants*, but there are also more general terms to describe arrivers and leavers. *Emigrants* refer specifically to international migrants who leave a place, and *immigrants* are international migrants who arrive to a place.

Demographers measure both the *stock* and the *flow* of migration. Stock refers to the number of immigrants in a place at a certain time, regardless of when they entered, whereas flow refers to the number of immigrants arriving or departing a place during a certain window. The most basic measure of stock is the number of immigrants living in a country. This number is obtained through census or survey data asking about location of birth; in the United States, it is obtained through the American Communities Survey. In 2010, the United States had the largest stock of immigrants of all countries worldwide, with 39 million immigrants.[14] The *percent foreign born* adjusts the stock of immigrants by the population in the receiving country; thus it is not a true rate because the denominator does not capture the population at risk of migrating:

$$\text{Percent foreign born} = \frac{\text{immigrants}}{\text{population in receiving country}} \times 100$$

Figure 3–5 shows the foreign born stock and percent foreign born in the United States from 1850 to 2010.[14]

In the United States in 2010, 39 million immigrants represented 12.8% of the U.S. population.[13] Although the foreign born stock has increased more or less steadily over this period (with a plateau between 1930 and 1970, what scholars refer to as an "immigration pause" in U.S. history), the percent foreign born in 2010 is still smaller than it was when it peaked in

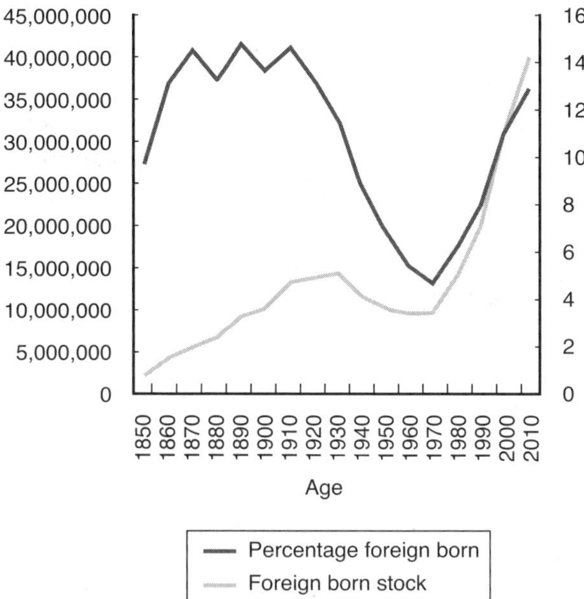

Figure 3–5 Foreign Born Stock and Percent Foreign Born in United States, 1850–2010

Data from U.S. Census Bureau. Historic data, 1850–2000: Gibson C, Jung K. *Historical Census Statistics on the Foreign-Born Population of the United States: 1850–2000.* Population Division Working Paper No. 81. Washington, DC: U.S. Census Bureau; 2006. http://www.census.gov/population/ www/documentation/twps0081/twps0081.pdf. The data point for 2010 came from the Census Bureau fact finder. http://factfinder2.census.gov/faces/nav/jsf/pages/index.xhtml.

1890 at 14.7%. This is because the U.S. population has grown more rapidly than the immigrant population over this period.

Globally, the United States is an outlier in foreign born stock, but not in percent foreign born. In 2009, Qatar had the highest percent foreign born, at 87%,[13] followed by a number of other Gulf states where migration to work in oil fields and services is very large in scale. Worldwide, 3.1% of the world's population consisted of immigrants—that is, people living outside their country of birth—in 2009.[13]

To measure migrant flows, demographers estimate *rates of in- and out-migration*. These rates simply adjust the number of people entering or leaving a population between the period x to $x + n$ via migration by the population size at time x, typically expressed per 1000 people:

$$\text{Rate of in-(or out-) migration} = \frac{\text{people migrating to (or out of) a population}_{x-x+n}}{\text{population}_x} \times 1000$$

The *rate of net migration* is the difference between the rate of in-migration and the rate of out-migration. Between 2005 and 2010, the rate of net migration to Qatar was 93 migrants per 1000 people.[10] Qatar thus gained more immigrants than it lost emigrants, but this rate alone does not say how many of each type. Rates of net migration are, in fact, more common than rates of in- or out-migration because direct counts of migrants, especially emigrants, are hard to come by. Instead, rates of net migration are often estimated with the demographic equation, using the change in population between two census dates, and counts of births and deaths in those periods, to estimate net migration via a residual analysis.

CONCLUSION

Although demographic measurement is an essential building block for social analyses of all sorts, it is important to use demographic estimates with caution and to understand the methodology underlying demographic measurement to be wary of potential error or bias. The collection of population data through censuses, vital registration systems, and surveys entails a number of decisions and constraints that may generate error. Because census data are collected at spaced intervals, most demographic estimates are based on indirect methods such as the demographic equation. Demographic measures can themselves be biased by basic demographic forces like age structure or period effects. Not understanding how the total fertility rate or life expectancy is constructed can lead to incorrect interpretations of those estimates, and understanding how they are measured can increase their utility for analysis. In Box 3–2, an example of the limits—as well as the strengths—of demographic analysis drives this point home. Nevertheless, measurement provides the basic building blocks of social analysis, and correct demographic measurement is a powerful tool for understanding and analyzing the social world.

Box 3–2 An Example of Demographic Data Analysis: Its Strengths and Weaknesses

A recent article by Emilio Parrado[15] in the journal *Demography* challenges the previously well-known demographic "fact" that Hispanics, and Mexican Americans in particular, have much higher fertility than other groups in the United States. This fact has been well documented using vital statistics data: in 2006, for example, the NCHS reported TFRs of 2.95 births per Hispanic woman, 3.1 births per Mexican American woman, and 1.9 births per non-Hispanic white woman. As Mexico and other Latin American countries experienced the demographic transition, and TFRs in those countries dropped to replacement or near-replacement levels, observers expressed concern that Hispanic fertility in the United States was higher not just than other U.S. groups but also higher than fertility in Hispanics' origin countries.

Parrado was not so easily convinced. He suspected that the elevated TFRs for U.S. Hispanics might have something to do with the construction of fertility rates, and in particular with the fact that the numerator and the denominator come from different sources. Whereas births—the numerator in the TFR—are counted every year through vital registrations, the population at risk of births—counts of women in various childbearing ages—come from census data. In inter-census years, these counts are estimated. For Hispanics in particular, counts of population may be incorrect if information about migration is lacking or misestimated. If estimates of the size of the Hispanic population underestimate the number of women who have immigrated between census years, the number of births per woman will be biased upward—that is, they will be higher than is actually the case.

Parrado demonstrated that a large portion of the difference in TFRs between Hispanics and other groups in the United States, as well as between Mexican Americans and Mexicans in Mexico, is, in fact, due to this data error. Using birth history data from surveys to estimate TFRs for Hispanics, he showed that if the numerator and the denominator come from the same source (as they do in survey data), much of the elevated fertility among Hispanics disappears. It also is driven by higher and misestimated fertility among immigrants, not among U.S.-born Hispanics, for whom this data error is less important.

This was not the only problem affecting Hispanic immigrant total fertility rates. As described in the section on the measurement of fertility, TFRs are sensitive to shifts in the timing of fertility—what demographers call "period effects." Parrado noted that migration is intricately linked to family formation: couples often postpone childbearing until after a migrant trip, spouses are reunited through migration after separations, and people may migrate for the purpose of marrying. As a result, TFRs can be artificially elevated for recent migrants as a result of a post-migration timing of childbearing. When Parrado estimated completed fertility rates, he found much smaller differences between groups. Specifically, using NSFG data from 2002, he estimated completed fertility rates for 35- to 39-year-olds of 2.1 births per Mexican American woman and 2.4 births per Mexican immigrant woman. These rates are much closer to the 1.9 TFR estimated for whites.

> Parrado's analysis is important because the artificially high fertility rates estimated for Hispanics have been used by anti-immigration scholars as an example of the lack of assimilation of Hispanic immigrants, and they are the basis for projections of the future racial/ethnic composition of the U.S. population. This example reveals the power of demographic measurement and mismeasurement, and serves as a reminder of the importance of understanding the data and methods of demography when using population measures as a tool for social analysis.

REFERENCES

1. Duncan GJ. When to promote, and when to avoid, a population perspective. *Demography*. 2008;45(4):763–784.
2. Snipp CM. Racial measurement in the American census: past practices and implications for the future. *Annu Rev Sociol*. 2003;29:563–588.
3. Hummer RA, Rogers RG, Nam CB, Ellison CG. Religious involvement and U.S. adult mortality. *Demography*. 1999;36(2):273–285.
4. PRB DataFinder [Internet]. Washington, DC: Population Reference Bureau; c2013. http://www.prb.org/DataFinder.aspx. Accessed March 4, 2013.
5. Razak N. The challenge of Muslim youth. *New York Times*. December 14, 2012: Opinion pages.
6. U.S. Census Bureau International Programs. *International Data Base*. Washington, DC: U.S. Census Bureau; 2013. http://www.census.gov/population/international/data/idb/informationGateway.php. Accessed March 4, 2013.
7. World Health Organization. Geneva: World Health Organization; 2013. http://www.who.int/healthinfo/statistics/mortality_life_tables/en/. Accessed March 4, 2013.
8. Human Mortality Database. Berkeley, CA: University of California, Berkeley; 2013. http://www.mortality.org. Accessed April 4, 2013.
9. Kristof ND. Unhealthy America. *New York Times*. November 4, 2009: Opinion pages.
10. Weeks JR. *Population: An Introduction to Concepts and Issues*. 11th ed. Belmont, CA: Wadsworth Cengage Learning; 2012.
11. Haupt A, Kane TT. *Population Handbook*. 5th ed. Washington, DC: Population Reference Bureau; 2004.
12. United Nations Population Division. Net reproduction rate by major area, region, and country, 1950–2100. http://esa.un.org/wpp/Excel-Data/fertility.htm.
13. United Nations Population Division. New York, NY: United Nations; 2013. http://www.un.org/esa/population/unpop.htm. Accessed March 4, 2013.
14. U.S. Census Bureau. Washington, DC: U.S. Census Bureau; 2013. http://www.census.gov/population/international/data/idb/informationGateway.php. Accessed March 4, 2013.
15. Parrado EA. How high is Hispanic/Mexican fertility in the United States? Immigration and tempo considerations. *Demography*. 2011;48(3):1059–1080.

Chapter 4

Measuring Reproductive Health

Sara Yeatman

INTRODUCTION

The World Health Organization (WHO) defines reproductive health as "a state of complete physical, mental and social well-being and ... not merely the absence of disease or infirmity, in all matters relating to the reproductive system and its functions and processes." This definition implies that all people are should have

- Satisfying and safe sex lives;
- The capability to reproduce as well as the freedom to decide if, when, and how often to do so; and
- Access to appropriate healthcare services that enable women to go safely through pregnancy and childbirth, thus providing couples the best chance of having a healthy infants.[1]

Promoting reproductive health is not just about minimizing illness and death but also about encouraging reproductive well-being. Given the breadth of the WHO definition, it is difficult to develop a comprehensive list of indicators of reproductive health. These should include not just what people do (e.g., contraceptive use) or suffer (e.g., maternal hemorrhage, sexually transmitted infections), but also their knowledge of and access to services like safe maternal care and contraceptive methods.

Faced with the challenges of implementing the ideals of the Millennium Development Goals, the International Conference on Population and Development, and other global summits, policy makers, program managers, and advocates need concrete indicators and measures to show that they are improving access to and quality of family planning and other sexual and reproductive health services, increasing skilled attendance at birth and strengthening referral systems, reducing recourse to abortion and improving the quality of existing abortion services, providing information and services that respond to young people's needs, and integrating the prevention and treatment of reproductive tract infections, including HIV/AIDS, with other sexual and reproductive health services.[2]

This chapter introduces key reproductive health indicators, moving from measures of sexual behavior (the precursor of most reproductive outcomes) to contraception to pregnancy and finally birth. However, before covering specific indicators (the *what*) and their measures (the *how*), it is important to discuss *why* we measure indicators of reproductive health. It is essential to measure aspects of reproductive health to identify problems in need of resources and attention, to highlight specific populations in greatest need of reproductive health services, and to monitor existing programs and assess which ones work.[3] Throughout the chapter, we will keep returning to the question of *why* with examples of how reproductive health indicators are used in practice.

Reproductive health can be challenging to measure for a number of reasons. First, it is related to sex—who is sexually active (e.g., age at first sex), how sexual activity is practiced (e.g., with or without condoms), and what the outcomes of sexual activity are (e.g., a pregnancy, a sexually transmitted infection). Sex usually takes place behind closed doors, and aspects of sexual activity and its consequences can be highly stigmatized. Researchers must rely on individuals to report honestly about private and sensitive behaviors or attitudes. Clearly, individual reports about sexual activity can present many obstacles for accurate measurement.

The second major measurement challenge for reproductive health is that the entire enterprise can be de-emphasized or ignored. While reproductive health is defined to include both men and women, many of its aspects specifically address women and children. In many parts of the world, women are second-class citizens; therefore, areas of health that predominantly affect women and their children are not considered important. As a result, data necessary for measuring reproductive health may not be systematically collected by governments or healthcare institutions.

A third challenge is the use of uniform definitions and comparable measures. To compare reproductive health across populations with diverse cultures, standardized measures and comparable data are needed. While international organizations, such as WHO, work to develop indicators that can be compared across contexts, this is not an easy task. Even something as seemingly simple as whether an individual has *had sex* presents measurement challenges because the definition of sex varies by context. Does it mean only vaginal sex? Is someone who has had oral sex considered to have had sex? What about someone who has had sex only with a same-sex partner? Without uniform definitions, the data collected are unlikely to allow for meaningful comparisons.

The fourth challenge is data collection. Most aspects of reproductive health, such as contraceptive use and pregnancy outcomes are not captured in vital registration systems like births and deaths can be. Moreover, only a small fraction of reproductive health outcomes occur in health facilities where they can be easily recorded. For example, 85% of births occurred outside a health facility in Bangladesh between 2002 and 2007.[4]

DATA SOURCES

Data for measuring reproductive health come from four main sources: vital statistics, censuses, service statistics, and population-based studies.

Censuses and *vital statistics* registries (births, deaths, marriage) can provide some important numerators and denominators for measures of fertility, mortality, and union formation (marriage). In much of the world, however, these data are of inconsistent quality. Even under ideal circumstances, they provide only the most basic information about reproductive health. For other indicators, service delivery statistics and surveys are required.

The clinics that provide contraceptive care or abortions, offer maternity services, or treat sexually transmitted infections are important sources of reproductive health data. Data generated from these venues, called *service statistics*, are generally limited to information routinely collected during service delivery (e.g., conditions treated or services provided). Usually, these data do not reflect patient behaviors, attitudes, or knowledge. In the United States, some reproductive health events require mandatory reporting as part of public health surveillance. The Centers for Disease Control and Prevention (CDC), for example, require that health facilities report anonymized data on assisted reproductive technology (see Figure 4–1), induced abortion, and new cases of gonorrhea.[5] A weakness of service statistics is that they are limited to the population who seek services, so they reveal little about those persons who do not use contraception, who do not have sexually transmitted infections (or do not show up to get diagnosed), or who give birth at home. In many contexts—particularly in developing countries—poor

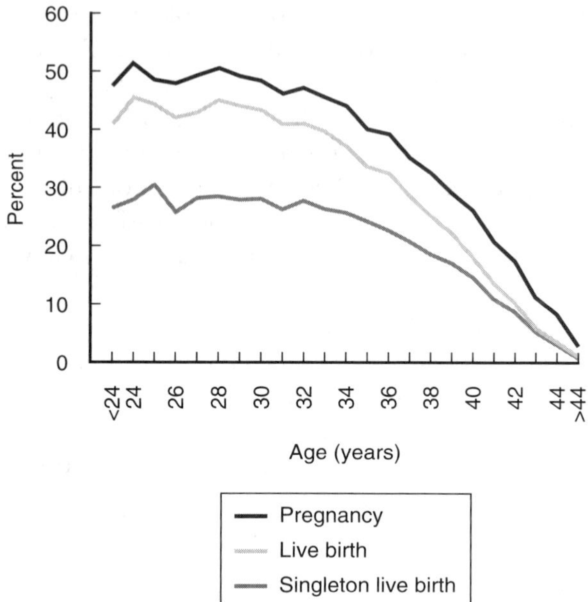

Figure 4–1 The CDC Requires Reporting of the Outcomes of Assisted Reproductive Technology (ART) in the United States

Reproduced from the National Center for Chronic Disease Prevention and Health Prevention. *2010 Assisted Reproductive Technology: National Summary Report.* Atlanta, Georgia: CDC; 2010. http://www.cdc.gov/art/ART2010/. Figure 14, p. 86.

and rural residents are less likely to use health services and are therefore likely to be missed by such statistics.[3]

Population-based studies (or sample surveys) use surveys of individuals to measure key reproductive health indicators. Using large representative samples of individuals, the data collected through the surveys can be generalized to larger populations of interest, such as counties or countries. Two examples of sample surveys that are widely used to measure aspects of reproductive health are highlighted next.

International: Demographic and Health Surveys

The Demographic and Health Surveys (DHS) are nationally representative studies from more than 90 countries, largely concentrated in Africa and Asia, funded by the U.S. Agency for International Development (USAID). DHS data have been critical for examining trends in a wide range of population and health topics in countries that generally have poor systems of vital registration. The DHS instrument includes reproductive health modules that measure sexual behavior, contraceptive use, birth spacing, maternal and infant mortality, knowledge and attitudes toward contraception, and fertility preferences.

DHS data allow researchers and policy makers to examine population and health trends over time, assess the impact of widespread interventions, and compare standardized health indicators across contexts. DHS data sets are publicly available (www.measuredhs.com) and the StatCompiler[6] tool offers easy access to key reproductive health indicators.[7]

Domestic: National Survey of Family Growth

Within the United States, one important source of reproductive health data is the National Survey of Family Growth (NSFG). The NSFG collects data on sexual and fertility-related behaviors (among other things) from a nationally representative sample of men and women between the key reproductive ages of 15 and 44. This survey uses a combination of in-person face-to-face interviews and audio computer-assisted self-interviewing (ACASI), where respondents listen to questions on headphones and enter their responses directly without having to say them to an interviewer. Since 2006, the NSFG has been collecting data continuously. Its results allow researchers and policy makers to examine trends in important indicators, such as the number of recent sexual partners or patterns of contraceptive use. This type of data collection also permits researchers to quickly detect significant changes in behavior.[8]

MEASURING SEXUAL BEHAVIOR

Sexual activity is a precursor to most reproductive health outcomes. Before someone delivers a baby or contracts a sexually transmitted infection, sexual activity is usually involved. Although hormonal contraception can be prescribed for nonsexual reasons, the use of contraception is usually associated with the desire to protect oneself from unwanted sexual outcomes. In measuring sexual activity, it is important to include behaviors not directly related to procreation. Sexual behaviors, such as oral and anal sex, do not lead to pregnancy, but they can transmit sexually transmitted infections, which have physical and mental health outcomes. Understanding sexual norms and behaviors in specific population is important for designing sexual education programs as well as for targeting reproductive and sexual health services.

While many indicators of sexual behavior exist, this section describes some of the most commonly used measures. Two of the most important describe the age at which the risk of reproductive and sexual outcomes start. *Age at menarche* describes the age at which a woman gets her first menstrual period and is therefore physiologically able to become pregnant. *Age at sexual initiation* (also called age at first sex or sexual debut) refers to the age at which sexual intercourse first occurs, thus exposing a woman not practicing contraception to the risk of fertilization. Other common measures of sexual behavior include the frequency and type of sexual activity, the number of recent and lifetime sexual partners, and the context of sexual activity (e.g., consensual or forced).

SEXUALLY TRANSMITTED INFECTIONS

Sexually transmitted infections (STI) are bacterial, viral, or parasitic infections that are primarily spread through interpersonal sexual contact. Some STIs, including the human immunodeficiency virus (HIV) and syphilis, can also be spread from mother to child during pregnancy and childbirth or between persons through blood products or tissue transfer.[9]

To understand the burden of disease imposed by various sexually transmitted infections as well as the risk of acquiring them, the concepts of prevalence and incidence are particularly important. Prevalence and incidence are key epidemiologic tools used to describe the distribution of a disease in a population. While they are related concepts, there are important distinctions between them.[10] *Prevalence*, the disease burden in a population, is most commonly presented as a percentage (although it can also be a count or a proportion):

$$\text{Prevalence (count)} = \text{number of current cases of a disease in a population}$$

$$\text{Prevalence (percent)} = \frac{\text{number of current cases of a disease}}{\text{population}} \times 100$$

Incidence, by comparison, refers to new cases of a disease over a given period of time and is a measure of disease risk. It can be presented as a count, proportion, or rate:

$$\text{Incidence (count)} = \text{number of new cases of disease during a specified time period}$$

Incidence proportion

$$= \frac{\text{number of new cases of disease during a specified time period}}{\text{population at start of time period}} \times 100$$

Incidence rate

$$= \frac{\text{number of new cases of disease during a specified time period}}{\text{person-years of exposure over time period}} \times 100$$

Human papillomavirus (HPV), for example, is the most common sexually transmitted infection in the United States. The CDC estimated that there were 14,100,000 people in the United States newly infected with HPV in 2008 (incidence) and 79,100,000 people living with this viral infection in the country.[10] Certain strains of HPV can cause cervical cancer, a fact that makes monitoring HPV trends particularly important.

Measuring HIV Prevalence and Incidence

HIV statistics are frequently cited to describe the burden and uneven distribution of HIV worldwide. For example, HIV prevalence is 11% in Malawi.[11] Sub-Saharan Africa is home to two thirds of all people in the world living with HIV, more than half of whom are women.[12] In the United States, HIV prevalence among African Americans is 10 times the prevalence among whites.[13] How do researchers know this? HIV tests are not part of most routine doctor well-health visits. Moreover, people can live with HIV for years before developing symptoms, let alone dying from acquired immunodeficiency syndrome (AIDS), the most severe form of HIV infection. So, how are these figures calculated?

In the United States, HIV cases are subject to mandatory reporting to the CDC. New cases of HIV are typically identified at health facilities such as an STI clinic or a facility that provides antenatal care. Once these cases are identified, they are reported confidentially to the CDC[14] (Figure 4–2).

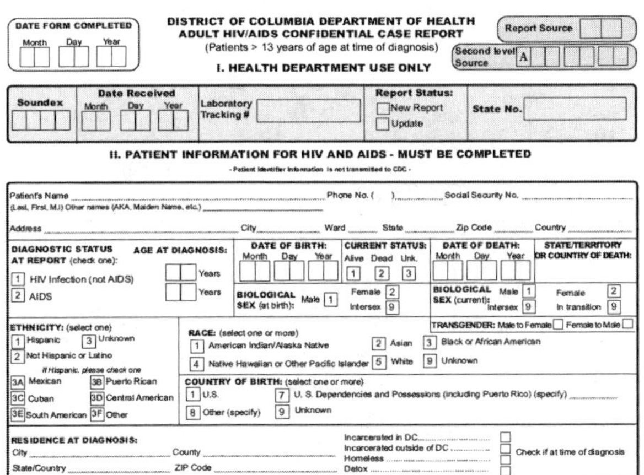

Figure 4–2 Form for Confidential Reporting of New HIV Diagnoses, District of Columbia

Reproduced from the District of Columbia Department of Health. *Adult HIV Case Report Form: Technical Assistance on Completing the Form.* Washington, DC: Author; 2013: p. 1 of 5-page form. http://doh.dc.gov/sites/default/files/dc/sites/doh/publication/attachments/Technical%20 Guidance%20for%20ACRF%202013_06_05.pdf.

Similarly, participating jurisdictions analyze blood samples to assess the recency of new infections so as to estimate HIV incidence.[15] In high-risk communities, HIV testing is often included as part of research studies, which helps contribute to monitoring. For example, since 2003 the CDC's National HIV Behavioral Surveillance (NHBS) has conducted rotating surveillance of three high-risk populations across the United States: men who have sex with men, injecting drug users, and heterosexuals at increased risk.[14]

These data are used to develop estimates of HIV prevalence and HIV incidence. In turn, these HIV estimates are used to monitor trends in transmission, allocate resources to populations at greatest risk, and assess the success of interventions. In 2009, there were approximately 1,148,200 persons older than the age of 12 living with HIV in the United States, representing 0.45% of the population.[16] The following year (2010), an estimated 47,500 people were newly infected with HIV.[15] As shown in Figure 4-3, the incidence rate varies widely by race/ethnicity and gender.

In the high HIV prevalence countries of sub-Saharan Africa, different approaches are used to monitor trends in HIV, often in combination with one another. First, population-based surveys such as the Demographic and Health Surveys (discussed earlier in this chapter) increasingly offer HIV testing to nationally representative samples of the reproductive-age population. Second, demographic surveillance sites in parts of sub-Saharan Africa collect detailed data on births, deaths, and often the HIV status of everyone living in a small geographic area, which can be used to estimate of HIV prevalence, AIDS-related deaths, and occasionally HIV incidence. Third, women are

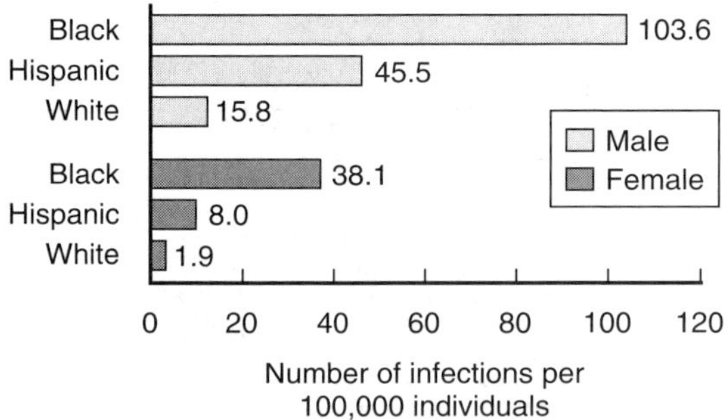

Figure 4-3 Incidence Rate of Newly Identified HIV Infections by Race/Ethnicity and Sex, 2010

Reproduced from the Centers for Disease Control and Prevention. New HIV Infections in the United States. CDC Fact Sheet. http://www.cdc.gov/nchhstp/newsroom/docs/2012/hiv-infections-2007–2010.pdf. Published December 2012.

routinely tested for HIV at antenatal clinics. In contexts where fertility is high and most pregnant women receive antenatal care, these service statistics are useful for monitoring trends in HIV over time and can be particularly sensitive to changes in risk in small geographic areas.

It is important to consider the denominator in HIV prevalence figures. Most HIV infections are sexually transmitted in sub-Saharan Africa, so HIV prevalence is usually calculated for the reproductive-age population, whose members are most likely to be sexually active. For example, as noted at the beginning of this section, HIV prevalence in Malawi is 11%. This statistic actually means that 11% of the population between the ages of 15 and 49 has contracted the HIV/AIDS virus. If prevalence were calculated for the entire Malawian population, it would be considerably lower than 11% because children and those living past the age of 50 are less likely to be infected.

DETERMINANTS OF FERTILITY

Demographers distinguish between the physiological capacity to reproduce, known as *fecundity*, and actual reproductive performance. *Parity* is a related term that is used to describe the number of live births a woman has experienced. For example, a woman who has two children is fertile (of parity 2) but may be infecund (or sterile) if she can no longer have children.[17,18] In contrast to parity, the obstetric term *gravida* refers to the number of pregnancies a woman has experienced, regardless of whether those pregnancies resulted in live births.

Fertility statistics are usually calculated for women of reproductive age, which is defined as either 15–44 or 15–49 years of age, depending on the data source. Because proportionately few births occur to women in their late 40s, 44 is often used as the cutoff point.

Proximate Determinants of Fertility

The fertility of a population is shaped by a variety of biological, social, economic, and cultural forces. These forces affect fertility through a series of so-called proximate determinants. Four proximate determinants explain most of the variation in fertility across populations:

- The proportion of women who are in sexual unions, and thus sexually active and at risk of pregnancy
- The proportion of women using contraceptives and the effectiveness of those contraceptives

- The level of induced abortion in the population
- The proportion of women who are fecund in the population[18,19]

We begin with the fourth proximate determinant, explaining the major reasons for infecundity in a population. Infecundity can be either short term or long term in nature. Short-term infecundity refers to postpartum periods of abstinence and lactational amenorrhea (i.e., the absence of a menstrual period after birth due to breastfeeding).[18,19] In some contexts, periods of abstinence and the fertility-reducing effect of breastfeeding can be considerable. Consider, for example, that in 2011 in Cameroon, the average period of abstinence after birth was 9 months and the average duration of breastfeeding was 11 months, which together result in a substantial period during which a postpartum woman could not get pregnant.[5]

In contrast, long-term infecundity refers to primary and secondary infertility. The definition of infertility varies across demographic and medical usage. In demography, primary infertility (also called primary sterility) describes the inability to bear any children. Secondary infertility, in contrast, describes the inability to bear another child after having earlier given birth. In medical terminology, however, infertility is defined as the inability to conceive (rather than give birth).[20] WHO, for example, defines infertility as "the inability to conceive a child. A couple may be considered infertile, if after two years of regular sexual intercourse, without contraception, the woman has not become pregnant (and there is no other reason, such as breastfeeding or postpartum amenorrhea."[21]

CONTRACEPTIVE USE

Family planning refers to the intentional effort of couples to regulate the number and spacing of births through the use of modern and traditional methods of contraception. *Contraceptives* are the devices that enable family planning by helping women and men reduce their likelihood of generating a pregnancy through sexual intercourse.

Most contraceptive statistics come from asking women in population-based surveys about their contraceptive behavior. Through a series of questions asked on these surveys, researchers can describe the contraceptive behavior and contraceptive needs of a population. Most often, surveys focus on the reported contraceptive behavior of women, because women experience pregnancies. Moreover, it is possible for women but difficult for men to use a contraceptive without their partner's knowledge.

One of the most frequently used indicators of contraceptive use is the contraceptive prevalence rate—a measure of the percentage of reproductive-age women in a population who are currently using contraception. It is usually calculated for specific populations, such as sexually active women, married women, or unmarried sexually active women.[18]

Contraceptive prevalence rate

$$= \frac{\text{number of women of reproductive age using contraception}}{\text{number of reproductive age women in the population}} \times 100$$

For example, 5% of married women in Chad were using contraceptives in 2010 compared to 79% of married women in Colombia.[22,23] Figure 4–4 shows contraceptive prevalence by sociodemographic characteristics in

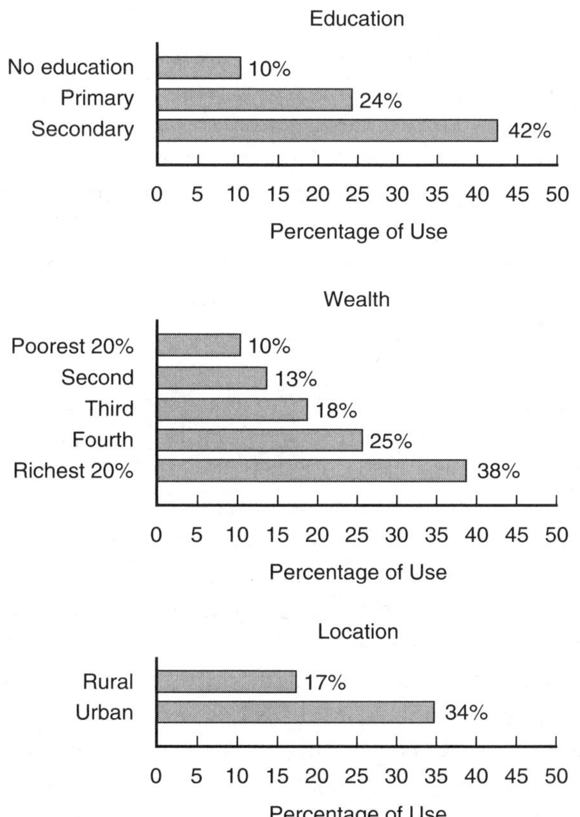

Figure 4–4 Contraceptive Prevalence in Sub-Saharan Africa by Sociodemographic Characteristics

Reproduced from Greene M, Joshi S, Robles O. *UNFPA State of the World Population 2012: By Choice, Not by Chance: Family Planning, Human Rights, and Development.* New York: United Nations Population Fund; 2012.

sub-Saharan Africa. Women who are more educated, are wealthier, and live in urban areas are the most likely to use contraceptives.

A common distinction made in surveys of contraceptive use is the percentage of women currently using a contraceptive and the percentage of women who have *ever used* contraception, or a particular contraceptive method. "Ever used" can be used to gauge the contraceptive willingness of a population, to describe the range of methods used, and to understand contraceptive discontinuation and the reasons for it.[24]

A wide variety of contraceptive methods exist today to meet the needs of sexually active people looking to prevent pregnancy or STIs (see the *Contraceptive History and Practice* chapter). Some contraceptive methods, such as male and female condoms and withdrawal, are coital dependent, meaning that action must be taken at the time of sexual intercourse to prevent pregnancy. A woman could have sex one day using a condom and then have sex the next day without a condom. Therefore, when measuring contraceptive use, it is important to know not just which type of contraceptive someone is using, but how consistently it is being used.

Other contraceptives are not coital dependent, such as the Depo-Provera injectable contraception. Within the 3-month period in which a Depo-Provera shot is effective, all acts of sexual intercourse will be protected from pregnancy. Not surprisingly, coital-dependent methods are less effective at preventing pregnancy in practice because there is more room for user error when a method must be used consistently and correctly for every act of sexual intercourse.

Couples can use more than one form of contraception, particularly when they want to protect themselves from both pregnancy (e.g., with a highly effective hormonal contraceptive device) and sexually transmitted infections (e.g., with condoms, which offer protection from most STIs). A couple is said to be using *dual protection* when they use a condom for protection from STIs in conjunction with a contraceptive method that is highly effective at preventing pregnancy. Perhaps the most important characteristic of a contraceptive device is its effectiveness in preventing pregnancy. *Contraceptive effectiveness* is measured by comparing the number of sexually active couples who would get pregnant in a year if they used a particular contraceptive. The effectiveness of any particular contraceptive method can be divided into pregnancy rates during perfect use and pregnancy rates during typical use.

The effectiveness of contraceptives varies widely by method. For example, with the hormonal (Levonorgestrel) intrauterine device (IUD), approximately 0.2% of couples will get pregnant after 1 year of perfect use.

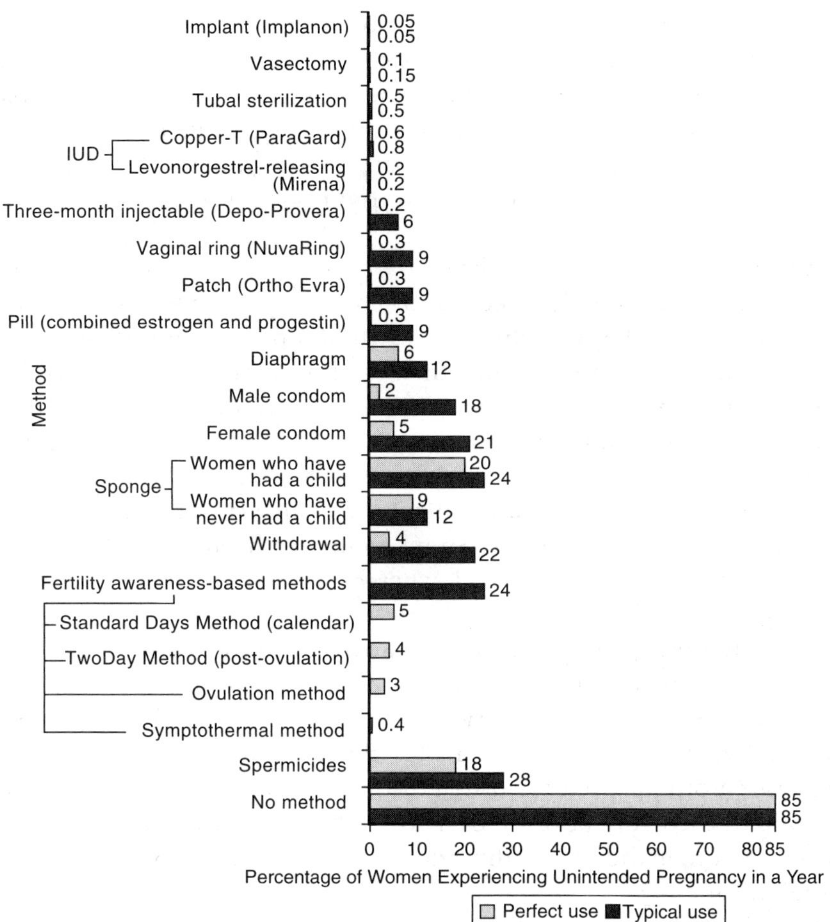

Figure 4–5 Proportion of U.S. Women Initiating Use of a Birth Control Method Who Will Become Pregnant During Their First Year of Use, by Method and Type of Use

Data from Trussell J, Guthrie KA. Choosing a contraceptive: efficacy, safety, and personal considerations. In: Hatcher RA, Trussell, J, Nelson AL, Cates, W, Kowal D, Policar MS. *Contraceptive Technology* Atlanta, GA: Ardent Media Inc. 2011: 24.

Thus this method is considered to be 99.8% effective (Figure 4–5). By comparison, with perfect use of the male condom, 2% of couples would be expected to get pregnant in year, so with perfect use the male condom is 98% effective. Of course, typical use is not always consistent and correct. Coital-dependent methods, in particular, have a larger discrepancy between perfect and typical use. With condoms, this means that pregnancy rates vary from 2% with perfect use to 18% with typical use (82% effectiveness).[25-27]

For methods that are not coital dependent, such as the IUD, injectable contraception, and female and male sterilization, there are almost no differences between correct and consistent use (i.e., perfect use) and typical use. For other hormonal methods that require more diligence on the part of the user (e.g., birth control pills, patch, or vaginal ring), there is considerable difference between these two figures because of inconsistent or incorrect use.

Another useful measure is the *contraceptive method mix* of a population. Having access to a wide selection of contraceptive methods is a positive indicator of the reproductive health of a population. When people have a range of methods from which to choose, they are more likely to find and use the method that is most appropriate for them. The contraceptive method mix of a population is the percentage of contraceptive use made up of different contraceptive methods. For a variety of historical, programmatic, and cultural reasons, the contraceptive mix of populations varies widely. Some countries have a lot of variety in the contraceptive methods used, whereas others rely on one dominant method.

Japan, for example, offers an illustrative, if extreme, case of major reliance on a single contraceptive method. Male condom use is higher in Japan than in any other country, accounting for 75% of all contraceptive use in 1997.[28] One of the reasons for this reliance is that birth control pills were not approved for distribution in Japan until 1999, compared to 1960 in the United States. Even after the pill was approved, however, Japanese women did not readily adopt it, as using male condoms was the social norm.

Figure 4–6 shows variation in the contraceptive method mix by region.[29] Clearly, there is considerable variation across the world. The birth control pill is a dominant method in every region except Asia, where the IUD is particularly popular. Condom use makes up a relatively smaller proportion of contraceptive use in Africa than in the rest of the world and is most popular in Europe. Also notable is the ratio of male to female sterilization. In regions that have more gender inequality, male sterilization (a relatively uncomplicated outpatient procedure) is considerably less common than female sterilization, which is more complicated and costly[30] (e.g., Asia, Latin America and the Caribbean, Africa). In contrast, in more-developed and gender-equitable contexts, male sterilization approximates the level of female sterilization.

Beyond contraceptive behavior, surveys also frequently capture women's and men's knowledge of, attitudes toward, and access to different types of contraceptives.

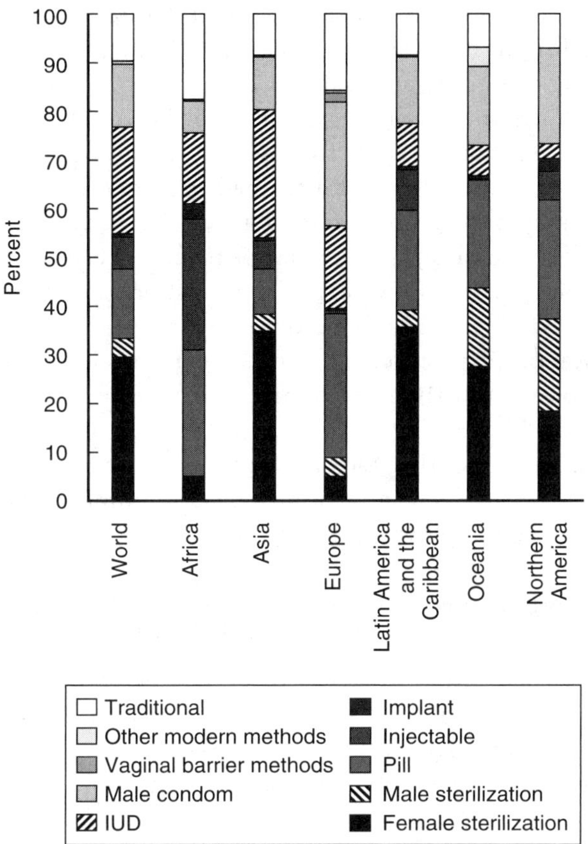

Figure 4–6 Contraceptive Method Mix (as Share of Overall Use) by Region

Reproduced from the United Nations Population Division. *World Contraceptive Patterns 2013.* New York: United Nations; 2013. http://www.un.org/en/development/desa/population/publications/family/contraceptive-wallchart-2013.shtml.

ABORTION

Technically, the term *abortion* refers to the premature expulsion or loss of embryo. This includes both induced abortion (hereafter referred to as *abortion*) and spontaneous abortion, conventionally referred to as *miscarriage.* Among women who know they are pregnant, the miscarriage rate ranges from 15% to 20%.[31]

Abortion can be particularly challenging to measure. Statistics on legal abortion are collected through service statistics (i.e., clinics performing abortion) or population-based surveys. In many countries, however, abortion is either illegal or difficult to acquire safely. In such contexts, most abortions are considered unsafe abortions, and these are considerably more difficult to measure accurately. WHO defines an unsafe abortion as any procedure to terminate an unintended pregnancy done either by people lacking the necessary skills or in an environment that does not conform to minimum medical standards, or both.[32] When an abortion has to be carried out in secret for legal, cultural, or social reasons, it is often unsafe, even if done by a trained provider, because emergency backup care may not be immediately available. Such abortions are far more difficult to measure, as there may be no clinic record and women are likely to underreport their own experience of abortion, particularly where abortion-related stigma is high. To develop estimates of unsafe abortion, a combination of hospital records from abortion-related complications, surveys of women and their confidantes, and key informant interviews are conducted.[33-35]

The two most frequently cited abortion indicators are the abortion rate and the abortion ratio. The abortion rate is the number of abortions per 1000 women of reproductive age in a given year:

$$\text{Abortion rate} = \frac{\text{number of abortions}}{\text{women of reproductive age}} \times 1000$$

For example, in 2008, Russia had one of the highest rates of abortion, at 38 abortions per 1000 women aged 15–44. In contrast, the Netherlands had a rate of 8 abortions per 1000 women aged 15–44.[36]

The abortion ratio is the number of abortions per 100 live births in a given year:

$$\text{Abortion ratio} = \frac{\text{number of abortions}}{\text{number of live births}} \times 100$$

The main interpretative difference between the two statistics is that the abortion ratio factors in the level of fertility in the population, while the abortion rate does not. Thus, for example, despite abortion rates being comparable between developing and developed countries (or somewhat higher in developing countries), the abortion ratio is lower in developing countries because the birth rate is higher, which leads to a lower abortion to live birth ratio.[37] The abortion ratio is also closely associated with another indicator: the percentage of pregnancies ending in abortion.

In 2009, the United States had an abortion ratio of 23 per 100 live births and an abortion rate of 15.1 abortions per 1000 women aged 15–44.[38] Figure 4–7 shows that the abortion rate in the United States has been declining over the last three decades.

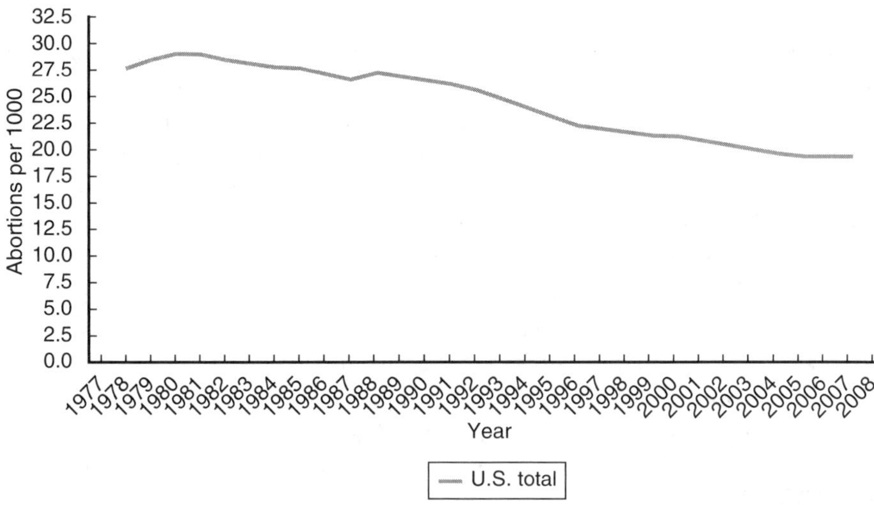

Figure 4–7 Abortion Rate (per 1000 Women Aged 15–44) in the United States, 1977–2008

Data for 1978–2004: Henshaw SK, Kost K. *Trends in the Characteristics of Women Obtaining Abortions, 1974–2004.* New York, NY: Guttmacher Institute; 2008. Data for 2005: Centers for Disease Control and Prevention's *2005 Abortion Surveillance Report* and unpublished tabulations of data from Guttmacher Abortion Provider Surveys. Data for 2007: Centers for Disease Control and Prevention's *2007 Abortion Surveillance Report* and unpublished tabulations of data from Guttmacher Abortion Provider Surveys.

FERTILITY PREFERENCES AND INTENTIONS

You may hear a friend say, "I'm going to wait until I finish graduate school before having children." Your friend is stating a fertility intention—that she intends to have children but does not want them in the near future. Fertility preferences and intentions are important tools for forecasting changes in fertility (as they are reasonable predictors of behavior) and for assessing reproductive health needs (as they indicate who—if sexually active—should be using contraception to prevent an unintended pregnancy).

Fertility preferences and fertility intentions are actually distinct constructs, although they are often conflated. Fertility preferences are feelings or desires related to having children, whereas intentions involve planning.[39] A number of survey tools have been developed to measure fertility preferences and intentions, with the most commonly used being introduced here.

Ideal family size is a measure of how many children an individual considers ideal if the matter were fully in his or her control. Because many men and women asked this question have already begun to have children, and some may already have more than their ideal, this question is usually asked hypothetically. For example, the Demographic and Health Surveys ask people who already have children: "If you could go back to the time you did not have any children and could choose exactly the number of children to have in your whole life, how many would that be?"[40] At the individual level, ideal family size is used to gauge desires for children in the future and to retrospectively assess the intentionality of children already born. Mostly ideal family size is used in the aggregate to gauge social norms around fertility (e.g., what is considered the ideal number of children in a particular community, and how that perception has changed over time).

Current fertility preferences are measured in surveys by asking men and women whether they want to have a child (or another child) and if so, how long they would like to wait. Retrospective fertility preferences (or intentions) are more difficult to measure. One approach is to ask women how they felt about a pregnancy when it occurred: Was it wanted (or intended) then, wanted (or intended) later, or not wanted (or intended) at all? This helps to identify pregnancies that were unwanted (or unintended/unplanned) at the time of conception and pregnancies that were mistimed (i.e., wanted at a later time). This approach is not without its challenges because mothers are considerably less likely to refer to one of their children as unwanted once they have had the child.

Among other uses such as forecasting fertility behavior, reproductive health advocates and policy makers use fertility preferences to calculate unmet need for family planning[41] (i.e., sexually active women who do not want to get pregnant but are not using contraception):

Percent unmet need for family planning

$$= \frac{\text{women not using any method of contraception}}{\text{woman wanting no more children or to delay next birth}} \times 100$$

A related indicator is the proportion of demand for contraceptives satisfied, which is a measure of how well contraceptive supply within the health system is meeting demand. It is also an indicator of how well women's stated desires regarding fertility and contraception are being fulfilled and can highlight inequities in access to appropriate and acceptable contraceptive methods.[41]

Proportion of demand for contraceptives satisfied

$$= \frac{\text{women using any method of contraception}}{\left[\begin{array}{l}\text{women using any method of contraception} + \begin{array}{l}\text{women not using contraception} \\ \text{and wanting no more children or} \\ \text{wanting to delay the next birth}\end{array}\end{array}\right]}$$

STILLBIRTHS AND CHILD MORTALITY

Pregnancies can end in four ways—through live birth, miscarriage, abortion, or stillbirth. Definitions of stillbirth (also called fetal death) vary across contexts but are usually determined by the size and length of gestation of a fetus that has died in utero. For purposes of international comparison, WHO defines a stillbirth as a death in utero of a fetus at or after 28 weeks' gestation,[42,43] but many countries' definitions have earlier cutoffs (commonly 20 weeks).

Stillbirths are largely preventable, as evidenced by the relatively low 2009 ratio of stillbirths in developed countries (3 per 1000 births) compared to 28 per 1000 births in sub-Saharan Africa and 27 per 1000 births in South Asia. Indeed, 98% of all stillbirths occur in low- or middle-income countries.[43]

Infant deaths describe deaths within the first year of life. The first days of life are particularly hazardous, and large numbers of children die shortly before or just after birth. The causes and determinants of deaths that occur before birth, immediately following birth, and later within the first year differ. The first two types of deaths are closely associated with quality of care during pregnancy and upon delivery, while the third has more to do with risks unrelated to birth and has been more successfully reduced worldwide through public health measures such as improved sanitation and immunizations.[42] *Perinatal* deaths include stillbirths and deaths within the first 7 days of life. *Neonatal* deaths, in contrast, are those that occur in the first month (28 days) of life.

$$\text{Stillbirth rate} = \frac{\text{number of stillbirths}}{\text{total births (live + stillbirth)}} \times 1000$$

$$\text{Perinatal mortality rate} = \frac{\text{number of stillbirths} + \text{deaths in first 7 days}}{\text{total births}} \times 1000$$

$$\text{Neonatal mortality rate} = \frac{\text{number of deaths in first 28 days of life}}{\text{live births}} \times 1000$$

$$\text{Infant mortality rate} = \frac{\text{deaths}_{0-1} \text{ (deaths under age 1)}}{\text{live births}} \times 1000$$

One of the most important measures of child health in a population is the under-5 mortality rate, which describes the probability of a child dying before his or her fifth birthday:

$$\text{Under-5 mortality rate} = \frac{\text{deaths}_{0-5} \text{ (deaths under age 5)}}{\text{live births}} \times 1000$$

You might notice that although these statistics are called "rates," most are not technically rates because the denominator is births rather than the population at risk of death during a certain period of time. Instead, these indicators represent probabilities of death.

MATERNAL MORTALITY

In parts of the world, women hardly think about the possibility of death when they become pregnant. And for good, as maternal mortality has fallen dramatically in much of the world such that it hardly ever occurs. For example, in 2010 in Austria, approximately 4 women died per 100,000 births.[44] Elsewhere, however, pregnancy and childbirth remain real threats to women's health. That same year in Somalia, 1000 women died for every 100,000 births.[44] In other words, for every 100 births, 1 woman died. Approximately 99% of maternal deaths occur in less developed countries, and this mortality is concentrated in sub-Saharan Africa and South Asia (Figure 4–8).[44] Indeed, approximately 800 women die every day from preventable causes related to pregnancy or childbirth.[45]

A maternal death is the death of a woman while pregnant or within 42 days of the termination of the pregnancy (through birth, miscarriage, or abortion) due to causes related to or aggravated by pregnancy.[44] The main causes of maternal death are severe bleeding (hemorrhage), indirect causes (diseases that complicate pregnancy or are aggravated by it such as HIV,

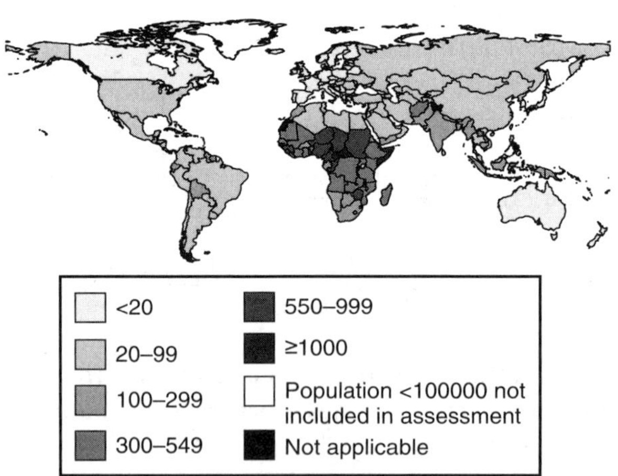

Figure 4–8 The Uneven Global Distribution of Maternal Mortality

Reproduced from World Health Organization. *Trends in Maternal Mortality: 1990 to 2010.* Geneva, Switzerland: WHO; 2012:23.

malaria, and anemia), infections, unsafe abortion, eclampsia, and obstructed labor.[46]

Maternal mortality is particularly difficult to measure with much precision because it requires information about the deaths of women, including their age, their pregnancy status at or near the time of death, and the medical cause of the death. Additionally, even in countries with high maternal mortality, it remains a relatively rare event; thus large quantities of data are necessary to produce reliable and precise estimates. To overcome these challenges, multiple sources of data such as vital registration systems, censuses, household surveys, and statistical models are used in tandem with record review and verbal autopsies.[47]

Two statistics are often calculated to compare the risk of pregnancy and childbirth across populations. The most frequently used indicator is the maternal mortality ratio, which estimates a woman's risk of dying per 100,000 live births:

$$\text{Maternal mortality ratio} = \frac{\text{number of maternal deaths}}{100,000 \text{ live births}}$$

The second indicator is the adult lifetime risk of maternal death, which combines the risk of dying from a given birth with the fertility level of the population by estimating the probability that a 15-year-old woman will eventually die from maternal causes.[44] In societies with high fertility, a woman is at greater risk of dying from pregnancy or childbirth both because she is at risk more often and because higher-order and closely spaced pregnancies carry additional risks.

In Yemen in 2010, the maternal mortality ratio was 200 maternal deaths per 100,000 live births and the lifetime risk of maternal death was 1 in 90. In contrast, Namibia had the same maternal mortality ratio but a relatively lower lifetime risk of maternal death, 1 in 160.[44] A major contributing factor explaining this difference is the fact that the total fertility rate in Yemen was double that of Namibia.

ANTENATAL CARE AND CHILDBIRTH

Key determinants of perinatal and maternal mortality are the care a woman receives while pregnant and the situation in which the baby is delivered. Antenatal care is important to monitor the pregnancy, detect and treat complications early, prevent diseases through immunization and nutrition, and communicate health messages about the delivery and newborn care. The coverage and quality of antenatal care in a population are assessed by measuring the prevalence of antenatal care, the training of the personnel providing antenatal care, when women first receive care, how frequently they receive care during their pregnancy, and what was included in the care. In most developing countries, this information comes from household surveys such as the Demographic and Health Surveys.[48]

Birthing systems vary widely by context—both because of issues of access and infrastructure and because of cultural preferences. For example, in 2009, midwife-attended home births accounted for approximately 30% of births in the Netherlands and fewer than 1% of births in the United States.[49,50] In 2011, Bangladesh had lower levels of births delivered in health facilities (27%) compared to its South Asia neighbors. However, this was a dramatic improvement from 1993, when fewer than 4% of Bangladeshi births occurred in such settings.[51]

Although there is no one universally accepted definition for safe childbirth, some generally accepted characteristics clearly improve the safety of delivery for mother and baby. These include being attended by trained personnel (i.e., physician, midwife, or skilled birth attendant) and delivering in a health facility or other sanitary location where safe transfer to a health facility is possible should complications arise. One reason deliveries in health facilities can be safer than home-based ones is access to cesarean section should the need arise. WHO suggests that between 10% and 15% of births should be delivered by cesarean section. Too few of these procedures indicate that women and babies who would benefit from cesarean deliveries do not have access to them and, therefore, that maternal and perinatal mortality are unnecessarily high. In contrast, where cesarean rates substantially exceed 15%, the costs to the health system are unnecessarily high without achieving gains in maternal and infant health.[52,53] Worldwide, there is considerable variation in the cesarean section rate: approximately 1% of births were delivered by cesarean section in Niger in 2006, compared to 46% that same year in Brazil.[53]

WOMEN'S STATUS

Gender, according to WHO, refers to the socially constructed roles, behavior, activities, and attributes that a particular society considers appropriate for men and women. These can give rise to gender inequalities that favor one group over the other. In most societies, women have a lower status than men, and this factor can contribute to inequities in health access and outcomes that favor men at the expense of women.

The status of women in a community is closely related to reproductive health outcomes. In general, where women have greater autonomy, contraceptive use increases, fertility declines, and maternal and infant mortality falls.[54,55]

No single indicator measures women's status. Rather, a number of different indicators are typically combined to develop a picture of the status of women in a society. Age at first marriage and at first child's birth are markers of the opportunities available to women and girls. Where girls marry early and start having children as adolescents, women are mostly valued for their reproductive roles and have few opportunities for education or employment outside the home.[41] They are also exposed to the additional health risks imposed by early and frequent childbearing.

Since 2000, most Demographic and Health Surveys have included special modules designed to measure the gender norms of a society, gender inequities, and women's empowerment. In addition to the variables mentioned previously, these modules measure women's education (e.g., average attainment, educational attainment relative to men's), economic autonomy (e.g., participation in labor force, involvement in household decision making, ability to control own earnings),[56] health autonomy (e.g., ability to make health-related decisions for themselves and their children, obstacles to accessing health care), proxies for relationship power such as spousal differences in age and education, and attitudes toward wife-beating and a wife's ability to refuse sex to her husband. Additionally, the surveys measure the prevalence of emotional, physical, and sexual violence and female genital cutting in communities.[7,57]

CONCLUSION

Reproductive health is broadly defined to include sexual activity, reproductive choices and behavior including contraceptive practice, childbirth outcomes for both women and children, miscarriages and induced abortions, and sexually transmitted infections. Measuring reproductive health is often challenging because of sparse data collection, the association of reproductive health with sexual activity, not; stigmatization in certain cultures, and the lack of uniform definitions and measures. Extant reproductive health data emanate from vital statistics, censuses, service statistics, and population-based studies.

Sources of reproductive health vary geographically. In the United States, for example, the National Survey of Family Growth gathers data about fertility behavior and intentions as well as contraceptive practice. In developing countries, these data are often collected through Demographic and Health Surveys (DHS). Because of the relationship between the women's status and fertility, data measuring gender norms and women's empowerment are often collected through the DHS. In every setting, more than one data source is required to provide an accurate view of the broad field of reproductive health.

DISCUSSION QUESTIONS

1. What are some of the possible repercussions of the difficulty in measuring reproductive health? Consider maternal mortality in particular when answering this question.
2. The countries that are in most need of improved reproductive health tend to be the ones with little data on this topic. How can this situation be improved? What could be accomplished with more accurate data?
3. Because there is a greater difference between "correct" and "consistent" use for methods that are coital dependent or require more diligence on the part of the user, should there be a greater emphasis on IUDs and other non-coital-dependent forms of contraceptives?
4. Although reproductive health as a whole includes men and women, discuss why it is often more concerned with women and children.
5. Why is it difficult to measure the status of women across the globe?

REFERENCES

1. United Nations. *Report of the International Conference on Population and Development, Cairo, 5–13 September 1994.* New York: United Nations; 1995: Sales No. 95.XIII.18. para 7.2.
2. World Health Organization. *The WHO Strategic Approach to Strengthening Sexual and Reproductive Health Policies and Programmes.* Geneva, Switzerland: WHO Division of Reproductive Health and Research; 2007. http://www.who.int/reproductivehealth/publications/strategic_approach/RHR_07.7/en/index.html. Accessed May 7, 2013.
3. Blanchard K, Elul B, Ramarao S. *Reproductive Health Indicators: Moving Forward.* New York: Population Council; 1999.
4. National Institute of Population Research and Training (NIPORT), Mitra and Associates, Macro International. *Bangladesh Demographic and Health Survey 2007.* Dhaka, Bangladesh/Calverton, MD: National Institute of Population Research and Training, Mitra and Associates, and Macro International; 2009.
5. Centers for Disease Control and Prevention. *Incidence, Prevalence, and Cost of Sexually Transmitted Infections in the United States.* CDC Fact Sheet. Atlanta, GA: CDC; February 2013.
6. ICF International. MEASURE DHS STATcompiler. 2002. http://www.statcompiler.com. Accessed February 21, 2013.
7. Measure DHS. Demographic and Health Surveys. http://www.measuredhs.com. Accessed May 8, 2013.
8. Groves RM, Mosher WD, Lepkowski J, Kirgis NG. Planning and development of the continuous National Survey of Family Growth. National Center for Health Statistics. *Vital Health Stat.* 2009;1(48).

9. World Health Organization. *Sexually Transmitted Infections Fact Sheet.* Geneva, Switzerland: WHO; August 2011:110.

10. Centers for Disease Control and Prevention. Data and statistics. http://www.cdc.gov/reproductivehealth/data_stats/index.htm. Accessed February 15, 2013.

11. National Statistical Office (NSO) and ICF Macro. *Malawi Demographic and Health Survey 2010.* Zomba, Malawi/Calverton, MD: NSO and ICF Macro; 2010.

12. UNAIDS. *Report on the Global AIDS Epidemic.* Geneva, Switzerland: Joint United Nations Programme on HIV/AIDS (UNAIDS). 2012. http://www.unaids.org/en/resources/publications/2012/name,76121,en.asp. Accessed March 17, 2014.

13. McQuillan GM, Kruszon-Moran D. *HIV Infection in the United States Household Population Aged 18–49 Years: Results from 1999–2006.* NCHS Data Brief No. 4. Hyattsville, MD: National Center for Health Statistics; 2008.

14. Centers for Disease Control and Prevention. HIV surveillance supported by the Division of HIV/AIDS Prevention. 2012. http://www.cdc.gov/hiv/topics/surveillance/resources/factsheets/pdf/HIV_surveillance.pdf. Accessed February 28, 2013.

15. Centers for Disease Control and Prevention. Estimated HIV incidence among adults and adolescents in the United States, 2007–2010. *HIV Surveill Suppl Rep.* December 2012;17(4). http://www.cdc.gov/hiv/topics/surveillance/resources/reports/#supplemental. Accessed March 11, 2013.

16. Centers for Disease Control and Prevention. Monitoring selected national HIV prevention and care objectives by using HIV surveillance data—United States and 6 U.S. dependent areas—2010. *HIV Surveill Suppl Rep.* June 2012;17(3 pt A). http://www.cdc.gov/hiv/topics/surveillance/resources/reports/. Accessed March 11, 2013.

17. Leridon H. Human fecundity: situation and outlook. *Pop Soc.* October 2010:471.

18. Population Reference Bureau. *Population Handbook.* 6th Ed. Washington, DC: Population Reference Bureau; 2011. http://www.prb.org/Publications/Reports/2011/prb-population-handbook-2011.aspx. Accessed March 17, 2014.

19. Bongaarts J. A framework for analyzing the proximate determinants of fertility. *Pop Develop Rev.* 1978;4(1):105–132.

20. Rutstein SO, Shah IH. *Infecundity, Infertility, and Childlessness in Developing Countries.* DHS Comparative Reports No. 9. Calverton, MD: ORC Macro and World Health Organization; 2004.

21. World Health Organization. Health topics: infertility. http://www.who.int/topics/infertility/en/. Accessed February 5, 2013.

22. ICF Macro, Profamilia. *Colombia Demographic and Health Survey 2009–2010.* Fairfax, VA: ICF Macro; 2011.

23. République du Tchad, Institut National de la Statistique, UN Population Fund, UNICEF. *Enquête par Grappes à Indicateurs Multiples: TChad 2010.* New York: UNFPA and UNICEF; May 2011.

24. Daniels K, Mosher WD, Jones J. *Contraceptive Methods Women Have Ever Used: United States, 1982–2010.* Atlanta, GA: Centers for Disease Control and Prevention; February 14, 2013:62.

25. Trussell J. Contraceptive failure in the United States. *Contraception.* 2011;83(5):397–404.
26. Trussell J. Estimates of contraceptive failure from the 1995 National Survey of Family Growth. *Contraception.* 2008;78:85.
27. Kost K, Singh, S, Bankoleet A. Estimates of contraceptive failure from the 2002 National Survey of Family Growth. *Contraception.* 2008;77(1):10–21.
28. Sato R, Iwasawa M. Contraceptive use and induced abortion in Japan: how is it so unique among the developed countries? *Japan J Pop.* 2006;4(1):33–54.
29. United Nations Population Division. *World Contraceptive Patterns 2012.* New York: United Nations; 2013.
30. Smith GL, Taylor GP, Smith KF. Comparative risks and costs of male and female sterilization. *Am J Public Health.* 1985;75(4):370–374.
31. United States National Library of Medicine. Miscarriage. In: *ADAM Medical Encyclopedia.* PubMed Health. http://www.ncbi.nlm.nih.gov/pubmedhealth/PMH0002458/. Accessed May 8, 2013.
32. World Health Organization. *The Prevention and Management of Unsafe Abortion: Report Of a Technical Working Group.* Geneva, Switzerland: WHO; 1992.
33. Sedgh G, Singh S, Shah IH, et al. Induced abortion incidence and trends worldwide from 1995–2008. *Lancet.* 2012;6736(11):61786–61788.
34. Singh S, Fetters T, Gebreselassie H, et al. Estimated incidence of induced abortion in Ethiopia, 2008. *Int Persp Sex Reprod Health.* 2010;36(1):16–25.
35. Sedgh G, Rossier C, Kaboré I, et al. Estimating abortion incidence in Burkina Faso using two methodologies. *Stud Fam Plan.* 2011;42(3):147–154.
36. Sedgh G, Singh S, Henshaw SK, Bankole A. Legal abortion worldwide in 2008: levels and recent trends. *Persp Sex Reprod Health.* 2011;43(3):188–198.
37. Sedgh G, Henshaw S, Singh S, et al. Induced abortion: estimated rates and trends worldwide. *Lancet.* 2007;370(9595):1338–1345.
38. Centers for Disease Control and Prevention. Abortion surveillance—United States, 2009. *MMWR.* 61(8):1–48.
39. Thomson E. Couple childbearing desires, intentions, and births. *Demography.* 1997;34(3):343–354.
40. Westoff CF. *Desired Number of Children: 2000–20008.* DHS Comparative Reports No. 25. Calverton, MD: ICF Macro; 2010.
41. United Nations Population Fund. *How Universal Is Access to Reproductive Health? A Review of the Evidence.* New York: UNFPA; 2010.
42. World Health Organization. *Neonatal and Perinatal Mortality: Country, Regional and Global Estimates.* Geneva, Switzerland: WHO; 2006.
43. Cousens S, Blencowe H, Stanton C, et al. National, regional, and worldwide estimates of stillbirth rates in 2009 with trends since 1995: a systematic analysis. *Lancet.* 2011;377(9774):1319–1330.
44. World Health Organization. *Trends in Maternal Mortality: 1990 to 2010.* Geneva, Switzerland: WHO; 2012.
45. World Health Organization. *Maternal Mortality Factsheet No. 348.* Geneva, Switzerland: WHO; May 2012.

46. World Health Organization. *The World Health Report: Make Every Mother Count.* Geneva, Switzerland: WHO; 2005.

47. World Health Organization. *Maternal Mortality in 2000: Estimates Developed by WHO, UNICEF and UNFPA.* Geneva, Switzerland: WHO; 2004.

48. World Health Organization. *Antenatal Care in Developing Countries: Promises, Achievements and Missed Opportunities: An Analysis of Trends, Levels and Differentials.* Geneva, Switzerland: WHO; 2003.

49. Hendrix, MJ, Evers SM, Basten MC, et al. Cost analysis of the Dutch obstetric system: low-risk nulliparous women preferring home or short-stay hospital birth: a prospective non-randomised controlled study. *BMC Health Serv Res.* 2009;9:211.

50. Macdorman MF, Mathews TJ, Declercq E. *2012 Home Births in the United States, 1990–2009.* NCHS Data Brief 84. Hyattsville, MD: National Center for Health Statistics; 2012.

51. ICF International. Measure DHS STATcompiler. http://www.statcompiler.com. Accessed February 21, 2013.

52. World Health Organization. Appropriate technology for birth. *Lancet.* 1985;2(8452):436–437.

53. Gibbons L, Belizán JM, Lauer JA, Betrán AP, Merialdi M, Althabe F. The global numbers and costs of additionally needed and unnecessary caesarean sections performed per year: overuse as a barrier to universal coverage. *World Health Report (2010): Background Paper No. 30.* Geneva, Switzerland: WHO; 2010:3–32.

54. Shen C, Williamson JB. Maternal mortality, women's status, and economic dependency in less developed countries: a cross-national analysis. *Soc Sci Med.* 1999;49:197–214.

55. Kishor S, Subaiya L. *Understanding Women's Empowerment: A Comparative Analysis of Demographic and Health Surveys (DHS) Data.* DHS Comparative Reports No. 20. Calverton, MD: Macro International; 2008.

56. Schuler SR, Hashemi SM, Riley AP. The influence of women's changing roles and status in Bangladesh's fertility transition: evidence from a study of credit programs and contraceptive use. *World Development.* 1997;25(4):563–575.

57. Schatz E, Williams J. Measuring gender and reproductive health in Africa using demographic and health surveys: the need for mixed-methods research. *Culture Health Sex.* 2012;14(7):811–826.

Chapter 5

Population Theories and Dynamics

John R. Weeks

INTRODUCTION

In many respects, the modern world was created in the 18th century with the rise of the Age of Enlightenment in Europe. This revolution in the way we think about and act toward one another and our environment is still in progress today. An important part of this perspective is that the rights of individuals supersede the demands of a monarchy or dictatorship—an idea that inspired both the American and French Revolutions. The Enlightenment ushered in an era of critically questioning traditional ideas and authority that continues to reverberate around the world.[1]

For many centuries, people had pondered the issue of population growth, but the world had seen relatively little of it until Europe's population began to grow, especially in the 17th and 18th centuries. It is not clear that the Enlightenment had anything directly to do with the initial drop in death rates that started Europe's population growing. Indeed, the causality may have worked in the other direction, in that we know that a growing population forces societies to adapt or die (an evolutionary concept that, of course, gained currency only as part of the scientific revolution taking place within the Enlightenment). What is arguably the most famous book on population—that of Thomas Robert Malthus—was born in the midst of the Enlightenment, as Europe was changing the world in historic ways.

In France, these new ideas were well expressed by Marie Jean Antoine Nicolas de Caritat, marquis de Condorcet, a member of the French aristocracy who forsook a military career to pursue a life devoted to mathematics and philosophy. His ideas helped to shape the French Revolution—although despite his inspiration for and sympathy with that cause, he died in prison at the hands of revolutionaries. Published in 1795 after his death, Condorcet's *Sketch for an Historical Picture of the Progress of the Human Mind*[2] revealed that he "saw the outlines of liberal democracy more than a century in advance of his time: universal education; universal suffrage; equality before the law; freedom of thought and expression; the right to freedom and self-determination of colonial peoples; the redistribution of wealth; a system of national insurance and pensions; equal rights for women."[3(px)]

Condorcet's optimism was based on his belief that technological progress has no limits: "With all this progress in industry and welfare which establishes a happier proportion between men's talents and their needs, each successive generation will have larger possessions, either as a result of this progress or through the preservation of the products of industry, and so, *as a consequence of the physical constitution of the human race, the number of people will increase.*"[2(p188)] Condorcet then asked whether it might not happen that eventually the happiness of the population would reach a limit. If that happens, he concluded, "we can assume that by then men will know that ... their aim should be to promote the general welfare of the human race or of the society in which they live or of the family to which they belong, rather than foolishly to encumber the world with useless and wretched beings."[2(p189)] Condorcet thus saw prosperity and population growth as increasing hand in hand, and if the limits to growth were ever reached, the final solution would be birth control, keeping in mind that his book was written well before the invention of modern methods of birth control.

In England at about the same time, William Godwin (father of Mary Wollstonecraft Shelley, author of *Frankenstein*, and father-in-law of the poet Percy Bysshe Shelley) was having similar thoughts. His *Enquiry Concerning Political Justice and Its Influences on Morals and Happiness* appeared in its first edition in 1793, revealing Godwin's ideas that scientific progress would enable the food supply to grow far beyond the levels of his day, and that such prosperity would not lead to overpopulation because people would deliberately limit their sexual expression and procreation. Furthermore, he believed that most of the problems of the poor were due not to overpopulation but rather to the inequities of the social institutions, especially greed and accumulation of property.[4] Today we might think of this as a Marxian idea, although of course Godwin's book came out 25 years before Marx was born.

Figure 5–1 Thomas Robert Malthus (1766–1834)

Courtesy of the National Library of Medicine

Thomas Robert Malthus (Figure 5–1) was a young man when Godwin's book was published. He had recently graduated from Cambridge University and was a country curate and a nonresident fellow of Cambridge University when he read and contemplated the works of Godwin, Condorcet, and others who shared this emerging enlightened worldview. Although he wanted to be able to embrace such an openly optimistic philosophy of life, Malthus felt that intellectually he had to reject it. In doing so, he unleashed a controversy about population growth and its consequences that rages to this very day. He was, however, also a product of the Enlightenment, despite being a clergyman. Malthus was a critical thinker as well as a moralist, and he did not think that the evidence of life around him warranted the conclusions being drawn by Godwin and Condorcet.

THE MALTHUSIAN PERSPECTIVE

Malthus's *Essay on the Principle of Population as It Affects the Future Improvement of Society; with Remarks on the Speculations of Mr. Godwin, M. Condorcet, and Other Writers* was published anonymously in 1798. Its author was not trying to become famous for his views on population; instead, he simply wanted to show that unbounded optimism about the future was misplaced.

He introduced his essay by commenting, "I have read some of the specula-tions on the perfectibility of man and society, with great pleasure. I have been warmed and delighted with the enchanting picture which they hold forth. I ardently wish for such happy improvements. But I see great, and, to my understanding, unconquerable difficulties in the way to them."[5(p7)]

These "difficulties" were the problems posed by Malthus's now-famous principle of population. In addressing these issues, he included two postulates. First, food is necessary for the existence of humanity. Second, "the passion between the sexes is necessary" and will continue "in its present state."[5(p11)] Given these postulates, Malthus suggested, "the power of popu-lation is indefinitely greater than the power in the earth to produce subsis-tence for man." Accordingly, "population, when unchecked, increases in a geometrical ratio," while subsistence "increases only in an arithmetical ratio." These laws of nature imply a strong and consistent check on population, but unfortunately, this check "must necessarily be severely felt by a large portion of mankind." Malthus concluded that his argument against the perfectibility of mankind or humanity prevailed over previous ideas presented by Godwin and Cordorcet.[5(p11)]

In other words, Malthus somewhat smugly assumed that he had demol-ished the utopian optimism by suggesting that the laws of nature, operating through the principle of population, essentially prescribed poverty for a certain segment of humanity. He was apparently taken aback by the reaction to his book. However, after owning up to its authorship, he proceeded to document his population principles and to respond to critics by publishing a substantially revised version in 1803, slightly but importantly retitled to read *An Essay on the Principle of Population; or a View of Its Past and Present Effects on Human Happiness; with an Inquiry into Our Prospects Respecting the Future Removal or Mitigation of the Evils Which It Occasions*. In all, seven editions of Malthus's essay on population were published, and as a whole they have undoubtedly been the single most influential work relating population growth to its social consequences. Although Malthus relied heavily on the work of earlier writers such as Hume[6] and Wallace,[7] he was the first to lay out a theory linking the consequences of population growth to its causes in a systematic way.

Causes of Population Growth

Malthus believed that human beings, like plants and subhuman ("nonrational," in his words) animals, are "impelled" to increase the popu-lation of the species by what he called a powerful "instinct," the urge to reproduce. Further, if there were no checks on population growth, human beings would multiply to an "incalculable" number, filling "millions of worlds in a few thousand years."[8(p6)] But why has this outcome failed to be realized? Because high fertility is checked by high mortality. Little was known in his time about causes of death, but in Malthus's view, the ultimate check to growth is lack of food (the "means of subsistence," as he called it), which would cause people to weaken and die. In turn, the means of subsistence are limited by the amount of land available, the "arts" or technology that can be applied to the land, and "social organization" or land ownership patterns. Lacking the perspective that we have today, Malthus believed that each society had its own particular level of technology and organization, and he did not foresee them changing over time. This was clearly one way in which he was wrong.

A cornerstone of Malthus's argument is that populations tend to grow more rapidly than the food supply does, because the population has the potential for growing geometrically—2 parents could have 4 children, 16 grandchildren, and so on—while he believed (incorrectly, as Darwin very importantly pointed out a half century later) that food production could be increased only arithmetically, by adding one acre at a time, as Malthus did not envision scientific advances that allow farmers to grow more food per acre. He argued, then, that in the natural order, population growth will outstrip the food supply, and the lack of food will ultimately put a stop to the increase of people. It is no wonder that he was called the "gloomy cleric."

Malthus was aware that starvation rarely operates directly to kill people; that is, something else usually intervenes to kill them before they actually die of starvation. This "something else" represents what Malthus calls "positive checks," which today we would call disease or degeneration that causes death. There are also "preventive checks"—limits to birth. In theory, the preventive checks would include all possible means of birth control, including abstinence, contraception, and abortion. Here, however, the moralist in Malthus overcame his critical thinking. He believed that the only acceptable means of preventing a birth was to exercise "moral restraint"—that is, to postpone marriage, remaining chaste in the meantime, until a man feels "secure that, should he have a large family, his utmost exertions can save them from rags and squalid poverty, and their consequent degradation in the community."[8(p13)] Any other means of birth control, including contraception (either before or after marriage), abortion, infanticide, or any "improper means," was viewed as a vice that would "lower, in a marked manner, the dignity of human nature." Moral restraint was a very important point with Malthus, because he believed that if people were allowed to prevent births by "improper means" (that is, contraception, abortion, or sterilization), then they would expend their energies in ways that are, so to speak, not economically productive. He was a clergyman in the Church of England, not the Catholic Church, but you can appreciate that his point of view on birth control would fit in easily in many communities of the world today.

Consequences of Population Growth

Malthus believed that a natural consequence of population growth was poverty. This is the logical end result of his arguments that (1) people have a natural urge to reproduce and (2) the increase in the food supply cannot keep up with population growth. In his analysis, Malthus turned the argument of Adam Smith[9] upside down. Instead of population growth depending on the demand for labor, as Smith had argued, Malthus believed that the urge to reproduce always forces population pressure to precede the demand for labor. Thus "overpopulation" (as measured by the level of unemployment) would force wages down to the point where people could not afford to marry and raise a family. At such low wages, with a surplus of labor and the need for each person to work harder just to earn a subsistence wage, cultivators could employ more labor, put more acres into production, and thus increase the means of subsistence. Malthus believed that this cycle of increased food resources leading to population growth leading to too many people for available resources leading then back to poverty was part of a natural law of population. Each increase in the food supply simply meant that eventually more people could live in poverty. In essence, he blamed poverty on the poor themselves.

Avoiding the Consequences

All was not despair for Malthus, no matter how gloomy his outlook may have seemed. From another influential Enlightenment thinker, John Locke, Malthus borrowed the idea that "the endeavor to avoid pain rather than the pursuit of pleasure is the great stimulus to action in life."[5(p359)] Malthus suggested that the well-educated, rational person would perceive in advance the pain of having hungry children or being in debt and would postpone marriage and sexual intercourse—exercise "moral restraint"—until the individual was sure that he or she could avoid that pain. If that motivation existed and the preventive check was operating, then the miserable consequences of population growth could be avoided. You will recall that Condorcet had suggested the possibility of birth control as a preventive check, but Malthus objected to this solution: "To remove the difficulty in this way, will, surely in the opinion of most men, be to destroy that virtue, and purity of manners, which the advocates of equality, and of the perfectibility of man, profess to be the end and object of their views."[5(p154)]

To Malthus, material success is a consequence of the human ability to plan rationally—to be educated about future consequences of current behavior—and he was a man who practiced what he preached. He planned his family rationally, waiting to marry and have children until he was 39, shortly after getting a secure job in 1805 as a professor of history and political economy at East India College in Haileybury, England (north of London). He and his wife, 11 years his junior, had only three children.[10,11]

Why Is Malthus So Important?

There are several reasons why Malthus is so important, although most of them have little to do directly with his Principle of Population. From a historical perspective, he is probably most important for having inspired others to think differently about the world. Charles Darwin, for one, acknowledged his debt to Malthus. The crucial part of Malthus's ratio of population growth to food increase was that food (including both plants and nonhuman animals) would not grow exponentially. Yet when Charles Darwin acknowledged that his *Origin of the Species* was inspired by Malthus's essay, he implicitly rejected this central tenet of Malthus's argument. "Darwin described his own theory as 'the doctrine of Malthus applied with manifold force to the whole animal and vegetable kingdoms; for in this case there can be no artificial increase of food, and no prudential restraint from marriage.' Thus plants and animals, even more than men, would increase geometrically if unchecked."[12(p128)] Put another way, without Malthus, we might not have had Darwin thinking about evolution.

Malthus's idea that too many children caused poverty among the poor led the English Parliament to change its thinking about welfare. During Malthus's lifetime, the number of people on welfare in England had been increasing, straining local budgets, and Parliament was trying to figure out what to do. Although the Poor Laws were not abolished, they were reformed largely because Malthus had given legitimacy to public criticism of the entire concept of welfare payments,[12] which in his view perpetuated misery. According to Malthus, such a system permitted poor people to be supported by others and thus not feel that great pain, the avoidance of which might lead to birth prevention. The Malthusian perspective that blamed the poor for their own poverty endures to this day, contrasted with the equally enduring view of Godwin and Condorcet that poverty is the creation of unjust human institutions.

THE MARXIAN PERSPECTIVE

Karl Marx (Figure 5–2) and Friedrich Engels were both teenagers in Germany when Malthus died in England in 1834. By the time they had met and independently moved to England, Malthus's ideas already were politically influential in their native land, not just in England. Several German states and Austria had responded to what they believed was overly rapid growth in the number of poor people by legislating against marriages in which the applicant could not guarantee that his family would not wind up on welfare.[13] As it turned out, that scheme backfired on the German states, because people continued to have children, but out of wedlock. Thus the welfare rolls grew as the illegitimate children had to be cared for by the state.[14] The laws were eventually repealed, but they had an impact on Marx and Engels, who saw the Malthusian point of view as an outrage against humanity. Their demographic perspective thus arose in reaction to Malthus. Would there have been a *Das Capital* without Malthus? Perhaps not.

Figure 5–2 Karl Marx (1818–1883)

© Photos.com/Thinkstock

Neither Marx nor Engels ever directly addressed the issue of why and how populations grew. They seem to have had little quarrel with Malthus on this point, although they were in favor of equal rights for men and women and saw no harm in preventing birth. Nonetheless, they were skeptical of the eternal or natural laws of nature as stated by Malthus (i.e., that population tends to outstrip resources), preferring instead to view human activity as the product of a particular social and economic environment. The basic Marxian perspective is that each society at each point in history has its own law of population that determines the consequences of population growth. Of course, this view was as unenlightened as Malthus's idea that each nation had its own technology and social organization that was fixed and unchangeable. Marx believed that under capitalism, the consequences of population growth are overpopulation and poverty, whereas for socialism, population growth is readily absorbed by the economy with no side effects. This line of reasoning led to Marx's vehement rejection of Malthus's argument, because if Malthus was right about his "pretended 'natural law of population,'"[15(p680)] then Marx's theory would be wrong.

Marx and Engels especially quarreled with the Malthusian idea that resources could not grow as rapidly as population: they saw no reason to suspect that science and technology could not increase the availability of food and other goods at least as quickly as the population grew. Engels argued in 1865 that whatever population pressure existed in society was really pressure against the means of employment rather than against the means of subsistence.[16] Thus Marx and Engels flatly rejected the notion that poverty can be blamed on the poor. Instead, they said, poverty is the result of a poorly organized society, especially a capitalist society. Implicit in the writings of Marx and Engels is the idea that the normal consequence of population growth should be a significant increase in production. After all, each worker obviously was producing more than he or she required—how else would all the dependents (including the wealthy manufacturers) survive? In a well-ordered society, if there were more people, there ought to be more wealth, not more poverty.[17]

Not only did Marx and Engels feel that poverty, in general, was not the end result of population growth, but they argued specifically that even in England at that time there was enough wealth to eliminate poverty. Engels had himself managed a textile plant owned by his father's firm in Manchester, and he believed that in England more people had meant more wealth for the capitalists rather than for the workers because the capitalists were skimming off some of the workers' wages as profits for themselves. Marx argued that they did that by stripping the workers of their tools and then, in essence, charging the workers for being able to come to the factory to work. For example, if you do not have the tools to make a car but want a job making cars, you could get hired at the factory and work eight hours a day. But, according to Marx, you might get paid for only four hours, with the capitalist (owner of the factory) keeping part of your wages as payment for the tools you were using. The more the capitalist keeps, of course, the lower your wages and the poorer you will be.

Furthermore, Marx argued, capitalism worked by using the labor of the working classes to earn profits to buy machines that would replace the laborers, which in turn would lead to unemployment and poverty. Thus the poor were not poor because they overran the food supply, but rather because capitalists had first taken away part of their wages and then taken away their very jobs and replaced them with machines. The consequences of population growth that Malthus discussed, then, were really the consequences of capitalist society, not of population growth per se. Overpopulation in a capitalist society was thought to be a result of the capitalists' desire for an industrial reserve army that would keep wages low through competition for jobs and, at the same time, would force workers to be more productive to keep their jobs. To Marx, however, the logical extension of this perspective was that the growing population would bear the seeds of destruction for capitalism, because unemployment would lead to disaffection and revolution. If society could be reorganized in a more equitable (that is, socialist) way, then population problems would disappear. History, of course, has generally shown that Marx was as wrong in his ideas as Malthus was in his. Both men were as dogmatic as they were scientific.

It is noteworthy that Marx, like Malthus, practiced what he preached. Marx was adamantly opposed to the notion of moral restraint, and his life repudiated that concept. He married at the relatively young age (compared with Malthus) of 25, proceeded to father eight children, including one illegitimate son, and was on intimate terms with poverty for much of his life.

THE NEO-MALTHUSIAN PERSPECTIVE

Even Malthus's friends were skeptical of his insistence on moral restraint as the only reasonable preventive check to population growth. But it took until the mid-20th century for those who criticized Malthus's insistence on the value of moral restraint, while accepting many of his other conclusions, to have a label applied to them: neo-Malthusians. Specifically, neo-Malthusians favor contraception (as a "preventive check") rather than simple reliance on moral restraint. Without question, the most famous neo-Malthusian is Paul Ehrlich (Figure 5–3). In the 1960s, the world became

Figure 5–3 Paul Ralph Ehrlich (born 1932)

Courtesy of Paul R. Ehrlich

keenly aware of the population crisis through his writings. His *Population Bomb*[18] was an immediate sensation when it came out in 1968, and to this day it often sets the tone for public debate about population issues. In the second edition of his book, Ehrlich[19] phrased the situation in three parts: "too many people," "too little food," and, adding a wrinkle not foreseen directly by Malthus, "environmental degradation" (Ehrlich called earth "a dying planet").

In 1990, Ehrlich, in collaboration with his wife Anne, followed with an update titled *The Population Explosion*,[20] reflecting their view that the bomb they worried about in 1968 had detonated in the meantime. The level of concern about the destruction of the environment has grown tremendously since 1968. Ehrlich's book inspired the first Earth Day in the spring of 1970 (an annual event ever since in most communities across the United States and elsewhere in the world).

Ehrlich argues that Malthus was right—dead right. But the death struggle is more complicated than that foreseen by Malthus. To Ehrlich, the poor are dying of hunger, while rich and poor alike are dying from the by-products of affluence—pollution and ecological disaster. A few benefit, all suffer. What does the future hold? Ehrlich suggests that there are only two solutions to the population problem: the birth rate solution (lowering the birth rate—the "preventive check") and the death rate solution (a rise in the death rate—the "positive check"). He views the death rate solution as being the most likely to happen, because, like Malthus, he has little faith in the ability of human-kind to pull its act together. The only way to avoid that scenario, he argues, is to bring the birth rate under control, perhaps even by force. That idea has generated death threats against him and his wife, but of course over time the world has responded with lower birth rates. Would there have been a modern environmental movement without a neo-Malthusian like Paul Ehrlich? Perhaps not.

THE DEMOGRAPHIC TRANSITION THEORY

The population-growth controversy, initiated by Malthus and fueled by Marx, emerged into a series of 19th-century and early-20th-century reformulations that have led directly to prevailing theories in demography. Although it has dominated demographic thinking for the past half century, the demographic transition theory actually began as a description of the demographic changes that had taken place in advanced nations over time. In particular, it described the transition from high birth and death rates to low birth and death rates, with an interstitial spurt in growth rates leading to a larger population at the end of the transition than there had been at the start.

This idea emerged as early as 1929, when Warren Thompson gathered data from "certain countries" for the period 1908–1927 and showed that the countries fell into three main groups, according to their patterns of population growth:

- *Group A (northern and western Europe and the United States):* From the latter part of the 19th century to 1927, these countries had moved from having very high rates of natural increase to having very low rates of increase "and will shortly become stationary and start to decline in numbers."
- *Group B (Italy, Spain, and the "Slavic" peoples of central Europe):* Thompson saw evidence of a decline in both birth rates and death rates but suggested that "it appears probable that the death rate will decline as rapidly or even more rapidly than the birth rate for some time yet. The condition in these Group B countries is much the same as existed in the Group A countries thirty to fifty years ago."[21(p968)]
- *Group C (the rest of the world):* In the rest of the world, Thompson saw little evidence of control over either births or deaths.

As a consequence of the relative lack of voluntary control over births and deaths at that time in world history, Thompson believed that the Group C countries (which included about 70%–75% of the population of the world at the time) would continue to have their growth "determined largely by the opportunities they have to increase their means of subsistence. Malthus described their processes of growth quite accurately when he said 'that population does invariably increase, where there are means of subsistence.'"[21(p971)]

Thompson's work, however, came at a time when there was little concern about overpopulation. The Group C countries had relatively low rates of growth because of high mortality and, shortly thereafter during the Great Depression, birth rates in the United States and Europe were so low that Enid Charles published a widely read book called *The Twilight of Parenthood*, which was introduced with the comment that "in place of the Malthusian menace of overpopulation there is now real danger of underpopulation."[22(pv)] A few years later, however, as the Depression had ended and World War II was coming to a close, Frank Notestein[23] picked up the threads of Thompson's thesis and provided labels for the three types of growth patterns that had been labeled simply A, B, and C. Notestein called the Group A pattern "incipient decline," the Group B pattern "transitional growth," and the Group C pattern "high growth potential."

That same year, Kingsley Davis[24] edited a volume of *The Annals of the American Academy of Political and Social Sciences* titled *World Population in Transition*. In the lead article (titled "The World Demographic Transition"), he noted that "viewed in the long-run, earth's population has been like a long, thin powder fuse that burns slowly and haltingly until it finally reaches the charge and explodes."[24(p1)] Thus was born the term *demographic transition*, meaning the process of moving from high birth and death rates to low birth and death rates, from high growth potential to incipient decline.

At that point in the 1940s, the demographic transition was merely a picture of demographic change, not a theory. But each new country studied fit into the picture, and it seemed as though some new universal law of population growth—an evolutionary scheme—was being developed. There was also an apparent historical uniqueness to the demographic transition, since all known cases have occurred within the last 200 years. Between the mid-1940s and the late 1960s, rapid population growth became a worldwide concern, and demographers devoted a great deal of time to the demographic transition perspective. By 1964, George Stolnitz was able to report that "demographic transitions rank among the most sweeping and best-documented trends of modern times ... based upon hundreds of investigations, covering a host of specific places, periods and events."[25(p20)] As the pattern of change took shape, explanations were developed for why and how countries pass through the transition. These explanations tended to be cobbled together in a somewhat piecemeal fashion from Enlightenment thinkers of the 19th and early 20th centuries, but overall they were derived from the concept of "modernization."

Modernization theory is based on the idea that in premodern times human society was generally governed by "tradition," and that the massive economic changes wrought by industrialization forced societies to alter traditional institutions: "In traditional societies fertility and mortality are high. In modern societies fertility and mortality are low. In between, there is demographic transition."[26(p502)] In the process, behavior has changed and the world has been permanently transformed. Modernization theory is a macro-level theory that sees human actors as being buffeted by the winds of changing social institutions. Individuals did not deliberately lower their risk of death to precipitate the modern decline in mortality; rather, society-wide increases in the standard of living and improved public health infrastructure brought about this change. Similarly, people did not just decide to move from the farm to town to take a job in a factory; instead, economic changes took place that created those higher-wage urban jobs while eliminating many agricultural jobs. These same economic forces improved transportation and communication and made it possible for individuals to migrate in previously unheard of numbers.

Modernization theory provided the vehicle that moved the demographic transition from a mere description of events to a demographic perspective. This theory drew on the available data for most countries that had gone through the transition. Death rates decline as the standard of living improves, and birth rates almost always decline a few decades later, eventually dropping to low levels. It was argued that the decline in the birth rate typically lags behind the decline in the death rate because it takes time for a population to adjust to the fact that mortality really is lower, and because the social and economic institutions that favor high fertility require time to adjust to new norms of lower fertility that are more consistent with the lower levels of mortality. Because most people value the prolongation of life, it is not difficult to lower mortality, but the reduction of fertility is contrary to the established norms of societies that have required high birth rates to keep pace with high death rates. Such norms are not easily changed, even in the face of poverty.

Birth rates eventually decline, it was argued, as the importance of family life is diminished by industrial and urban life, thus weakening the pressure for large families. Large families are presumed to have been desired because they provided parents with a built-in labor pool, and because children provided old-age security for parents.[27] The same economic development that lowered mortality is theorized to transform a society into an urban industrial state in which compulsory education lowers the value of children by removing them from the labor force, and people come to realize that lower infant mortality means that fewer children need to be born to achieve a certain number of surviving children.[28] Finally, as a consequence of the many alterations in social institutions, "the pressure from high fertility weakens and the idea of conscious control of fertility gradually gains strength."[29(p421)]

Reformulating the Demographic Transition Theory

As it turned out, modernization theory alone could not capture all of the aspects of the demographic transition. This was made clear by the results of the European Fertility Project, directed by Ansley Coale at Princeton University. In one of the first analyses, focusing on the fertility decline in Spain, data for that nation's 49 provinces revealed that the history of fertility change in Spain was not explained by a simple version of the demographic transition theory. Fertility in Spain declined in contiguous areas that were culturally similar, even though the levels of urbanization and economic development might be different there.[30] At about the same time, other students began to uncover similarly puzzling historical patterns in European data.[31] A systematic review of the demographic histories of Europe was thus begun to establish exactly how and why the transition occurred.

With the discovery that the decline of fertility in Europe occurred in the context of widely differing social, economic, and demographic conditions, it became apparent that economic development was a sufficient cause of fertility decline, but not a necessary one.[32] For example, many provinces in Europe experienced a rapid drop in their birth rates even though they were not very urban, infant mortality rates were high, and a low percentage of the population was engaged in industrial occupations. The data suggest that one of the more common similarities in those areas that have undergone fertility declines is the rapid spread of "secularization." Secularization is an attitude of autonomy from otherworldly powers and a sense of responsibility for one's own well-being.[1,33,34] It is associated with an enlightened view of the world—a break from traditional ways of thinking and behaving.

It is difficult to determine exactly why such attitudes arise when and where they do, but we do know that industrialization and economic development are almost always accompanied by secularization. Secularization, however, can occur independently of industrialization. It might be thought of as a modernization of thought (i.e., part and parcel of the Enlightenment), distinct from a modernization of social institutions. Some theorists have suggested that secularization is part of the process of Westernization.[35] In all events, when it pops up, secularization often spreads quickly, being diffused through social networks as people imitate the behavior of others to whom they look for clues to proper and appropriate conduct.

Education has been identified as a key stimulant to such altered attitudes, especially mass education, which tends to emphasize modernization and secular concepts.[36] Education facilitates the rapid spread of new ideas and information, which would perhaps help explain another of the important findings from the Princeton European Fertility Project—namely, that the onset of long-term fertility decline tends to be concentrated in a relatively short period of time.[37] The data from Europe suggest that once marital fertility has dropped by as little as 10% in a region, the decline spreads rapidly. This "tipping point" occurs whether or not infant mortality has already declined.[38] The data from the Princeton European Fertility Project revealed that some areas of Europe that were similar with respect to socioeconomic development did not experience a fertility decline at the same time, whereas other provinces that were less similar socioeconomically experienced nearly identical drops in fertility.

This riddle may be solved by examining cultural factors, not just socio-economic ones. Areas that share a similar culture (same language, common ethnic background, similar lifestyle) are more likely to share a decline in fertility than areas that are culturally less similar.[39] The principal reason for this is that the idea of family planning seems to spread quickly until it runs into a barrier to its communication. Language is one such barrier; social and economic inequality in a region is another.[40] Social distance between people turns out to effectively inhibit communication of new ideas and attitudes.

Overall, then, the principal ingredient in the reformulation of the demographic transition perspective was to add "ideational" factors to "demand" factors as the likely causes of fertility decline.[41] The original version of the theory suggested that modernization reduces the demand for children and so fertility falls—if people are rational economic creatures, then this is what should happen. But the real world is more complex, and diffusion of ideas can shape fertility behavior along with, or even in the absence of, the usual signs of modernization. People may generally be rational, but Massey[42] has reminded us that much of human behavior continues to be powered by emotional responses that supersede rationality. We are animals and though we may have vastly greater intellectual capacities than other species, we are still influenced by a variety of nonrational forces, including our hormones.[43–45]

One strength of the effort to reformulate the demographic transition is that nearly all other perspectives can find a home here. Malthusians note with satisfaction that fertility first declined in Europe primarily as a result of a delay in marriage, much as Malthus would have preferred, although this was largely because of the scarcity of effective means of birth control. Neo-Malthusians can take heart from the fact that rapid and sustained declines occurred simultaneously with the spread of knowledge about family planning practices. Marxists also find a place for themselves in the reformulated demographic transition perspective, because its basic tenet is that a change in the social structure (modernization) is necessary to bring about a decline in fertility. This is only a short step away from agreeing with Marx that there is no universal law of population, but rather that each stage of development and social organization has its own law, and that cultural patterns will influence the timing and tempo of the demographic transition—when it starts and how it progresses.

The Theory of Demographic Change and Response

The work of the European Fertility Project focused on explaining regional differences in fertility declines. This was a very important theoretical development, but not a comprehensive one because it only partially dealt with a central issue of the demographic transition theory: How (and under which conditions) can a mortality decline lead to a fertility decline? To answer that question, Kingsley Davis[46] asked what happens to individuals when mortality declines. The answer is that more children survive through adulthood, putting greater pressure on family resources, and people have to reorganize their lives in an attempt to relieve that pressure; that is, people respond to the demographic change. Nevertheless, their response will be couched in terms of personal goals, not national goals. It rarely matters what a government wants. If individual members of a society do not stand to gain by behaving in a particular way, they probably will not behave that way. Indeed, that was a major argument made by the neo-Malthusians against moral restraint. Why advocate postponement of marriage and sexual gratification rather than contraception when you know that few people who postpone marriage will actually postpone sexual intercourse, too? In fact, Ludwig Brentano[47] quite forthrightly suggested that Malthus was insane to think that abstinence was the cure for the poor.

Davis (Figure 5–4) argued that the response that individuals make to the population pressure created by more members joining their ranks is determined by the means available to them. A first response, nondemographic in nature, is to try to increase resources by working harder—longer hours perhaps, a second job, and so on. If that effort is not sufficient or there are no such opportunities, then migration of some family members (typically unmarried sons or daughters) is the easiest demographic response. This

Figure 5-4 Kingsley Davis (1908–1997)

Courtesy of The American Sociological Association

is, of course, the option that people have been using forever (undoubtedly explaining in large part why human beings have spread out over the planet).

But what will be the response of that second generation—the children who now have survived when previously they would not have, and who have put the greater pressure on resources? Davis argues that if social or economic improvement is indeed possible, then people will try to take advantage of those opportunities by avoiding the large families that caused problems for their parents. He suggests that the most powerful motive for family limitation is not fear of poverty or avoidance of pain, as Malthus argued; rather, it is the prospect of rising prosperity that will most often motivate people to find the means to limit the number of children they have. Davis argues that the desire to maintain one's relative status in society may lead to an active desire to prevent too many children from draining away one's resources. Davis's analysis is important in reminding us of the crucial link between the everyday lives of individuals and the kinds of population changes that take place in society. Many of these changes are taking place in the richer countries of the Global North, not simply in the less rich countries of the Global South.

The Second Demographic Transition

One of the assumptions underlying most discussions of the demographic transition is that societies are seeking a situation of zero population growth—where births equal deaths and the age structure remains stable over time. So, the transition is hypothesized to be one from zero population growth at high birth and death rates to zero population growth at low birth and death rates. However, this assumption has been called into question by the dramatic changes that have taken place in family and household structure since World War II, especially in Europe, but also more recently in much of East Asia, leading van de Kaa,[48] Lesthaeghe and Neels,[49] and others to talk about the "second demographic transition." A demographic centerpiece of this change in the richer countries has been a decline in fertility to below-replacement levels, associated with the new personal freedom to do what one wants, especially among women. So, rather than grow up, marry, and have children, this transition is associated with a postponement of marriage, a rise in single living, cohabitation, and prolonged residence in the parental household.[49,50] There has also been an increasing lack of permanence in family relationships, leading to "rising divorce rates and high separation rates of cohabitants. And the forms of family reconstitution are shifting away from remarriage in favor of postmarital cohabitation as well."[51(pp5–6)]

Billari and Liefbroer[52(p1)] have suggested that the second demographic transition "constitutes 'the' mainstream concept among population scholars dealing with demographic change in European societies." For his part, van de Kaa[53(p4)] has noted, "in my view it is really impossible to understand the demographic changes that have occurred in Europe, and in many other industrialised countries as well, since the mid-1960s, without accepting the idea that the many and very varied changes we have observed in a whole series of demographic variables are interrelated and may in their totality be indicative of, and represent, the manifestation of a change in demographic regime." By this he means that instead of assuming that the end result of the demographic transition is population stability (zero population growth), this new demographic regime is one in which that assumption is thrown out the window, and replaced by the idea that young people make decisions about having children on the basis of self-fulfillment, without regard to biological replacement, knowing as they do that, at least in the rich European countries in which this occurring, migrants can make up the difference if too few children are being born.

DEMOGRAPHIC DYNAMICS AND THE AGE AND SEX STRUCTURE

When the population increases (or decreases) in size, it never does so with an equal increase or decrease at each age. Some ages are affected more than others, and it is the changing number and percentage of people at each group that force societies to respond. The key ingredient in how the demographic transition impacts society, then, is how it impacts the age structure. As it turns out, fairly predictable changes take place over time. Thus we can refer to the age transition as the shift from a very young population in which there are slightly more males than females to an older population in which there are more females than males. In between, bumps and dents in the age and sex structure represent powerful forces for social, economic, and political change.

In general, it is the interaction of fertility, mortality, and migration that produces the age and sex structure, which can be viewed as a key to the life of a social group—a record of past history and a portent of the future. Population processes not only produce the age and sex structure, but also are affected by it. A high birth rate does not simply mean more people: it means that a few years from now, there will be more children entering school than before; that 18 years from now, there will be more new job hopefuls and college freshmen than before. An influx of young adult immigrants this year means a larger-than-average number of older people 30 to 40 years from now (and it may mean an immediate sudden rise in the number of births, with all the attendant consequences). The very low birth rates in Europe and East Asia do not imply just that there will be fewer people in the future: they mean that schools will be closing, jobs will go begging, businesses will close, houses will be vacant, and life will be very different than it once was.

For most of human history, populations were very young. They had a high proportion of people in the younger ages, a modest fraction in the middle adult ages, and very few older people. Reaching an advanced age was truly an exceptional circumstance, and it is easy to see why the elderly would be revered and maybe even feared. The demographic transitions have changed all that in complex, but still decipherable ways. The end of the demographic transition, if such a thing really occurs, is assumed in the ideal to be a population with essentially the same number of people at each age until the very older ages, when people die off fairly quickly.

In between the historical pattern and this expected future pattern is where it gets messy, and that is where we are in the world today. Figure 5-5 illustrates this with examples from three different countries—a very young country (Nigeria), a rapidly aging country (Japan), and a country in the middle (Mexico). The graphs themselves are called *population pyramids* because the "classic" picture is of a high-fertility, high-mortality society with a broad base built of numerous births, rapidly tapering to the top (the older ages) because of high death rates in combination with the high birth rate. Nigeria's age and sex structure still reflects the classic look of the population pyramids, as you can see in Figure 5-5. Until fairly recently, Mexico also looked like that, but the decline in fertility in Mexico since the 1970s

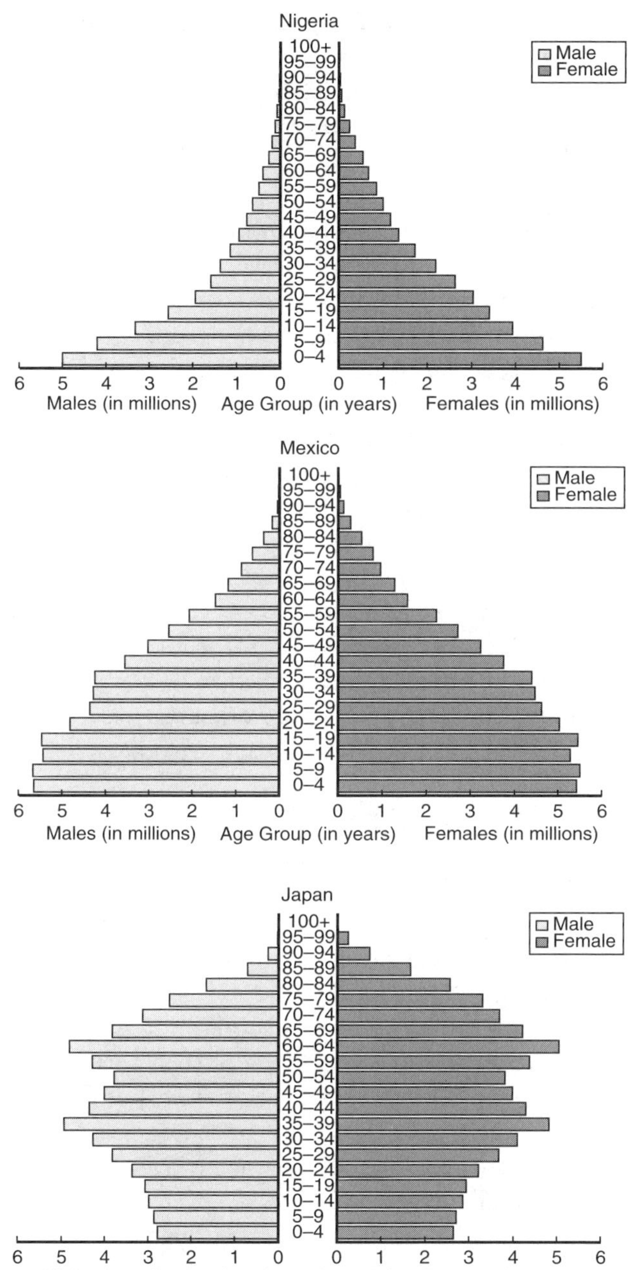

Figure 5–5 Population Pyramids for Nigeria, Mexico, and Japan, 2010

Data from the United Nations Population Division. World population prospects: the 2010 revision. http://esa.un.org/unpd/wpp/unpp/panel_indicators.htm. Accessed January 7, 2014.

has narrowed the base of the pyramid rather noticeably, whereas Japan is somewhat barrel-shaped because of its very low fertility. No matter the shape, we still call the graph a population pyramid.

The population pyramids shown in Figure 5–5 do not happen by chance. Each of the three population processes has predictable impacts on the age structure. Mortality declines affect every age and both sexes, but in virtually all societies the very youngest and the very oldest ages are most susceptible to death, and in modern societies (where maternal mortality is fairly low), males are more likely than females to die at any given age. As a result, the early drop in mortality increases the proportion of children who survive and, therefore, serves to increase the fraction of the population at the younger ages.

As mortality declines without any change in fertility, the age structure actually becomes more pyramidal than it was before, as a result of the greater survivability of children, which means that the younger population is growing faster than the population at the older ages; thus the average age actually declines in such a scenario. While the *average age* is one way to summarize an age structure, another index commonly used to measure the social and economic impact of different age structures is the *dependency ratio*—the ratio of the dependent-age population (the young and the old) to the working-age population. The higher this ratio is, the more people each potential worker has to support; conversely, the lower it is, the fewer people dependent on each worker:

$$\text{Dependency ratio} = \frac{(\text{population } 0\text{–}14) + (\text{population } 65+)}{\text{population } 15\text{–}64}.$$

Both mortality and migration can affect all ages and differentially affect each sex. However, the impact of fertility is not quite the same as that of mortality or migration, and it is for this reason that changes in fertility tend to have the most dramatic long-term impact on the age structure (migration has the biggest short-term impact, as noted later in this section). Fertility obviously adds people only at age zero to begin with, but that effect persists for the population age after age. Thus, if the birth rate were to drop suddenly in one year, then as those people get older, there will always be fewer of them than there are people of surrounding ages. Conversely, if fertility goes up, then there will be more people in each successive birth cohort. In general, the impact of fertility levels is so important that with exactly the same level of mortality, just altering the level of fertility can produce age structures that run the gamut from those that might characterize primitive to highly developed populations.

Migration has its own unique impact on the age and sex structure. In the short run, immigration adds people mainly into the young adult ages in the host area, and of course those people have been taken out of the age structure in the donor area. In the long run, however, the impact of migration is felt indirectly through its influence on reproduction, because these young adult immigrants are of prime reproductive ages. Thus, over the long term, the impact of immigration is largely to increase the population at each age. The migrants have babies, who grow up through each age, and of course the immigrants themselves proceed through each successive age. Conversely, emigration has the effect of dramatically slowing population growth: those young adults take not only themselves out the community, but also their potential future children. Likewise, those individuals will not be around to contribute to the aging process within the community they left behind.

Although the age transition means that every society's age and sex structure will be uniquely determined by its combination of mortality, fertility, and migration rates, two aspects of the age transition seem particularly troublesome: *youth bulges* and *aging populations*. Both are problematic because of the issue of dependency. Youth bulges call into question the resources that societies can devote to creating jobs and fulfilling lives for its youth and young adults, whereas aging populations call into question the ability of the younger generation to support an increasingly dependent older generation.

Youth bulges are not the same thing as a young population, which characterized most human societies for most of history. Rather, such a bulge is the result of a quick drop in fertility that reduces the very young population without affecting the older population. It can have a dramatic and potentially positive impact on a society, increasing the economic productivity of the adult population, with few children to deal with and also few older people to worry about.[54] The size and timing of such bulges—of which there are many in the world today—in combination with society's response to them, will help tell the tale of how dramatic the changes are in the context of the age transition and whether those changes will be used for good or evil. The "good" reaction relates especially to the use of the young population to spur economic development and lift people out of poverty. The "evil" reaction relates to the use of young people to promote violence and terrorism.

No matter how you have defined a youth bulge, the underlying reason why it matters is that the young adult ages are unsettled, even tumultuous, in every society, especially for men. It has been said that the "dogs of war" (with no disrespect meant to dogs) are young and male.[55(p7)] Moreover, because in most traditional societies males are routinely accorded higher status than females, an increase in the number and proportion of young men in a population creates conditions for change. When that change is revolutionary and violent, young men are almost invariably involved.[56]

Crenshaw and his associates[57(p974)] examined the pattern of age-specific growth rates and economic development for the period 1965–1990 and concluded that an increase in the child population (the impact of declining mortality in a high-fertility society) does, indeed, hinder economic progress in less-developed nations. On the more positive side, an increase in the adult population relative to other ages (the delayed effect of a decline in fertility) fosters economic development, producing what these authors call a "demographic windfall effect whereby the demographic transition allows a massive, one-time boost in economic development as rapid labor force growth occurs in the absence of burgeoning youth dependency." The demographic "windfall" has also been called the *demographic dividend*,[58] the "demographic bonus," and the "window of opportunity,"[59] and China has been the poster child.

The challenge of the youth bulge is that it represents a potential conflict between successive cohorts. Ryder emphasizes the fact that you cannot understand the impact of a younger cohort without also understanding the older cohorts.[60] Accordingly, it is not just numbers that matter, but also the differences or inequalities that crop up between cohorts. In societies undergoing modernization, the younger people may be better educated, for example, than their parents. This might lead to different and better paying occupations, which may give younger people more economic power than their parents would have enjoyed at the same age. The drop in the death rate will almost certainly propel other changes in society. In particular, as the population grows and society responds to that growth, the traditional family structure may no longer suffice for the tasks of socialization—that is, for preparing children to be the adults needed by the changing society.[61]

The aging of society also leads to a clash of cohorts, but in this case it is the older population that makes demands on the younger cohorts: the dependency ratio increases because the older population is increasing as a fraction of the total, while the working-age population is declining as a percentage of the total population. In the United States in the 1960s, there were nearly four workers for every Social Security retiree, but by 2030 that number will have dropped to only two. The burden on the younger generation will obviously be intense. Some of this will be alleviated by immigrants and their children—"replacement migration," as it is sometimes known. The United States accepts more immigrants than any other country in the world, which is one reason why the U.S. fertility rate is just at replacement level, rather than being well below replacement level, as is true in many European nations.

Immigrants to Europe have created generational issues between aging Europeans (who are largely Christian) and the younger immigrants from Africa, the Middle East, and Asia (especially those who are Muslim). The fundamental changes that have been discussed in Europe as means to deal with the aging population include especially the following ideas:

- Workers will have to stay in the labor force longer, although this would not necessarily imply a shorter retirement if, for example, retirement age increased in tandem with increases in life expectancy.
- An even larger fraction of women in the labor force would help to increase the number of people paying into the system.
- Reducing unemployment among young people will get them into the labor force and paying into the system.
- Increasing productivity per person could raise wages and, in turn, the payments into the system.
- Pension plans will have to be restructured.[62]

Why do we have to think about all of these societal changes? The answer is simple: the age transition has forced us into it.

CONCLUSION

Modern population theories hearken back especially to the work of Thomas Robert Malthus, but his influence is less on how we view the world demographically, and more on the bigger questions of the causes and consequences of population growth. Although many people accepted his arguments during his lifetime, Malthus is more famous today for the arguments that arose in opposition to his belief that poverty was an inevitable result of population growth and for his insistence that the only acceptable means of preventing population growth was through moral restraint. The most dramatic changes in the demography of the world have occurred subsequent to Malthus's death, so we will never know how he might have interpreted the massive decline in mortality, almost equally huge declines in fertility, and the incredible volume of international migration that characterize the world more than 200 years after he first published his *Essay on Population*.

Modern population theory revolves around the demographic transition, which began as a description of the population changes occurring everywhere in the world, but subsequently evolved into a mid-range theory of how mortality, fertility, and migration are inextricably linked to changes in families and societies, especially changes related to modernization in its broadest sense. At the same time, society is importantly impacted by demographic change, and this influence is felt most notably through changes in the age and sex structure of society. The age transition forces communities and societies to adjust, and the means by which societies deal with transitions such as youth bulges and rapidly aging populations are key components of current and future world affairs.

DISCUSSION QUESTIONS

1. Why is Thomas Malthus so important?
2. How did Karl Marx's views of population and poverty differ from those of Malthus? Which insights do Marx and Engels add to population theory?
3. How do the neo-Malthusians differ from the Malthusians?
4. Discuss the demographic transition both as a description and as a theory.
5. Does age structure really matter for a population? In which ways?

REFERENCES

1. Norris P, Inglehart R. *Sacred and Secular: Religion and Politics Worldwide.* New York: Cambridge University Press; 2004.
2. Condorcet MJANdC. *Sketch for an Historical Picture of the Progress of the Human Mind.* Baraclough J, trans. London: Weidenfield and Nicholson; 1795 [1955].
3. Hampshire S. Introduction. In: Condorcet MJANdC, ed. *Sketch for an Historical Picture of the Progress of the Human Mind.* London: Weidenfield and Nicholson; 1955: vii–xii.
4. Godwin W. *Enquiry Concerning Political Justice and Its Influences on Morals and Happiness.* Toronto, Canada: University of Toronto Press; 1793 [1946].
5. Malthus TR. *An Essay on Population.* New York: Augustus Kelley; 1798 [1965].
6. Hume D. Of the populousness of ancient nations. In: Hume D, ed. *Essays: Moral, Political and Literary.* London: Oxford; 1752 [1963]: 377–464.
7. Wallace R. *A Dissertation on the Numbers of Mankind, in Ancient and Modern Times.* 1st and 2nd eds., revised and corrected. New York: Kelley; 1761 [1969].
8. Malthus TR. *An Essay on the the Principle of Population.* 7th ed. London: Reeves & Turner; 1872 [1971].
9. Smith A. *An Inquiry into the Nature and Causes of the Wealth of Nations.* 1776. http://www.socsci.mcmaster.ca/~econ/ugcm/3ll3/smith/wealth/wealbk01. Accessed 2003.
10. Nickerson J. *Homage to Malthus.* Port Washington, NY: National University Publications; 1975.
11. Petersen W. *Malthus.* Cambridge, MA: Harvard University Press; 1979.
12. Himmelfarb G. *The Idea of Poverty: England in the Early Industrial Age.* New York: Alfred A. Knopf; 1984.

13. Glass DV. *Introduction to Malthus*. New York: Wiley; 1953.

14. Knodel J. Two and a half centuries of demographic history in a Bavarian village. *Pop Stud.* 1970;24:353–369.

15. Marx K. *Capital: A Critique of Political Economy*. Moore S, Aveling E, trans. Engels F, ed. New York: Modern Library; 1890 [1906].

16. Meek R. *Marx and Engels on the Population Bomb*. Berkeley, CA: Ramparts Press; 1971.

17. Engels F. *Outlines of a Critique of Political Economy*. Reprinted in: Meek RL. *Marx and Engels on Malthus*. London: Lawrence and Wishart; 1844 [1953].

18. Ehrlich P. *The Population Bomb*. New York: Ballantine Books; 1968.

19. Ehrlich P. *The Population Bomb*. 2nd ed. New York: Sierra Club/Ballantine Books; 1971.

20. Ehrlich P, Ehrlich A. *The Population Explosion*. New York: Simon & Schuster; 1990.

21. Thompson W. Population. *Am J Sociol.* 1929;34(6):959–975.

22. Charles E. *The Twilight of Parenthood*. London: Watt's and Co.; 1936.

23. Notestein FW. Population: the long view. In: Schultz TW, ed. *Food for the World*. Chicago: University of Chicago Press; 1945: 36–57.

24. Davis K. The world demographic transition. *Ann Am Acad Polit Soc Sci.* 1945;237:1–11.

25. Stolnitz GJ. The demographic transition: from high to low birth rates and death rates. In: Freedman R, ed. *Population: The Vital Revolution*. Garden City: Anchor Books; 1964:Chapter 2; 30–46.

26. Demeny P. Early fertility decline in Austria–Hungary: a lesson in demographic transition. *Daedalus.* 1968;97(2):502–522.

27. Hohm C. Social security and fertility: an international perspective. *Demography.* 1975;12(4):629–644.

28. Easterlin R. The economics and sociology of fertility: a synthesis. In: Tilly C, ed. *Historical Studies of Changing Fertility*. Princeton, NJ: Princeton University Press; 1978; 57–133.

29. Teitelbaum M. Relevance of demographic transition for developing countries. *Science.* 1975;188:420 425.

30. Leasure JW. *Factors Involved in the Decline of Fertility in Spain: 1900–1950*. Doctoral dissertation, Princeton University, Department of Economics, Princeton, NJ, 1962.

31. Coale A. Preface. In: Coale A, Watkins SC, eds. *The Decline of Fertility in Europe*. Princeton, NJ: Princeton University Press; 1986: xix–xxii.

32. Coale A. The demographic transition. In: IUSSP, ed. *Proceedings of International Population Conference*, Vol. I. Liege, Belgium: IUSSP; 1973:53–72.

33. Lesthaeghe RJ. *The Decline of Belgian Fertility, 1800–1970*. Princeton, NJ: Princeton University Press; 1977.

34. Leasure JW. L' baisse de la fecondité aux États-Unis de 1800 a 1860. *Population.* 1982;3:607–622.

35. Caldwell J. Toward a restatement of demographic transition theory. *Pop Develop Rev.* 1976;2(3–4):321–366.

36. Caldwell J. Mass education as a determinant of the timing of fertility decline. *Pop Develop Rev.* 1980;6(2):225–256.

37. van de Walle E, Knodel J. Europe's fertility transition: new evidence and lessions for today's developing world. *Pop Bull.* 1980;34(6).

38. Watkins SC. Conclusion, Chapter 11. In: Coale A, Watkins SC, eds. *The Decline of Fertility in Europe.* Princeton, NJ: Princeton University Press; 1986: 420–449.

39. Watkins SC. *From Provinces into Nations: Demographic Integration in Western Europe, 1870–1960.* Princeton, NJ: Princeton University Press; 1991.

40. Lengyel-Cook M, Repetto R. The relevance of the developing countries to demographic transition theory: further lessons from the Hungarian experience. *Pop Stud.* 1982;36(1):105–128.

41. Cleland J, Wilson C. Demand theories of the fertility transition: an iconoclastic view. *Pop Stud.* 1987;41:5–30.

42. Massey DS. A brief history of human society: the origin and role of emotion in social life. *Am Sociol Rev.* 2002;67:1–29.

43. Udry JR. The nature of gender. *Demography.* 1994;31(4):561–574.

44. Udry JR. Biological limits of gender construction. *Am Sociol Rev.* 2000;65: 443–457.

45. Shermer M. *The Believing Brain: From Ghosts and Gods to Politics and Conspiracies: How We Construct Beliefs and Reinforce Them as Truths.* New York: St. Martin's Griffin; 2012.

46. Davis K. The theory of change and response in modern demographic history. *Pop Index.* 1963;29(4):345–366.

47. Brentano L. The doctrine of Malthus and the increase of population during the last decade. *Econ J.* 1910;20:371–393.

48. van de Kaa DJ. Europe's second demographic transition. *Pop Bull.* 1987;42(1).

49. Lesthaeghe RJ, Neels K. From the first to the second demographic transition: an interpretation of the spatial continuity of demographic innovation in France, Belgium and Switzerland. *Eur J Pop.* 2002;18:325–360.

50. McLanahan S. Diverging destinies: how children are faring under the second demographic transition. *Demography.* 2004;41(4):607–627.

51. Lesthaeghe RJ. On theory development: applications to the study of family formation. *Pop Develop Rev.* 1998;24(1):1–14.

52. Billari FC, Liefbroer AC. Is the second demographic transition a useful concept for demography? Introduction to a a debate. *Vienna Yearbook Pop Res 2004.* 2004:1–3.

53. van de Kaa DJ. Is the second demographic transition a useful research concept: questions and answers. *Vienna Yearbook Pop Res 2004.* 2004:4–10.

54. Weeks JR, Fugate D, eds. *The Youth Bulge: Challenge or Opportunity.* New York: IDEBATE Press; 2012.

55. Weeks JR. *Population: An Introduction to Concepts and Issues.* 11th ed. Belmont, CA: Wadsworth Cengage Learning; 2012.

56. Mesquida CG, Wiener NI. Male age composition and severity of conflicts. *Polit Life Sci.* 1999;18(2):181–189.

57. Crenshaw EM, Ameen AZ, Christenson M. Population dynamics and economic development: age-specific population growth rates and economic growth in developing countries, 1965 to 1990. *Am Sociol Rev.* 1997;62:974–984.

58. Bloom DE, Canning D, Sevilla J. *The Demographic Dividend: A New Perspective on the Economic Consequences of Population Change*. Santa Monica, CA: RAND; 2003.

59. Adioetomo SM, Beningisse G, Guitiano S, et al. *Policy Implications of Age-Structural Changes*. Paris: CICRED; 2005.

60. Ryder N. The cohort as a concept in the study of social change. *Am Sociol Rev.* 1965;30(6):843–861.

61. Courbage Y, Todd E. *A Convergence of Civilizations: The Transformation of Muslim Societies Around the World*. New York: Columbia University Press; 2011.

62. Nyce SA, Schreiber SJ. *The Economic Implications of Aging Societies: The Costs of Living Happily Ever After*. New York: Cambridge University Press; 2005.

III *Part*

HEALTH

Chapter 6

Contraceptive History and Practice

Deborah R. McFarlane
Richard Grossman

INTRODUCTION

Contraception is a generic term referring to the intentional avoidance of pregnancy through the use of various devices, agents, drugs, sexual practices, or surgical procedures.[1] Another name for contraception is *birth control*. Figure 6–1 illustrates that contraception is used to prevent conception.[2]

The desire to regulate fertility is universal. In every known culture, past and present, people have attempted to control their family sizes.[3] The Hutterites, the world's most famous high-fertility group, may offer an exception to this generalization, with the average woman giving birth to at least 11 children. However, this level of fertility occurred only in certain time intervals and within specific regions.[4(pp202-203),5] In other periods, the Hutterites, like other human populations, did not reproduce up to their biological capacity.

Myriad means and behaviors have been used to limit fertility, including delayed marriage, celibacy, infanticide, sexual taboos, contraception, and induced abortion.[6] Historically, those societies with a later average age at marriage have had lower fertility. Where early or child marriage is still practiced, higher than average levels of fertility continue—for example, among traditional populations in sub-Saharan Africa and south Asia.[4]

Obviously, avoiding or postponing sexual activity remains an effective way for a population, couple, or woman to delay childbearing. However, raising the age of marriage itself does not negate the need for contraception. In many cultures throughout the world, delaying marriage does not preclude sexual activity. Table 6-1 shows that in the United States and Western Europe, the average age of sexual initiation is between 16 and 19,[7] years before the average age of first marriage. In the United States, for example, the average age of first marriage is 27 for women and 29 for men.[8] Interestingly, the average age of first birth in the United States is 25.6.[9]

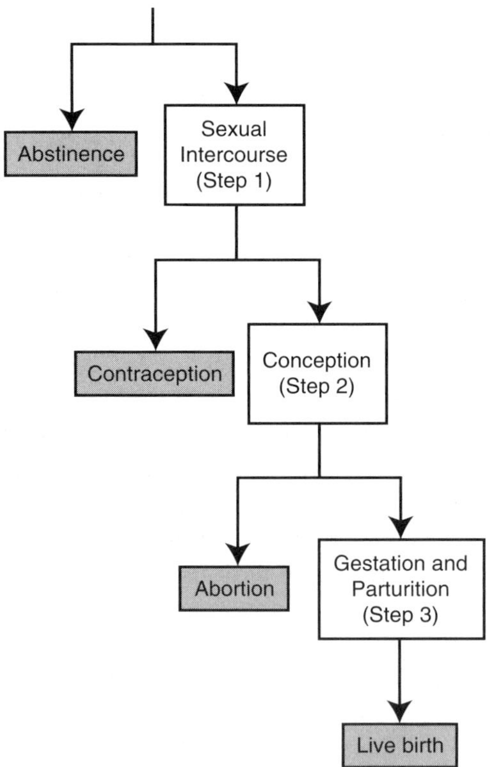

Figure 6–1 The Reproductive Process

Data from Davis K, Blake J. Social structure and fertility: an analytic framework. *economic development and cultural change.* 1956;4:211-235.

Table 6–1 Median Age of Sexual Initiation by Selected Countries (Years)

Country	Men	Women
United Kingdom	16.5	17.5
France	17.5	18.5
Italy	17.5	18.5
Norway	18.5	17.5
Switzerland	18.5	18.5
United States	17.3	17.5

Data from Wellings K, Collumbien M, Slaymaker E, et al. Sexual behavior in context: a global perspective. *Lancet.* 2006;365(9460):702–710.

HISTORY OF CONTRACEPTION

Ancient Egyptians

As Figure 6–2 illustrates, birth control has been practiced for a very long time. The oldest known recipes for contraceptives are from an ancient Egyptian papyrus dating back to 1850 BCE. Known as the Kahun papyrus, this document gave instructions for three contraceptive methods. The first described a vaginal diaphragm concocted from crocodile dung and other pasty substances. (The later substitution of elephant dung, which is more acidic, would have improved the diaphragm's contraceptive effectiveness.) The second Kahun recipe instructed women to douche with honey and natron (a native sodium carbonate), and the third mentioned a gum.[6(pp59–63)] Some 300 years later, the Ebers papyrus described a medicated lint tampon designed to prevent conception. This recipe called for the tips of the acacia, "which when compounded produced anhydride, the lactic acid used in modern contraceptive jelly."[2,3(p31)]

Ancient Hebrews

The ancient Hebrews practiced what many consider to be the oldest contraceptive method, coitus interruptus—that is, withdrawal of the penis prior to ejaculation.[10] In ancient times, many rabbis recommended coitus interruptus for two years following a birth, so that lactation could continue. Later, however, coitus interruptus was widely interpreted to be a mortal sin.[11] To prevent pregnancy, early Hebrew women also placed spongy elements in their vaginas. Other birth control attempts included magical potions and twisting and performing violent abdominal movements following coitus.[2,6]

Figure 6–2 Birth Control Has Been Around for 4000 Years. Isn't It Time You Knew All About It?

Courtesy of The Family Planning Association, London

Ancient Greeks

Egyptian knowledge of birth control was probably known to ancient Greeks, at least to prominent physicians and other intellectuals. Ancient Greeks, in turn, made their own contributions to contraceptive knowledge. As early as the seventh century BCE, they imported silphium, a giant fennel with contraceptive properties, from a southern colony, now known as Libya. This plant could not be cultivated on the Aegean peninsula, but demand for silphium became so great in ancient Greece, and later the Roman Empire, that by the third or fourth century AD, silphium became extinct. Other plants that did survive—asafoetida, pennyroyal, artemisia, myrrh, and rue—were also used as contraceptives. Modern analyses suggest that many of these afforded ancient Greeks effective methods of birth control.[2,3(pp30–31)]

Soranus (AD 98–138), a pre-eminent Greek physician, practiced medicine in the first half of the second century. He described about 30 formulas for contraceptive medications having ingredients such as old oil, honey, cedar gum, alum (a metallic astringent), and fruit acids, which were to be used to medicate wool suppositories.[6] Soranus also recommended four oral contraceptives.[2,3]

Other Ancient Cultures

In the ancient Islamic world, physicians and scientists recorded valid information on contraception.[12] While some ineffective potions were recommended, early Islamic literature addressing contraception was generally more accurate than early Christian writings.[6] Early Christians, few in number and widely scattered geographically, tended to follow regional customs and practices.[12]

Scant evidence exists of Islamic or Christian condemnation of early contraceptive practice. In AD 416, St. Augustine (AD 354–430) did denounce birth control practice. However, it is unclear to whom this critique was directed, and there is no evidence that it influenced religious adherents of the day.[2,13]

"In ancient India, rock salt was used as a spermicide."[2(p24)] Some Indian women placed honey and oil in their vaginas, hoping to prevent pregnancy. "Tampons soaked in these substances or in ghee (melted butter) also served as contraceptives."[2(p24)] Completely ineffective birth methods, such as passiveness or holding one's breath during coitus, were commonly practiced as well.[2,6]

Ancient Japanese made condoms made from tortoise shell or horn. In later times, Japanese couples would use leather condoms. (Rubber sheaths did not come into use until the second half of the 19th century.) "Prostitutes in ancient Japan used disks made of oiled tissue paper to prevent pregnancy.[2(p24),6]

Birth Control in Medieval Europe

Contraceptive practice was widespread during the Middle Ages. Many birth control methods existed, and apparently Europeans were quite successful in controlling their fertility. For example, the average household in medieval Italy had only 2.44 children. In the German territories, that figure is estimated to be 2.36. Lay midwives, serving as healers and childbirth assistants, transmitted much of the knowledge about fertility control.[2,14]

Before 1350, birth control was openly discussed in both canonical and secular documents. A classic medieval textbook, *Canons of Healing,* describes birth control methods. Clerical scholar Albert the Great (1200–1280) explained the contraceptive effects of certain plants and herbs in his *Summa de Creaturis* "with no indication that he was addressing a delicate or tabooed subject."[14(p24)] Chaucer also referred to birth control, another sign of its prevalence. Many church records of this time also document contraceptive practice.[2,6,10]

Disappearance of Birth Control

By the 15th century, however, hardly any Western European documents refer to birth control—a void that has long perplexed scholars. Because midwives were their major target, the witch hunts that occurred between 1360 and 1700 offer a plausible explanation of the disappearance of herbal contraceptives in that region. The witch hunts began shortly after the Black Plague (1347–1353) devastated the European population. England was particularly hard hit, losing an estimated 60% of its population. Throughout

Europe, the plague produced a massive labor shortage, which threatened the position of the landed aristocracy that depended on having enough people to serve as laborers, soldiers, and civil servants. At the time, the Catholic Church was the largest land owner in Europe.[2,14]

The idea of a deliberate state-organized system to supply labor was not new; the Roman Caesars had formulated such a concept, including the prohibition of birth control. Up until the 1300s, contraception had been commonly practiced despite a nearly 1000-year-old written Christian ethic prohibiting birth control (authored by St. Augustine). After the Black Plague, however, midwives, who had been the sources of birth control knowledge, were hunted down and persecuted as witches for the next two and a half centuries.[2,14]

The witch massacres were massive and widespread. Reliable estimates of the number of killings range from 1 million women to multiples of that figure (Figure 6–3). The pervasiveness of these massacres can be better appreciated when one considers that in 1600 the entire population of northwestern Europe, the center of witch hunting, was less than 50 million.[2,14]

Figure 6–3 Woodcut of Witch Being Burned at the Stake

Given the scale of these executions and those who were targeted, it is reasonable to assume that by 1700, when these killings ceased, birth control knowledge had been almost obliterated in Western Europe. While birth control practice did not completely cease, contraceptive information certainly was not as widespread or as sophisticated as it had been in the 14th century. Medieval midwives had been replaced by an emerging, largely male medical profession, whose members were not as knowledgeable of or as interested in birth control techniques as their predecessors.[14]

The Puritans

Among the other groups who opposed birth control were the Puritans (Figure 6–4), who emerged as a religious sect during the Elizabethan era (1558–1603). The Puritans disapproved of birth control for several reasons. They believed that children were a blessing from God, and that by being fruitful and multiplying, they could participate in the Creation. In addition, they worried that birth control might reduce the number of the Elect. Moreover, they were especially concerned about unmarried people practicing birth control. "Because many Puritans immigrated to the English colonies, their ideas became part of the American heritage.[2(p26),12(pp661-665)]

European Birth Control Rebounds

By 1500, Europe was experiencing rapid population growth, largely due to declining mortality.[4] Europe's population increased greatly for the next four centuries, despite massive emigration to all parts of the globe, particularly

Figure 6–4 Puritan Family, 1563
© Photos.com/Thinkstock

the Americas and Australia.[14] By the end of the 18th century, Europeans had begun to recognize both population growth and its concomitant poverty as social problems.

Among those concerned about the population explosion was the Rev. Thomas Malthus (1766–1834; see also the *Population Theories and Dynamics* chapter). In 1798, he published *An Essay on the Principle of Population as It Affects the Future Improvement of Society*. This work is important for several reasons, including Malthus's warnings about the likelihood of unchecked population growth. Because of his religious beliefs, Malthus considered birth control immoral, and he did not foresee the widespread use of contraceptives. His concern about the growing masses gained a great deal of attention, not only in his lifetime, but in also the 20th and 21st centuries.[2,15(pp53–59)]

Despite Malthus's misgivings, birth control practice was already rebounding. Condoms came into widespread use in 18th-century Europe (Box 6–1).[6(p195)] By 1700, shops in central London sold condoms made from animal membranes. (Vulcanized rubber condoms were not developed until the mid-19th century.[16]) This commercial activity was not clandestine; competition in the condom trade spurred widespread advertising, including jingles and leaflets announcing the availability of prophylactics.[2]

Francis Place and the English Birth Control Movement

Francis Place, a London leather breeches maker and an influential member of the reform movement of his day, is widely credited with launching the English birth control movement. In 1822, he published *The Principle of*

Box 6–1 Male Condoms: Origins Unknown but Documented by Casanova

While the exact origin of condoms is unknown, the first mention of a condom in the scientific literature appeared in a 1564 work of the great Italian anatomist Gabriel Fallopius. This work, published two years after the author's death, describes an experiment in which he instructed 1100 men in the use of a linen condom. Fallopius reported that not one of these men contracted syphilis.[6,11]

During the 18th century, prostitutes in brothels reportedly used condoms. This association with sexually transmitted diseases and prostitution has stigmatized this contraceptive method through current times. In the 1700s, Casanova, the Italian adventurer, reported his personal use of condoms. He documented his use of membrane prophylactics, not only to prevent infection, but also "to avoid impregnating his women." Casanova even described how he tested them for imperfections by inflating them with air.[2,6(p195)]

Population, Including an Examination of the Proposed Remedies of Thomas Malthus. In this monograph, he demonstrated the utter futility of deferred marriage or complete sexual abstinence as means for controlling population growth.[2,17(p139),18]

"Even more courageous than his rebuttal of Malthus's ideas about birth control was Place's publication of contraceptive brochures in 1823."[2(p27)] Widely distributed in London as well as in the industrial districts of northern England, the first of these leaflets was dubbed "The Diabolical Handbook." Advocating the use of coitus interruptus and tampons, Place instructed working-class people in practical ways to plan their families. For Francis Place, father of 15 children, this advice was not an abstract intellectual or theological argument; he knew the difficulty of supporting such a large brood.[17] Beyond his own family, Place's activities were socially groundbreaking for at least two reasons: his emphasis on socioeconomic desirability of birth control and his attempt to organize the masses.[2]

Contraception in Early-19-Century America

"Although the birth control movement did not take hold in the United States until the 20th century, the 19th century was an important time for American birth control practice. In 1800, a native-born white woman bore an average of 7.04 children. By 1860, that number was 5.21, and by 1900, it had plummeted to 3.56."[2(p28)] (Here it should be noted that statistics for women of color and immigrant women are far less complete and accurate than those for native-born, white women.[16]) To effect such dramatic change, Americans had to have been using birth control.[2]

At the beginning of the 19th century, the major contraceptive methods were coitus interruptus and prolonged lactation. Coitus interruptus became much more widespread among certain communities during this century, but how contraceptive information was diffused is not well understood. The Puritans' disapproval of birth control and other social conventions conspired against keeping records about contraceptive practices and other intimate behavior.[2,16,12]

During the first half of the 19th century, contraception was discussed in increasingly wider circles. Puritan ideas did not disappear from American life, but other ideas began to surface. "Two important books in disseminating contraceptive information were Dr. Charles Knowlton's *Fruits of Philosophy* (1831), also titled *The Private Companion of Young Married People*, and *Moral Physiology* by Robert Dale Owens. *Fruits of Philosophy* was the best medical presentation of its time and contained an excellent discussion of the reasons for family planning. Proof that in the 1830s the Puritan ethic still held sway is the fact that Knowlton served time in prison for writing and circulating this book."[2(p28)]

As the 19th century progressed, additional contraceptive methods became available, including douching and vaginal sponges (with and without spermicide), cervical caps and vaginal diaphragms (Box 6–2 and Figure 6–5), and periodic abstinence or the rhythm method. (Periodic abstinence was not very effective at this time because the correct time of ovulation had not been identified.) After the Civil War, a thriving commercial market for contraceptive devices developed. In 1872, for example, the Fuller and Fuller Company of Chicago sold seven kinds of troches (small, circular medicinal lozenges) for use in douching.[2,16]

Comstock Laws in America

In 1873, passage of the federal Comstock Law brought the development of American birth control to a standstill (Box 6–3). This legislation prohibited interstate trade in obscene literature and articles of immoral use, including contraceptives. Over the next two decades, this legislation spawned 22 "little Comstock" state laws, many of which went further in prohibiting birth control practice.[16(p257)] These laws stayed on the books for nearly a century, thwarting the dissemination of contraceptive information as well as chilling public and professional attitudes toward birth control.[2]

Box 6–2 Contraceptive Diaphragms and Cervical Caps

"A German physician, Wilhelm Peter Mesinga, is usually credited with inventing the diaphragm in 1882, but American inventions based on the same principle had been patented in the 1840s and were circulating by midcentury. Cervical caps, first advertised as "womb veils" in 1864, were probably derived from an obscure 1838 treatise written by German gynecologist, Friedrich Adolph Wilde. In 1823, Wilde had noticed that the farm families in a particular area numbered only two or three children. Investigating why their fertility was so much lower than average, he eventually traced the secret back to a midwife, who had discovered that placing an object in front of the cervix prevented impregnation. Wilde decided the best way to develop this idea was to take a wax imprint of the cervix and then make a rubber cap of the same size.[2(p29),16(p217),11(p11)] Knowledge of contraceptive pessaries that German immigrants brought with them to the United States may also have contributed to the development of the cervical cap.[2]

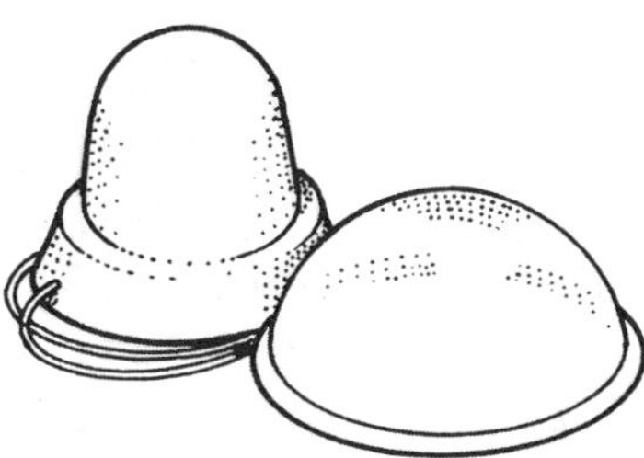

Figure 6–5 Illustration of Cervical Cap and Latex Diaphragm Contraceptive Barrier Methods

© Dorling Kindersley/Thinkstock

Box 6–3 Little Comstock Laws Follow the 1873 Federal Comstock Legislation

The passage of the 1873 federal Comstock Law prohibited interstate trade in obscene literature and materials, "including any article whatever for the prevention of conception, or for causing unlawful abortion." During the next 15 years, 22 states passed similar legislation. Known as "little Comstock laws," these state laws often went considerably further than federal law in trying to thwart contraceptive trade and practice.[16(p257)]

Fourteen states (Colorado, Indiana, Iowa, Massachusetts, Minnesota, Mississippi, Missouri, Montana, Nevada, New Jersey, New York, Pennsylvania, Washington, and Wyoming) prohibited the verbal transmission of information about contraception or abortion. Eleven states (Colorado, Indiana, Iowa, Minnesota, Mississippi, New Jersey, New York, North Dakota, Ohio, Pennsylvania, and Wyoming) made possession of instructions for the prevention of pregnancy a criminal offense. Four states (Colorado, Idaho, Iowa, and Oklahoma) authorized the search and seizure of contraceptive instructions. Connecticut alone outlawed the act of controlling conception.[2,16]

"One of the most striking legacies of Comstockery can be found in Margaret Sanger's account of her fruitless six-month search for contraceptive information in 'the best libraries in America.'"[2(p31),16(p283)] Fearing legal prosecution, the authors of early-20th-century medical textbooks did not even mention contraception. Indeed, it was 1959 before an American television show dared to mention birth control.[19] Comstockery underlay the context in which change occurred and portended current ideological battles.

Margaret Sanger

Margaret Sanger (1879–1966), the most influential birth control crusader of the 20th century, found her inspiration in the desperate conditions of the poor (see Figure 6–6). Leading a campaign from 1914 to 1937, she was successful in removing the stigma of obscenity from contraception.[20(p67)] After being jailed for distributing contraceptive literature in 1914, she fled to Europe to avoid a trial. When she returned in 1915, the prosecutor refused to try the case, although Sanger would have preferred another judicial test of the Comstock Law. She was jailed again in 1916 after opening a birth control clinic in Brooklyn—the first of its type in the country. In 1923, Sanger opened the Birth Control Clinical Research Bureau in New York City, which dispensed contraceptives and studied their effectiveness.[2]

On the political front, the persistent efforts of Sanger, the American Birth Control League, and others resulted in the gradual weakening of the Comstock Law. In 1918, the New York Court of Appeals ruled that legally practicing physicians could prescribe contraceptives for married couple "to cure or prevent disease."[21] "By 1937, the birth control movement had orchestrated and won the *U.S. v. One Package* case (86 F. 2d 737), which largely invalidated the 1873 federal Comstock Law."[2(p32-33)]

Figure 6–6 Margaret Sanger (1879–1966), American birth control activist.
Courtesy of Library of Congress, Prints and Photographs Division [LC-USZ62-29808].

In the *One Package* decision, a federal court of appeals ruled that had those legislators who enacted the Comstock Law known the facts about the dangers of pregnancy and the utility of contraception, they might not have outlawed all contraception as obscene. More than six decades after its passage, the court went on to say that the federal Comstock Law was not intended "to prevent the importation, sale, or carriage of mail of things which might intelligently be employed by conscientious and competent physicians for the purpose of life or the well-being of their patients."[22(p112)] Although the *One Package* decision opened the mails to contraceptive materials intended for physicians, "the right of individuals to bring such devices into the country was not established until 1971."[20(p121)] Box 6–4 explains that the majority of Americans at the time agreed with the *One Package* decision, but legislative action was still decades away.[2]

Developments in American Contraceptive Policies, 1940–1960

Although *One Package* tacitly permitted married couples to use contraception, diaphragms and cervical caps—the most effective methods of the day—were not widely available. These methods required a medical prescription and a clinical fitting by one's physician. Low-income women, in particular, had little access to these methods and limited information about them. In 1942, at the urging of First Lady Eleanor Roosevelt (1884–1962), Dr. Thomas Parran Jr. (1892–1968), the U.S. Surgeon General, *permitted* the states to use federal maternal and child health funds for birth control.[2]

Box 6–4 Public Opinion, the Court, and AMA Far Ahead of Legislative Action

"In 1937, at the time of the *One Package* decision, a national survey showed that 71% of Americans were in favor of birth control and 70% wanted to revise the federal Comstock Law. In the same year, the American Medical Association (AMA) recognized birth control as an integral part of medical practice and education."[2(p33),21] Nevertheless, the birth control movement's experience in both the U.S. Congress and state legislatures showed that legislators were not willing to take on such a controversial subject.[22] In concurring with the majority in the *One Package* decision, Justice Learned Hand commented upon the reticence of legislative bodies toward birth control, writing that legislative action might come, but "long after a majority would repeal birth control restrictions if a poll were taken." (U.S. v. One Package [86 F. 2d 737])

Birth control got another boost in 1958. After a prolonged controversy, physicians in New York City's municipal hospitals were permitted to prescribe and fit contraceptive diaphragms. This change marked a turning point in U.S. contraceptive policy because it served as a catalyst for other cities and states to change their practices.[19,20]

Contraception generated more publicity in 1959, when a presidential committee (formally the President's Commission to Study the United States Military Program) chaired by General William H. Draper (1894–1974) recommended that support for family planning be given to foreign governments requesting such assistance. Catholic bishops denounced this suggestion, and President Dwight Eisenhower initially opposed it (a position he publicly rescinded just a few years later); thus real policy change had to wait. Nonetheless, this recommendation spawned generated widespread discussion of contraception.[2,23]

"More effective birth control methods soon became available. In 1960, the U.S. Food and Drug Administration (FDA) approved the sale of oral steroid pills for contraception."[2(p34)] This development was the culmination of a decade-long effort by activist Margaret Sanger, philanthropist and biologist Katharine Dexter McCormick,[24] scientists Gregory Pincus and M. C. Chang, and obstetrician-gynecologists John Rock and Celso-Ramon Garcia.[25-29] A detailed timeline of the pill's development is available online.[25] Plastic intrauterine devices (IUDs; also known as intrauterine contraceptives [IUCs]) came on the global market in the 1960s.[30,31] Because of their effectiveness and separation from coitus, these methods ushered in a new era for birth control.[31]

For more than a half century, both birth control pills and IUDs have been widely used. Among married women globally, the IUD is the most commonly used reversible method of modern contraception, followed by oral contraceptives (OCs).[31] In more developed countries, oral contraceptive use exceeds use of other reversible methods, and among sexually active unmarried women worldwide, OCs are the most widely used modern contraceptive method.[32(p2),33]

CURRENT CONTRACEPTIVE METHODS

A wide variety of contraceptive methods exists today. Among these are reversible contraceptives and permanent methods. Reversible family planning methods include intrauterine methods, hormonal methods, barrier methods, fertility awareness-based methods, lactational amenorrhea, and coitus interruptus (withdrawal). Permanent methods include both male and female sterilization.[34]

Intrauterine Methods

IUCs or IUDs (Figure 6–7) are among the safest and most effective methods of contraception available today. They are also the most commonly used reversible method of contraception worldwide."[31(p147),33] Although their precise mechanism of action is not known, IUCs are suspected to work by preventing sperm from fertilizing ova. Therefore, an IUC is not an abortifacient; it does not interrupt an implanted pregnancy. "Pregnancy is prevented by a combination of the 'foreign body effect' of the plastic or metal frame and the specific action of the medication (copper or levonorgestrel) that is released."[31(pp147–152)]

Two types of IUCs are available in the United States at present: a copper T-shaped IUC and another T-shaped device that releases progestin. The copper T has been approved for 10 years' implantation, but data show that it is effective for at least 12 years. The approved life span of the levonorgestrel-containing IUC is 5 years, but some studies have shown efficacy up to 7 years.[31]

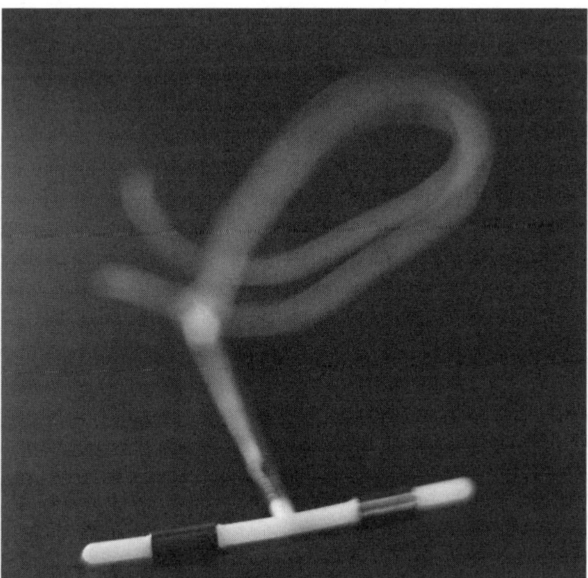

Figure 6–7 Intrauterine Contraceptive (IUC) or Device (IUD)
© Spike Mafford/Photodisc/Thinkstock

Hormonal Methods

Hormonal methods include the contraceptive implant, contraceptive injection, combined oral contraception (also called "the pill"; Figure 6–8), progestin-only pills, the patch, hormonal vaginal contraceptive ring, and emergency contraception.

Contraceptive implants consist of one or more thin rods or tubes that are inserted under the skin of a woman's upper arm. The rods contain progestin, which is released into the body for at least 3 years. Progestin suppresses ovulation; it also alters endometrial structure as well as changes "cervical mucus in a way that may impede sperm penetration."[34,35(p194)]

Contraceptive injections are similar to other hormonal methods in that they prevent pregnancy by inhibiting ovulation and changing cervical mucus. Usually given every 3 months, this birth control method is popular in China, Eastern Asia, and Latin America.[36]

Combined oral contraceptives (COCs) also suppress ovulation. The progestins in COCs "provide most of the contraceptive effect by suppressing ovulation and thickening cervical mucus, although the estrogens also make a small contribution to ovulation suppression."[37(p257)] Cycle control (e.g., reducing spotting) is enhanced by the estrogen.[37]

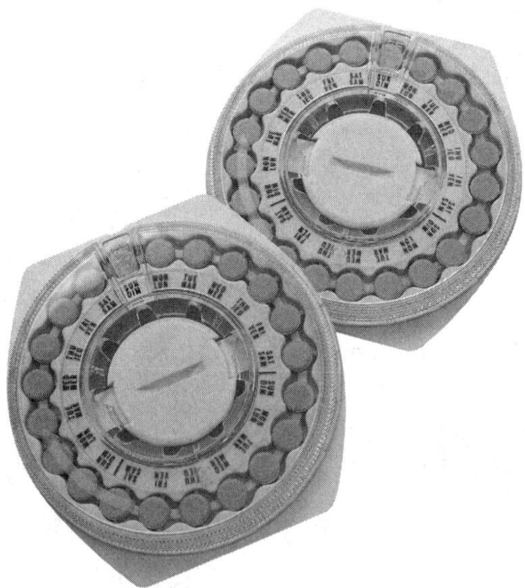

Figure 6–8 Birth Control Pills

© Stockbyte/Thinkstock

Progestin-only pills (sometimes called mini-pills) contain only one hormone, progestin, instead of both estrogen and progestin (as in COCs). Progestin-only pills may be a good option for women who should not take estrogen, particularly while breastfeeding. No adverse effects on lactation or infant health have been reported for this method.[34,38]

The *contraceptive patch* was first approved by the FDA in 2002. Like other combined hormonal methods, the patch prevents ovulation, thickens cervical mucus to prevent sperm penetration, and changes the endometrium, which "could affect implantation."[39(p344)]

The vaginal contraceptive ring also suppresses ovulation, thickens cervical mucus, and thins the endometrium. This method is designed to be placed in the vagina, where it releases the hormones progestin and estrogen, for 3 weeks. It is then removed for a week and a new ring placed in the vagina.[34,39]

Emergency contraceptives (EC) are methods that women can use after sexual intercourse to prevent pregnancy.[40] EC is not considered to be a regular method of birth control, but rather a backup if no birth control was used during sex or if the birth control method failed (e.g., a condom broke).[34] Emergency contraception is not considered equivalent to induced abortion because it is used before conception occurs. Indeed, EC greatly reduces a woman's chance of pregnancy after an episode of unprotected intercourse. In the United States, three ECs are approved for use; elsewhere in the world, other regimens are available. Because of the lack of awareness of EC and the politics surrounding it, EC is not used as widely as would be warranted by the incidence of unprotected coitus.[41,42]

Barrier Methods

Barrier methods include vaginal diaphragms, cervical caps, male condoms, female condoms, and spermicides.[34] All vaginal barriers and spermicides prevent fertilization by interfering with sperm transport into the female upper genital tract.[43(p392)] The male condom prevents pregnancy by blocking the passage of semen. Moreover, it prevents infection by covering the major portals of entry and exit for many sexually transmitted pathogens.[44(p372)]

A *vaginal diaphragm* is a flexible plastic dome with a soft spring in the rim. It folds in half so it can be inserted easily in the woman's vagina. Once in place, it springs gently into the dome shape. It is large enough to extend from the woman's pubic bone in front to the top of her vagina, and must cover the cervix. Although a diaphragm has some effectiveness without spermicide, this birth control method is much more effective when used with a spermicide.

It is placed in the diaphragm before insertion, so that the spermicide is held against the woman's cervix. Because of anatomic variations among women, diaphragms come in different diameters. In the United States, a prescription is required so that a healthcare professional can fit the diaphragm and instruct its user.[43]

A *cervical cap* is a thimble-shaped plastic dome with a brim that flares outward. "The concave side of the cap covers the cervix completely while the rim fits against the vaginal fornices."[43(p395)] To increase its effectiveness, spermicide is placed on the inside and the outside of the cap.

Male condoms are made from three types of material—latex, synthetic, and natural membrane. The first two provide prophylaxis against sexually transmitted diseases, while the third does not. Male condoms are available in a variety of shapes, colors, textures, and thicknesses as well as with or without lubricants or spermicides, and with or without reservoir-tip or nipple ends.[44(p372)]

Like male condoms, *female condoms* provide substantial protection against both unwanted pregnancy and sexually transmitted infections (Figure 6–9). One product is available for over-the-counter use in the United States. Other types of female condoms are available in about 100 different countries.[43]

Figure 6–9 Female Condom

© Keith Brofsky/Photodisc/Thinkstock

Fertility Awareness–Based Methods

"Fertility Awareness Based (FAB) methods of family planning depend on identifying the fertile window or the days in each menstrual cycle when intercourse is most likely to result in pregnancy. Some FAB methods may simply involve keeping track of cycle days to understand on which days of her cycle a woman is most likely to be fertile."[45(p417)] The success of any FAB method depends on three factors: "(1) the accuracy of the method in identifying the woman's actual fertile window, (2) the ability to correctly identify the fertile time," and (3) the ability to follow the method's regimen.[45(p419)]

Lactational Amenorrhea (Breastfeeding)

Breastfeeding extends postpartum infertility and depresses ovarian function. In terms of contraception, breastfeeding could be considered a hormonal method, because its effectiveness is due to the change in a lactating woman's hormones. Although partial protection against pregnancy lasts for many months, research has shown that it is most effective for the first 6 months of a newborn baby's life. Called the lactational amenorrhea method (LAM) of contraception, breastfeeding is approximately 98% effective if three criteria are met: the baby is younger than 6 months, the baby is exclusively or almost exclusively breastfed, and the mother has not had a menstrual period since the birth of the child.[46,47(p485)]

In both traditional societies and developing countries, lactation plays a major role in prolonging birth intervals, thereby reducing pregnancy. The contraceptive impact of breastfeeding is much less in developed countries because a far smaller fraction of infants are breastfed and they are weaned much earlier. "For example, in Indonesia, 96% of infants are breastfed, and those who are breastfed are not completely weaned until they are 2 years old on average. In contrast, in the United States, only 74% of infants are ever breastfed with only 21% still breastfeeding at 1 year."[47(p486)]

Coitus Interruptus (Withdrawal)

"Coitus interruptus is simply the practice of withdrawing the penis from the vagina and away from the external genital organs of the woman before ejaculation with the intention of avoiding pregnancy."[48(p409)] As explained earlier in the chapter, this method has been widely practiced throughout history, "playing a role in fertility declines occurring prior to the advent of the pill."[48(p409)] More than half of all sexually active woman in the United States say that they have relied upon coitus interruptus at some point, although less than 1% report it as their current family planning method. Clearly, this method is "unforgiving of incorrect or inconsistent use."[48(p411)]

Permanent Methods

Contraceptive sterilization is a permanent, safe, and highly effective approach for birth control. Sterilization is meant only for people who are certain that they do not desire a pregnancy in the future.[34] Current female methods include transabdominal and transcervical approaches. Sterilization for men, known as vasectomy, "blocks fertilization by cutting or occluding both vas deferens so that sperm can no longer pass out of the body in the ejaculate."[49(p460)] "Contraceptive sterilization (male and female) is the most widely used method of family planning in the world in both developed and developing countries."[49(p435)]

CONTRACEPTIVE EFFECTIVENESS AND SAFETY

Considerable variation exists in the efficacy of different contraceptive methods as well as in the possibilities of user failure. The real importance of any contraceptive method is how it improves on the likelihood of getting pregnant if no method is used.[4(p225)] If they do not use contraception, an estimated 85 out of every 100 sexually active couples will become pregnant within one year.[50(p780)] The effectiveness of any particular contraceptive method can be measured by pregnancy rates during perfect use and pregnancy rates during typical use.[51(p49)]

- Pregnancy rates during *perfect* or *theoretical use* show how effective contraceptives methods can be when used in ideal circumstances.
- Pregnancy rates during *typical use* reflect how effective a method is for the average person who does not always use the method correctly or consistently.[51]

Figure 6–10 shows that the difference between perfect (or theoretical) and typical use varies substantially among methods. For example, the percentage of women experiencing an unintended pregnancy within the first year of use of oral contraceptive pills is estimated to be 0.3% for perfect use and 9% for typical use. For the male condom, the pregnancy rate is 2% for perfect use and 18% for typical use. For some methods, such as sterilization, implants, levonorgestrel, and copper-T IUCs, there is little difference between the unintended pregnancy rates for perfect use and for typical use. The nature of the method itself simply precludes human error, because it does not require the user follow specific steps, especially around the time of coitus.[51] The typical use effectiveness of different methods is summarized in Figure 6–11.

Continuation rates are also important. Contraceptives are effective at preventing pregnancy only if women or couples continue to use them. Here again, there is wide variation among methods—ranging from a 47% continuation rate after the first year among women using fertility awareness methods to a 78% rate among copper-T IUC users to 100% among those who have undergone a sterilization procedure.

Over time, annual contraceptive effectiveness rates go up for birth control methods because women or couples find a method that works for them and continue to use it. This point is important because women spend long periods of their lives trying to avoid pregnancy. For example, the average woman in the United States spends about 39 years at potential biological

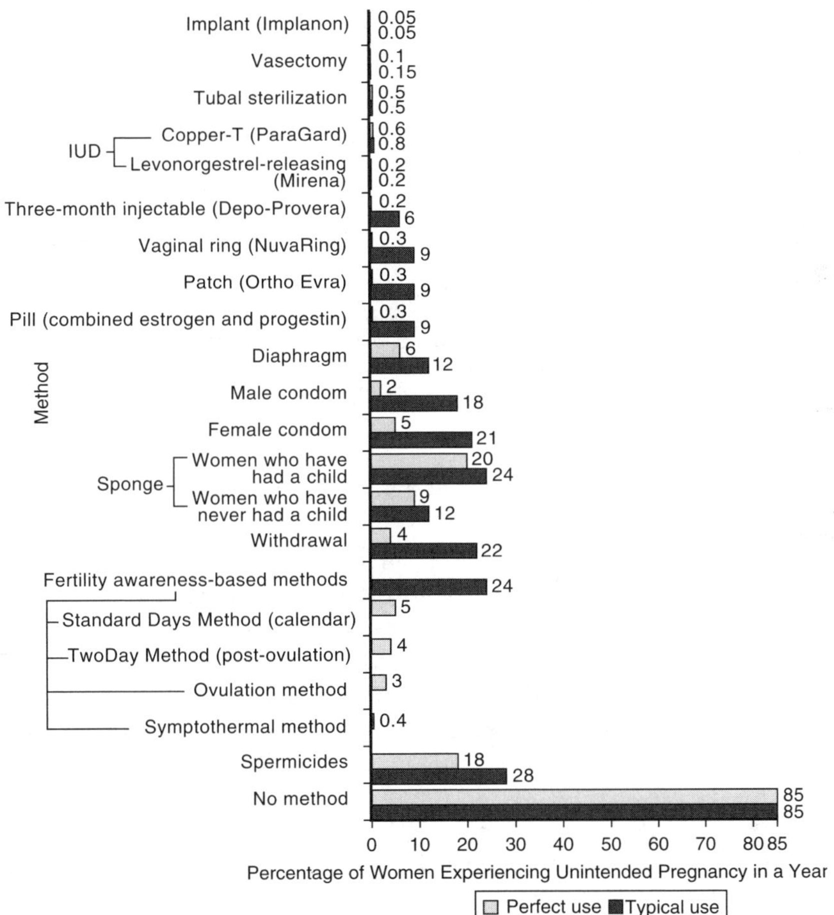

Figure 6–10 Unintended Pregnancy Rates During First Year of Perfect Use and Typical Use by Contraceptive Method

Data from Trussell J, Guthrie KA. Choosing a contraceptive: efficacy, safety, and personal considerations. In: *Contraceptive Technology.* Hatcher RA, Trussell, J, Nelson AL, Cates W, Kowal D, Policar MS, eds. Atlanta, GA: Ardent Media Inc.; 2011:24.

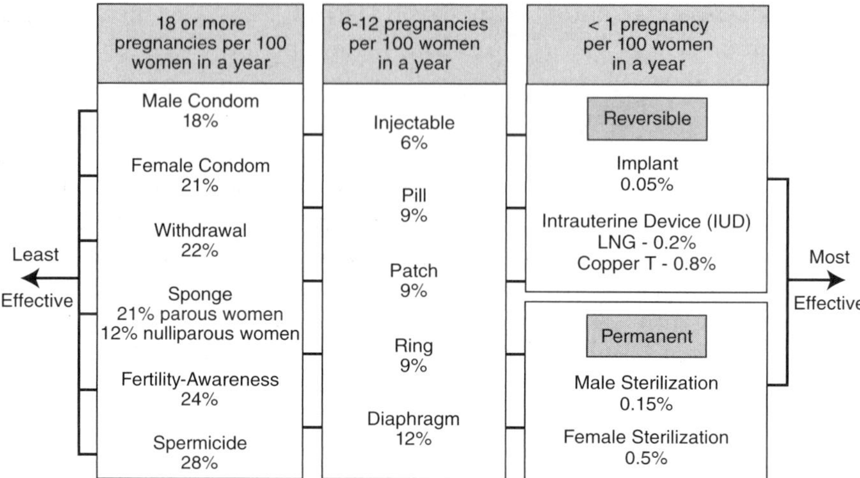

Figure 6–11 Incidence of Pregnancy for Typical Use of Different Contraceptive Methods (Estimates for First Year of Use)

Data from the Centers for Disease Control and Prevention, Contraception. 2013. http://www.cdc.gov/reproductivehealth/UnintendedPregnancy/PDF/Contraceptive_methods_508.pdf. Adapted from Hatcher RA, Trussell J, Nelson AL, Cates W, Kowal D, Policar MS. Atlanta, GA: Ardent Media Inc 2011 and World Health Organization (WHO) Department of Reproductive Health and Research, Johns Hopkins Bloomberg School of Public Health/Center for Communication Programs (CCP). Knowledge for health project. Family planning: a global handbook for providers (2011 update). Baltimore, MD; Geneva, Switzerland: CCP and WHO; 2011.

risk of pregnancy, from menarche at age 12.5 years to natural menopause at 51.3 years. More than half of that reproductive life span is spent trying to avoid further childbearing.[54(p68)]

The method that the couple uses and how it is used is vital. The typical woman in the United States who uses reversible methods of contraception continuously from age 15 to age 45 experiences 1.8 contraceptive failures. If both reversible methods and sterilization are practiced, the typical woman would experience only 1.3 contraceptive failures from age 15 to age 45.[51(p55)]

Contraceptive Safety

"In general, contraceptives pose few serious health risks to users. Moreover, the use of contraceptive methods is generally far safer than pregnancy."[51(p61)] In the United States, for example, the risk of death from pregnancy is 1 in 6900, while the risk of death for a nonsmoking woman aged 15–34 years using combined oral contraceptives is 1 in 1,667,000.[51(p62)]

Unintended pregnancy, therefore, unnecessarily place women at risk. "Women in many developing countries will experience an even greater advantage in using contraceptive methods—in comparison with pregnancy-related mortality—than women in the developing world."[51(p61)] "Increasing contraceptive use in developing countries has cut the number of maternal deaths by 40% over the past 20 years, merely by reducing the number of unintended pregnancies."[52(p149)]

Nevertheless, use of some contraceptive methods entails potential risks as well as benefits.[51] For women 35 and older who smoke and take COCs, the risks of death go up considerably (1 in 5200) and exceed the risk of pregnancy.[51(p63)] Certainly, other methods would be warranted in that situation.

Some birth control methods even provide noncontraceptive benefits. COCs, for example, protect against ectopic pregnancy, pelvic inflammatory disease (PID), and ovarian and endometrial and colorectal cancer, and they may improve bone health in some women with low estrogen levels.[51(p65,68)] However, throughout the world, women are more familiar with the health risks of birth control than the noncontraceptive benefits.[53]

CONTRACEPTIVE PREVALENCE

Contraceptive prevalence is a useful measure for tracking progress toward achieving universal access to reproductive health. The contraceptive prevalence rate is simply the number of reproductive-age women who are using contraception. It may be refined to include categories of methods or specific methods.[54]

Contraceptive methods are often classified as either modern or traditional. Modern methods of contraception include female and male sterilization, oral hormonal pills, IUDs, male condoms, injectables, implants (including Norplant), vaginal barrier methods, female condoms, and emergency contraception. Traditional methods of contraception include fertility awareness methods, coitus interruptus, prolonged abstinence, douching, LAM, and folk methods.[54] Table 6–2 shows that contraceptive prevalence varies greatly around the world. Countries differ both in the number of methods offered and the extent to which each is made available. Contraceptive prevalence is highest in countries where a full range of birth control methods is readily accessible.[55]

CONCLUSION

Contraception has been practiced throughout history. The first known recipes for birth control date back to 1850 BCE, and birth control has continued throughout the ages and across cultures. During that time, there have been efforts to limit contraceptive practice, notably during the

Table 6–2 Global Contraceptive Prevalence

World	Modern Contraceptive Prevalence	Traditional Contraceptive Prevalence
1. More-developed regions	61.3	11.0
2. Less-developed regions	55.2	5.9
a. Least-developed countries	25.0	6.4
b. Other developing countries	60.2	5.9

Data from United Nations Population Division. *World Contraceptive Use 2011*. New York: United Nations; 2011. http://www.un.org/esa/population/publications/contraceptive2011/contraceptive2011.htm.

infamous witch hunts in medieval Europe. More recent efforts to suppress birth control include the federal and state Comstock laws that made contraceptive practice illegal in some American states and turned interstate transport of contraceptive devices into a federal crime.

Twentieth-century activists, notably Margaret Sanger, have had a profound effort on women's and couples' ability to practice contraception. After nearly a century, the Comstock Laws were finally struck down. Through the efforts of Sanger and others, new hormonal contraception became available in 1960. Intrauterine devices were marketed shortly after. Both methods revolutionized global contraceptive practice.

The contraceptive options of today are multiple and largely woman centered. For the most part, modern contraceptive methods are highly effective and safe. However, their availability varies greatly around the world. Not surprisingly, contraceptive prevalence is lowest in the poorest countries.

DISCUSSION QUESTIONS

1. Discuss the legacy of the Comstock Law, both for the United States and the rest of the world.
2. Among countries, greater availability of contraceptive choices is associated with higher contraceptive prevalence. Explain why this is the case.
3. Why was developing a birth control pill so important?

4. Do you think that most people would find the data in Table 6-2 surprising? Why or why not?

5. Recently, there have been some promising developments in the development of male contraception. Do you think a male birth control pill would be as important as oral contraception has been for women?

REFERENCES

1. *The American Heritage Dictionary of the English Language.* 3rd ed. Boston, MA: Houghton Mifflin; 1992:409.
2. McFarlane DR, Meier KJ. *The Politics of Fertility Control: Family Planning and Abortion Policies in the American States.* New York: Chatham House Press 2001.
3. Riddle JM, Estes JW, Russell JC. Ever since Eve: birth control in the ancient world. *Archeology.* 1994;47:29–35.
4. Weeks JR. *Population: An Introduction to Concepts and Issues.* 11th ed. Belmont, CA: Wadsworth Cengage Learning; 2012.
5. Johnson JT. *Oral History Memoir, Population and Reproductive Health Oral History Project.* Northampton, MA: Sophia Smith Collection, Smith College; 2004. www.smith.edu/library/libs/ssc/prh/transcripts/johnson-trans.pdf. Accessed November 3, 2013.
6. Himes NE. *Medical History of Contraception* (reprint of 1936 edition). New York: Schoken Books; 1970.
7. Wellings K, Collumbien M, Slaymaker E, et al. Sexual behavior in context: a global perspective. *Lancet.* 2006;365(9460):702–710.
8. Hymowitz K, Carroll JS, Wilcox WB, Kaye K. *Knot Now: The Benefits and Costs of Delayed Marriage in America.* Charlottesville, VA: National Campaign to Prevent Teen and Unplanned Pregnancy, Relate Project, National Marriage Project at the University of Virginia; 2013. http://twentysomethingmarriage.org/summary/. Accessed December 2, 2013.
9. Martin JA, Hamilton EE, Ventura SJ, et al. *Births: Final Data for 2011.* National Vital Statistics Reports, Vol. 62, No. 1. Hyattsville, MD: National Center for Health Statistics; 2013. http://www.cdc.gov/nchs/fastats/unmarry.htm. Accessed December 7, 2013.
10. Tietze C. History of contraceptive methods. *J Sex Res.* 1965;1(2):69–85.
11. Suitters B. Contraception in ancient and modern society. *Roy Soc Health J.* 1968;88(1):9–11.
12. Schnucker RV. Elizabethan birth control and puritan attitudes. *J Interdiscipl Hist.* 1975;4:655–667.
13. Noonan JT. *Contraception: A History of Its Treatment by Catholic Theologians and Canonists.* Cambridge, MA: Harvard University Press; 1986.
14. Heinsohn G, Steiger O. The elimination of medieval birth control and the witch trials of modern times. *Int J Women's Stud.* 1982;5:193–214.
15. Thomlinson R. *Population Dynamics: Causes and Consequences of World Demographic Change.* New York: Random House; 1965.

16. Brodie JF. *Contraception and Abortion in Nineteenth Century America.* Ithaca, NY: Cornell University Press; 1994.

17. Finch BE, Green H. *Contraception Through the Ages.* London: Peter Owen; 1963.

18. *British Library, Francis Place Collection: History of the Collection and Its Collector.* London, British Library. n.d. http://www.bl.uk/reshelp/findhelprestype/microform/francisplace/index.html#history. Accessed December 4, 2013.

19. Jaffe FS. Public policy on fertility control. *Sci Am.* 1967;229(1):17–23.

20. Reed J. *From Public Vice to Private Virtue.* New York: Basic Books; 1978.

21. Planned Parenthood Federation of America. *Family Planning in America: Chronology of Major Events: Fact Sheet.* New York: Planned Parenthood of America; 1992.

22. Dienes CT. *Laws, Politics, and Birth Control.* Urbana: University of Illinois Press; 1972.

23. Piotrow PT. *World Population Crisis: The United States Response.* New York: Praeger; 1973.

24. Fields A, McCormick KD. *Pioneer for Women's Rights.* New York: Praeger; 2003.

25. Public Broadcasting Corporation. The pill: timeline. *The American Experience.* Boston, MA: PBS; 2002. http://www.pbs.org/wgbh/amex/pill/timeline/timeline2.html. Accessed December 12, 2013.

26. Halberstam D. *The Fifties.* New York: Villard Books; 1993.

27. Pincus G, Garcia C-R, Rock J, et al. Effectiveness of an oral contraceptive: effects of a progestin–estrogen combination upon fertility, menstrual phenomena, and health. *Science.* 1959;130:81–83.

28. Greep RO. *Min Chueh Chang, 1908–1991.* Washington, DC: National Academies Press; 1995.

29. Strauss JF, Mastroianni L. In memorium Celso Ramon Garcia, MD (1922–2004), reproductive medicine visionary. *J Exper Clin Assisted Reprod.* 2005;2:2. http://www.biomedcentral.com/1743-1050/2/2. Accessed December 12, 2013.

30. Mishell DR. Assessing the intrauterine device. *Fam Plan Persp.* 1975;7:103–111.

31. Dean G, Schwarz EB. Intrauterine contraceptives (IUCs). In: Hatcher RA, Trussell, J, Nelson AL, et al., eds. *Contraceptive Technology.* Atlanta, GA: Ardent Media; 2011:147–191.

32. Blackburn RD, Cunkelman JA, Zlidar VM. *Oral Contraceptives: An Update.* Population Reports, Series A, No. 9. Baltimore, MD: Johns Hopkins Bloomberg School of Public Health, Center for Communication Programs; 2000:1–40. http://www.k4health.org/toolkits/info-publications/oral-contraceptives-update. Accessed March 17, 2014.

33. United Nations Population Division. *World Population Use 2011.* New York: United Nations; 2011. http://www.un.org/esa/population/publications/contraceptive2011/contraceptive2011.htm.

34. Centers for Disease Control and Prevention. Contraception. 2013. http://www.cdc.gov/reproductivehealth/unintendedpregnancy/contraception.htm. Accessed December 9, 2013.

35. Raymond EG. Contraceptive implants. In Hatcher RA, Trussell, J, Nelson AL, et al., eds. *Contraceptive Technology.* Atlanta, GA: Ardent Media; 2011:193–207.

36. Bartz D, Goldberg AB. Injectable contraceptives. In Hatcher RA, Trussell, J, Nelson AL, et al., eds. *Contraceptive Technology.* Atlanta, GA: Ardent Media; 2011:209–236.
37. Nelson AL, Cwiak C. Combined oral contraceptives (COCs). In Hatcher RA, Trussell, J, Nelson AL, et al., eds. *Contraceptive Technology.* Atlanta, GA: Ardent Media; 2011:249–341.
38. Raymond EG. Progestin-only pills. In Hatcher RA, Trussell, J, Nelson AL, et al., eds. *Contraceptive Technology.* Atlanta, GA: Ardent Media; 2011:237–247.
39. Nanda K. Contraceptive patch and vaginal contraceptive ring. In Hatcher RA, Trussell, J, Nelson AL, et al., eds. *Contraceptive Technology.* Atlanta, GA: Ardent Media; 2011:343–369.
40. Trussell J, Bimla Schwarz E. Emergency contraception. In Hatcher RA, Trussell, J, Nelson AL, et al., eds. *Contraceptive Technology.* Atlanta, GA: Ardent Media; 2011:113–145.
41. Glasier A. Emergency contraception: clinical outcomes. *Contraception.* 2013;87(3):309–313.
42. McFarlane DR. Reproductive health policies in President Bush's second term: old battles and new fronts in the United States and internationally. *J Public Health Policy.* 2006;27(4):405–426.
43. Cates W Jr, Harwood B. Vaginal barriers and spermicides. In Hatcher RA, Trussell, J, Nelson AL, et al., eds. *Contraceptive Technology.* Atlanta, GA: Ardent Media; 2011:391–408.
44. Warner L, Steiner MJ. Male condoms. In Hatcher RA, Trussell, J, Nelson AL, et al., eds. *Contraceptive Technology.* Atlanta, GA: Ardent Media; 2011:371–389.
45. Jennings VH, Burke AE. Fertility awareness based methods. In Hatcher RA, Trussell, J, Nelson AL, et al., eds. *Contraceptive Technology.* Atlanta, GA: Ardent Media; 2011:418–434.
46. Labbok MH, Hight-Laukaran V, Peterson AE, et al. Multicenter Study of the Lactational Amenorrhea Method (LAM): I. Efficacy, Duration, and Implications for Clinical Application. *Contraception.* 1997;55(6):327–336.
47. Kennedy KI, Trussell J. Postpartum contraception and lactation. In Hatcher RA, Trussell, J, Nelson AL, et al., eds. *Contraceptive Technology.* Atlanta, GA: Ardent Media; 2011:483–511.
48. Kowal D. Coitus interruptus (withdrawal). In Hatcher RA, Trussell, J, Nelson AL, et al., eds. *Contraceptive Technology.* Atlanta, GA: Ardent Media; 2011:409–415.
49. Roncari D, Hou MY. Female and male sterilization. In Hatcher RA, Trussell, J, Nelson AL, et al., eds. *Contraceptive Technology.* Atlanta, GA: Ardent Media; 2011:435–482.
50. Trussell J. Contraceptive efficacy. In Hatcher RA, Trussell, J, Nelson AL, et al., eds. *Contraceptive Technology.* Atlanta, GA: Ardent Media; 2011:779–863.
51. Trussell J, Guthrie KA. Choosing a contraceptive: efficacy, safety, and personal considerations. In Hatcher RA, Trussell, J, Nelson AL, et al., eds. *Contraceptive Technology.* Atlanta, GA: Ardent Media; 2011:45–74.

52. Cleland J, Conde-Agudelo A, Peterson H, et al. Contraception and health. *Lancet.* 2012;380:149–156.
53. Edwards JE, Oldman A, Smith L, McQuay HJ, Moore RA. Women's knowledge of, and attitudes to, contraceptive effectiveness and adverse health effects. *Br J Fam Plan.* 2000;26:73–80. http://www.medicine.ox.ac.uk/bandolier/booth/sexhlth/womenoc.html. Accessed December 14, 2013.
54. United Nations Population Division. *World Contraceptive Use 2009: Contraceptive Prevalence.* New York: United Nations; 2009. http://www.un.org/esa/population/publications/WCU2009/Metadata/CPR.html. Accessed December 4, 2013.
55. Ross J, Hardee K, Mumford E, Eid S. Contraceptive method choice in developing countries. *Int Fam Plan Persp.* 2002;28(1):32–40. http://www.guttmacher.org/pubs/journals/2803202.html.

Chapter 7

Abortion and Reproductive Health

Andrzej Kulczycki

INTRODUCTION

Abortion is a generic term for pregnancies that do not end in a live birth or a stillbirth. Although spontaneous abortions occur, abortion invariably refers to the deliberate termination of pregnancy before viability, the point at which a fetus can survive independently outside of the uterus.* An abortion can be induced through a medical (pharmacologic) or surgical procedure, and it may bring relief from a pregnancy that is unwanted for a range of reasons.

Both in the United States and globally, more than one-fifth of all known pregnancies end in abortion. Abortion is now one of the safest procedures in medicine when performed by a trained professional in hygienic conditions using modern methods. Nevertheless, of the estimated 44 million abortions performed annually worldwide, about half are unsafe, causing some 47,000 maternal deaths, or 13% of all maternal mortality.[1] Millions more women are injured, some seriously and permanently, but these deaths and injuries are almost entirely preventable.

---

* Fetal viability may vary from 20 to 28 weeks' gestation, depending on fetal organ maturity, environmental conditions, and biomedical and technological capacities. Medical advances now imply that viability can be generally assumed at about 24 weeks' gestation in affluent countries, but there is no uniform gestational age that defines or predicts viability.

The incidence of abortion may be reduced through good access to a range of effective contraceptive methods, sex education, and appropriate support for women who want to have a child. In many countries, however, abortion is a controversial field of health and social policy due to fundamental differences in views on such issues as when human life begins, on women's roles and rights, and on the role of government in individuals' private lives.

Abortion has been with us since time immemorial. Historically, many women who underwent abortions risked their personal health and social standing. During the second half of the 20th century, this situation changed in many countries. Abortion procedures became safer, and efforts to legalize abortion gained momentum. In many countries, however, abortion remains illegal and/or inaccessible, resulting in unsafe procedures that contribute to maternal mortality and morbidity.

ABORTION IN HISTORY

Throughout recorded history, human beings have attempted to control family size through several steps: before conception by delaying marriage or through abstinence or contraception (the latter well understood and effective only in recent times); after birth by infanticide; or by inducing abortion, a step that falls temporally between these two extremes by preventing a conception from becoming a live birth. Infanticide was rarely used as a form of family limitation but, like abortion, it allowed fertility control after conception. An extensive pharmacopoeia of herbal abortifacients and menstrual-regulating practices evolved, many risky and ineffective.[2,3] These gradually became viewed with more suspicion by medical and pharmaceutical personnel keen to assert their professional role and interests. In the United States, for example, the medical establishment was far more instrumental than religious activism in pushing through the late-19th-century wave of restrictive abortion laws at the state level,[4] although it advocated for more permissive laws a century later.

Historically, human society has placed a high value on human life. Methods of fertility control were less likely to be sanctioned the further along they occurred in the continuum from heterosexual intercourse to a live child. Pregnancy was evidenced only when women could feel fetal movements (then known as quickening) which typically occurs naturally at about the middle of a pregnancy, or somewhat before (from 17–20 weeks gestation in most cases, but ranging from 13–25 weeks). Many customary laws condemned its interruption, not least due to the harm it could cause women. In the 19th century, most jurisdictions criminalized abortion. With advances in women's rights, abortion techniques, and safety, however, laws in the second half of the 20th century increasingly recognized women's needs and rights as the principal decision makers over their own reproduction. As attitudes slowly became more permissive through the 20th century, highly skilled laypersons and physicians increasingly provided safe abortion services to some affluent and well-connected women.[5,6]

Induced abortion was almost universally illegal at the beginning of the 20th century. The Soviet Union first made it legal and widely available. After World War II, the Scandinavian countries legalized abortion, as did most of Western and Eastern Europe, as well as Japan. In the United States, the 1973 U.S. Supreme Court ruling *Roe v. Wade* permitted abortions during the first three months of pregnancy and with increasing restrictions thereafter. The Court subsequently reaffirmed its landmark decision despite numerous legal challenges, although the 1976 Hyde Amendment passed by the U.S. Congress barred the use of federal funds for abortion except for all but the most extreme circumstances (rape, incest, or if the pregnant woman's life was threatened).

Both in the United States and elsewhere, abortion-related mortality fell greatly after nationwide legalization; moreover, the incidence of abortion has fallen with gains in contraceptive practice. However, conflict over abortion continues in many other countries, making it difficult to reach a necessary and practical societal consensus.[7,8] The International Conference on Population and Development Conference (ICPD), organized by the United Nations (UN) in Cairo in 1994, articulated and affirmed the concept of reproductive rights inclusive of safe, quality abortion services where allowed by national laws. At the beginning of the 21st century, abortion was legal in most of Europe and Asia, but illegal in much of Africa and South America. In northern America, Europe, Australasia, and Japan, most abortions in the last few decades of the 20th century were performed surgically using suction curettage. Since the turn of the century, more abortions have been induced by administering the pharmaceutical agents mifepristone and misoprosotol.[†] Mifepristone causes an abortion in almost all cases when used in combination with misoprostol, which is less effective when used by itself.

DATA

The number and characteristics of women who have abortions are known in detail only for those countries where abortion is permitted with few restrictions and statistics collected by government agencies are reasonably

---

[†] Known earlier as RU-486, mifepristone was approved in many European countries in the 1990s and in 2000 in the United States, where it is also known by its brand name, Mifeprex. Long used to prevent stomach ulcers, misoprostol is a prostaglandin analogue whose various gynecologic uses came to be known more recently.

complete. However, induced abortions are often underreported wherever the procedure remains a sensitive issue. A range of estimation methodologies has been developed for use in countries where abortion is outlawed and performed surreptitiously.[9,10] Nevertheless, the quality of data is not clear even in many countries where abortion is legal and widely accepted, such as in China and Vietnam. For such countries, however, estimates can be made by applying demographic techniques to reports from official national reporting systems, nationally representative surveys, published studies, and other data.

WHO provides the most up-to-date information regarding the epidemiology of abortion around the world, including estimates of the significant morbidity and mortality arising from unsafe procedures for countries in which they occur.[1] These estimates are based on national surveys of women, hospital records, and other data. The most widely cited estimates of abortion incidence for major world regions are presented by WHO and researchers from the Guttmacher Institute.[1,11] In the United States, the Centers for Disease Control and Prevention (CDC) is the federal agency that compiles annual numbers and basic characteristics of women obtaining abortions.[12] However, data are unavailable for some states that lack mandatory reporting and are of varying reliability for others where reporting is poorly enforced. A more complete count of the total number of abortions is available from the Guttmacher Institute based on its periodic census of abortion providers.[10] All these sources are cited in this chapter.

In addition, there is a need for much qualitative research to investigate multiple aspects of abortion, as well as different facets of abortion policy. The demography and epidemiology of induced abortion are discussed further in this chapter, following a brief description of spontaneous embryonic and fetal loss.

SPONTANEOUS ABORTIONS

Spontaneous abortion refers to a natural biological process by which an unknown but large percentage of pregnancies never become infants for reasons unrelated to human intervention. The uterus will typically expel a seriously defective fetus very early in the pregnancy, with the woman experiencing only a larger than usual blood flow around the time of her expected menstrual period and often not even being aware that she is pregnant. A spontaneous abortion occurring later in the pregnancy is commonly called a miscarriage and may be due to maternal health problems. It can also result in maternal death if left untreated. There is no definitive information about the number and rate of spontaneous abortions, although it is believed that these cases may account for up to one in four clinically recognized pregnancies.[14]

GLOBAL AND REGIONAL ABORTION INCIDENCE AND TRENDS

The numbers and rates of induced abortions can be estimated with more clarity, although these estimates are somewhat crude for most countries. Worldwide, the incidence of induced abortion is believed to have held steady in recent years after a long period of decline that was attributable to improved access to family planning education and services. In 2008, the global abortion rate was estimated at 29 abortions per 1000 women aged 15–44, a rate virtually unchanged since 2003, after having fallen from 35/1000 in 1995 (Table 7–1). The stall in the abortion rate is correlated with a plateau in the level of contraceptive use. Several factors account for the rise in the proportion of all abortions that occurred in developing regions (from

Table 7–1 Global and Regional Estimates of Induced Abortions, 1995, 2003, and 2008

Region	Number of Abortions (Millions)			Abortions Rate*		
	1995	2003	2008	1995	2003	2008
World	45.6	41.6	43.8	35	29	28
Developed countries	10.0	6.6	6.0	39	25	24
Excluding Eastern Europe	3.8	3.5	3.2	20	19	17
Developing countries	35.5	35.0	37.8	34	29	29
Excluding China	24.9	26.4	28.6	33	30	29
Africa	5.0	5.6	6.4	33	29	29
Asia	26.8	25.9	27.3	33	59	28
Europe	7.7	4.3	4.2	48	28	27
Latin America	4.2	4.1	4.4	37	31	32
Northern America	1.5	1.5	1.4	22	21	19
Oceania	0.1	0.1	0.1	21	18	17

*Abortions per 1000 women aged 15–44 years.

Data from Sedgh G, Singh S, Shah I, et al. Induced abortion: incidence and trends worldwide from 1995 to 2008. *Lancet.* 2012;379(9816):625–632; World Health Organization. *Unsafe Abortion: Global and Regional Estimates of the Incidence of Unsafe Abortion and Associated Mortality in 2008.* 6th ed. 2011. http://www.who.int/reproductivehealth/publications/unsafe_abortion/9789241501118/en/.

78% to 86%) over 1995–2008. These included declining fertility preferences and increases in the number of pregnancies terminated (up by 2.3 million to 38 million), as well as population growth. Overall, women living in the developing world have a higher likelihood of having an abortion than do their counterparts in more-developed countries, as reflected in their higher abortion rate (29 versus 24 per 1000 women of childbearing age, respectively).

The highest abortion rates in the world are found in many former Soviet Bloc republics. Family planning services have been severely lacking in these societies, whereas abortion has long been readily available because it was legalized in the Soviet Union and its allies well ahead of other countries. A 1999 survey documented that the average Georgian woman had 3.7 abortions during her lifetime, although this level had declined by 16% to 3.1 in 2005.[15] This was still the highest known total abortion rate in the world at the time. Levels of abortion fell elsewhere in Central Asia and Russia to a varying degree as the availability, accessibility, and quality of available contraceptive options improved.

Abortion is generally legal on broad grounds in Europe, but rates are estimated to vary from 12/1000 women in Western Europe to 43/1000 in Eastern European countries. The latter have a higher proportion of terminated pregnancies (about 30%) due to lower and less effective contraceptive use rates. The abortion rate is believed to have remained virtually unchanged in these regions over 2003–2008, having fallen significantly during the previous decade when greater gains in modern contraceptive use took place. The Netherlands, Belgium, Scandinavia (with the exception of Sweden), and Germany have had the lowest abortion rates in the world for many years (about 5–7/1000 women aged 15–44 years, approximately one-third the U.S. rate). These countries are characterized by greater contraceptive responsibility, very good access to family planning services, and widely available, safe, and inexpensive abortion services.

Among the world's major developing regions, Latin America has the world's highest abortion rates, although these are estimated to have fallen from 37 to 31 abortions per 1000 women between 1995 and 2008, with a steep decline occurring between the mid-1990s and the early 2000s. Subregional abortion rates are lowest in Central America (which includes Mexico) at 29/1000, rising to 32/1000 in South America and 39/1000 in the Caribbean.[16]

In Asia, estimates suggest that after a prolonged period of decline that coincided with gains in contraceptive use, abortion incidence may have risen in China in the first decade of the 20th century due to increases in premarital sexual activity and unintended pregnancy. Elsewhere in Asia, abortion rates showed no decline over 2003–2008 and are highest in Southeastern Asia (36/1000) and lowest in South Central Asia and Western Asia (26/1000).

In Africa, where the vast majority of abortions are clandestine and unsafe, the overall abortion rate was estimated to have fallen modestly between 1995 and 2003, and then remained mostly steady during 2003–2008 (at 29 abortions per 1000 women of childbearing age). Of all African subregions, Southern Africa has the lowest abortion rate (15/1000), attributable to the legalization of abortion in South Africa in 1997. Rates are much higher in Central and East Africa (36/1000 and 38/1000, respectively) than in West and North Africa (28/1000 and 18/1000, respectively).

STRUCTURAL AND INDIVIDUAL-LEVEL DETERMINANTS OF ABORTION

Unintended Pregnancy

Unintended pregnancies underlie nearly all abortions and have massive public health consequences. Unintended pregnancies, abortions, and unplanned births can be prevented through reducing unmet need for modern contraception among women and couples who want to avoid a pregnancy but who either are not using a method or are using traditional family planning methods that have relatively high user failure rates.

Family planning is hailed as one of the great public health achievements of the last century. By 2012, worldwide contraceptive use had risen to more than three-fifths (64%) of exposed couples.[17] In many countries, however, uptake of modern contraception is constrained by limited access and weak service delivery, and the burden of unintended pregnancy remains large. In 2008, 41% of pregnancies worldwide were estimated to be unintended, with the highest rates in eastern and middle Africa.[18] Worldwide, the annual increase in modern contraceptive use slowed in 1997–2009 to 0.2 percentage point per year in Africa, and to only 0.1 percentage point worldwide. In developing regions, approximately 222 million women were estimated to have an unmet need for modern contraception in 2012, accounting for about four-fifths of unintended pregnancies.[19] The United States has a higher abortion rate than other Western nations because a higher proportion of pregnancies are unintended (48%), about half of which are ultimately aborted.

Contraceptive failure also contributes substantially to unintended births and induced abortions, as the vast majority of contraceptive failures result in either one or the other outcome. Analysis of data from a range of low- and middle-income countries that conducted Demographic and Health Surveys (DHS) over 2002–2006 found that in the six countries studied that collected induced-abortion data (and that may have above-average abortion rates), 32% of all pregnancies ended in induced abortion and 53% of all induced abortions resulted from contraceptive failure.[20],‡ Most contraceptive failures occurred among traditional method users. Although nonuse of contraception constitutes a much larger problem, even widespread modern contraceptive use will not entirely eliminate the need for recourse to abortion, because no contraceptive method works perfectly every time. An additional complication in countries that lack quality family planning services is a frequent lack of contraceptive security. This particularly affects users of reversible methods, which demand regular resupply, and contributes further to contraceptive discontinuation, unintended pregnancy, and abortion.

Abortion over the Course of the Reproductive Transition

Evidence from a diverse set of countries shows that over time, abortion rates fall as contraceptive use rises.[21–23] The relationship between these rates may be mediated by the stability of levels of fertility. As the desirability of fertility (i.e., family size) declines, there is initially an increase in both contraceptive use and abortion. Abortion rates will be higher where desired family size and effective contraceptive use rates are lower, regardless of the legal status of abortion, cultural factors, and religious sanctions. In South Korea, average family size fell by more than half between the 1960s and 1980s. Both contraceptive prevalence and abortion rates increased alongside the sharp decrease in desired family size. Two decades later, the birth rate began to stabilize, the abortion rate plateaued, and contraceptive prevalence continued to rise.

Historically, Russia has had one of the world's highest abortion rates, largely due to the poor accessibility and quality of the limited available contraceptive options, as well as the ready availability of abortion (re-legalized in 1955). Sexually active women in this country tended to rely more on abortion rather than contraception to limit their family size. Indeed, as late as 1990, the Russian abortion rate was well over 100 per 1000 women of childbearing age. The situation finally changed in the early 1990s when the government began to promote contraception; over 1988–1998, modern contraceptive use increased 80% while the abortion rate fell by 53%.[24] However, continued progress has proved more difficult to achieve, not least because greater improvements in access and quality of family planning services are still needed.

---

‡ These results are not too dissimilar from levels in more affluent countries, which have higher contraceptive prevalence rates. Although rates vary by country, large proportions of unintended pregnancies and abortions in these countries result from contraceptive failure because the proportion of unintended pregnancies stemming from contraceptive failure is directly linked to the proportion of the population using contraception.

Individual-Level Determinants of Abortion

Women have abortions for many other reasons, including a desire to delay or end childbearing, concern over the interruption of work or education, perceived health concerns, and issues of financial or relationship stability. Most women who terminate their pregnancies do so because they feel unable in their current circumstances to fulfill their parental responsibilities as they would like, or to provide the kind of family support they believe their children deserve. Even if it is dangerous or forbidden, many women will resort to abortion to protect their family from poverty or to conceal an illegitimate pregnancy where it is stigmatized. Unwanted pregnancy is also frequently associated with sexual coercion and violence, particularly in younger women.

When asked about their reasons for having an abortion, three-fourths of recent U.S. abortion patients cited concern for or responsibility to other individuals (74%) and said they could not afford a baby then (73%). Almost as many (69%) said that having a baby would interfere with their employment, education, or ability to care for dependents. In addition, almost half of these women (48%) said they were having relationship problems and did not want to have to raise the child as a single parent; 38% did not wish to have another child because they had already completed their childbearing.[25]

ABORTION LEVELS AND DIFFERENTIALS IN THE UNITED STATES

In the United States as in other countries, decriminalization of abortion led to an initial increase in abortion rates that subsequently declined as contraceptive practice improved. The number of legal abortions increased from 1970 until 1990 and declined thereafter, and both the abortion rate and the abortion ratio had started to decline earlier. In 2008, 1.21 million abortions were performed, down from 1.31 million in 2000 and 1.61 million (an all-time high) in 1990, and the abortion rate declined from 27/1000 women aged 15–44 in 1990 to 21/1000 in 2000 (a level comparable to the mid-1970s) and then to 19/1000 in 2005. Despite these improvements, the U.S. rate remains higher than the rates in most countries of Western Europe, Australia, and New Zealand, where contraceptive use is more widespread and effective. Almost every second pregnancy among American women is unintended, with 4 in 10 of these pregnancies ending in abortion. In all, 22% of all viable pregnancies are terminated. At current rates, almost one in three U.S. women will have an abortion by age 45; each year, 2% of reproductive-age women have an abortion.

A broad cross-section of U.S. women have abortions, but unintended pregnancy and abortion rates are higher among specific groups of women, typically including those younger than age 30, in poverty, and from more disadvantaged racial and ethnic minorities. This situation is correlated with the rate of contraceptive nonuse, which is greatest among those who are young, poor, less educated, and non-white. This suggests the need for better contraceptive access and family planning counseling among these underserved groups. Women in their 20s account for 58% of abortions; at current rates, 1 in 10 women will have an abortion by age 20, 1 in 4 by age 30, and 3 in 10 by age 45.[26] Approximately 61% of abortions are obtained by women who are already a parent. Women who have never married and are not cohabitating account for 45% of all abortions. U.S. Hispanic women have slightly higher than average abortion rates, and African American women are more than twice as likely as women overall to have an abortion.

Overall, the U.S. unintended pregnancy rate remained constant over 1994-2006. Looking at subgroups, however, this rate increased 50% among poor women, but decreased 29% among higher-income women. Similarly, the abortion rate decreased 8% over 2000-2008, but increased by 18% among poor women, while decreasing by 28% among higher-income women. More than two-thirds (69%) of women obtaining abortions are economically disadvantaged; 42% have incomes below the federal poverty level[§] and 27% have incomes between 100% and 199% of this threshold.

SAFE AND UNSAFE ABORTION

Abortion is a very safe procedure for women if carried out early in a pregnancy by a trained provider in sanitary conditions. Estimates for the United States using diverse sets of data show that risks of maternal demise are significantly greater than risks from abortion.[27] The most recent estimate, based on national surveillance, survey, and birth certificate data for 1998-2005, shows that the risk of death associated with childbirth (8.8 deaths per 100,000 live births) is about 14 times higher than that associated with abortion (0.6 death per 100,000).[28] Similarly, the overall morbidity

---

[§] The U.S. Department of Health and Human Services calculates the federal poverty level, with poverty guidelines updated periodically. In 2008, this threshold stood at $10,830 for a single woman with no children—a level that many people would judge underestimates what truly constitutes absolute poverty.

associated with childbirth exceeds that with abortion. Less than 0.3% of U.S. abortion patients experience a complication that requires hospitalization.

As with most medical procedures, however, abortion becomes unsafe if performed by persons lacking the necessary skills or in an environment without the minimal medical standards, or both.[29] For many women with unwanted pregnancies, a safe abortion can be too expensive, unavailable, or illegal. These factors may lead them to delay getting an abortion until later in pregnancy, when the risk of complications increases, and to either attempt to induce abortions themselves or turn to providers outside the conventional medical care system. Unsafe procedures tend to be concentrated in low-income countries with weak health system capacity, underdeveloped postabortion care (PAC) services, and legal restrictions on the provision of safe abortion.

Potential complications of spontaneous and induced abortion include retained products that lead to sepsis and hemorrhage, toxic shock, injury to internal organs, psychological trauma, and long-term conditions such as chronic pelvic pain, pelvic inflammatory disease, and infertility. Along with the burden of care that falls on scarce hospital resources, significant negative consequences of such outcomes include increased costs for families, children left motherless, and losses in women's productivity. Poor, young, and uneducated women are far more likely to experience medical complications from unsafe abortions, because they are less able to pay for safe procedures and tend to have the least access to family planning services to begin with and to PAC when they need it. In many African countries especially, nonmarital adolescent pregnancies lead to expulsion from school, abandonment by the family, and sharply diminished life chances. Many such women will seek an abortion as a way out, but all too often they lack the resources to obtain a safe abortion. In cultures where a woman's status still depends on her ability to give birth to many children, infertility arising from pregnancy termination may be devastating.

Almost half of all abortions worldwide are unsafe, nearly all (98%) of which occur in developing countries (Table 7–2). The rate of unsafe abortions worldwide (14 abortions per 1000 women aged 15–44) and the proportion of abortion-related maternal deaths (13%) are believed to have remained steady between 1995 and 2008.[1] Nevertheless, the proportion of all abortions that were unsafe increased from 44% to 49%, while the number of maternal deaths due to unsafe abortion declined from approximately 68,000 to 47,000. The decline in the number of deaths, along with concurrent increases in the annual number of unsafe abortions performed (from about 19.9 million to 21.6 million over 1995–2008, essentially in poorer countries where abortion is illegal and populations are growing), indicate that the risks associated

Table 7-2 Global and Regional Estimates of Incidence and Mortality Due to Unsafe Abortions, 2008

Region	Number of Unsafe Abortions (1000s)	Unsafe Abortion Rate (per 1000 Women Aged 15–44 years)	Number of Deaths from Unsafe Abortion	Mortality from Unsafe Abortion per 100,000 Live Births	Percentage of Maternal Deaths
World total	21,600	14	47,000	30	13
Developed regions[a]	360	1	90	0.7	11
Developing regions	21,200	16	47,000	40	13
Africa	4200	28	29,000	80	14
Asia[b]	10,780	11	17,000	20	13
Europe	360	2	90	1	11
Latin America[c]	4230	32	1100	10	12
Oceania[b]	18	17	100	30	12

[a]Developed regions are taken to include northern America, Europe, Japan, Australia, and New Zealand. Developing regions include Africa, the Americas (excluding Canada and the United States), Asia (excluding Japan), and Oceania (excluding Australia and New Zealand).

[b]Australia, Japan, and New Zealand have been excluded from the regional estimates, but are included in the total for developed country regions.

[c]Includes Caribbean states.

Note: Figures may not add exactly to totals due to rounding.

Data from World Health Organization. *Unsafe Abortion: Global and Regional Estimates of the Incidence of Unsafe Abortion and Associated Mortality in 2008.* 6th ed. Geneva, Switzerland: WHO; 2011. http://www.who.int/reproductivehealth/publications/unsafe_abortion/9789241501118/en/.

with clandestine procedures may be decreasing. However, more deaths would be prevented if legal restrictions on providing safe abortions were removed, menstrual regulation and PAC made more widely available and accessible, and unwanted pregnancies reduced through increased and more effective contraceptive use.

African women experience the greatest risk of abortion-related death, with about one in three maternal deaths in parts of sub-Saharan Africa due to abortion. In 2008, more than 97% of abortions in Africa were unsafe, although the proportion is lower (58%) in Southern Africa; in that region, most people live in South Africa, which liberalized abortion in 1997. The proportion of abortions that are unsafe varies widely in Asia, from virtually none in Eastern Asia to 65% in South Central Asia. Although estimates by the WHO and the Guttmacher Institute continue to suggest that 95% of Latin American abortions are unsafe, many of these procedures are now performed in improved environments, and medication abortion (typically using misoprostol alone) has become a more common method in clandestine procedures.[16,30]

ABORTION TECHNIQUES AND CHANGING MEDICAL PRACTICE

Surgical methods of abortion include use of transcervical procedures such as vacuum aspiration, manual vacuum aspiration (MVA), and dilation and curettage (D&C). Vacuum aspiration is the preferred surgical method prior to 12 weeks' gestation and uses a plastic cannula or tube attached to either an electric pump or a hand-held vacuum syringe (MVA) which generates the suction mechanism. D&C (sharp curettage) remains common in many countries, as it is a standard gynecologic procedure performed for a variety of reasons, but it is less safe and is now recommended by WHO only when MVA is unavailable.[29,31]

Menstrual regulation (MR) refers to the evacuation of the contents of the uterus up to 10 weeks after the start of the last menstrual cycle. Such services are widely available as a backup to contraception in Bangladesh, Indonesia, and other countries before a pregnancy is confirmed, where they are often referred to in vernacular terms as "bringing the period down." This renders the procedure more acceptable to the larger culture and to many religious authorities otherwise critical of abortion. The provision of MR greatly reduces unsafe abortions occurring in the second and third trimesters of pregnancy and lessens demands on more expensive PAC.[32] Other technological improvements, including increased use of highly sensitive at-home urine pregnancy tests and transvaginal ultrasonography, have also enabled a shift toward earlier abortions, which reduce the risk of complications.

A medication abortion (also often referred to as medical abortion) uses pharmacologic drugs to end an early-term pregnancy, typically before 9 weeks' gestation. It involves a combination of mifepristone (an antiprogesterone) followed by a prostaglandin, usually misoprostol, that causes uterine contractions. This regimen is safe, effective, and acceptable to most women, albeit with some cases needing subsequent surgical intervention for incomplete abortion.** Use of mifepristone for this indication remains

---

** An incomplete abortion, in which parts of the fetus or placental tissue are retained in the uterus, can result in hemorrhage, intense pain, uterine infection, and death if left untreated.

limited to high-income countries where abortion is legal. The medication is given under the supervision of a physician, with the resulting abortion completed in the privacy of one's own home. After being first approved by France in 1988, a decade later mifepristone accounted for more than half of eligible early abortions in France, Scotland, and Sweden. A higher fraction of European women now have abortions at or before 9 weeks' gestation than did so before the drug's introduction, which has not been associated with any increase in the overall abortion rate. In the United States, integration of mifepristone into the healthcare system has been slow but its use has steadily increased, though mostly at sites that had already provided surgical abortion.

Misoprostol is a proven medication for a variety of obstetric and gynecologic uses. Although less effective when used alone, its adoption has increased throughout South America, where it has reduced incomplete abortions and hospitalizations.[16] The increased use of medication abortion in this region, Europe, and Northern America is ushering in a worldwide trend and has contributed to declines in the proportion of clandestine abortions that result in severe morbidity and maternal death.[33]

In the United States, about one in nine abortions is performed in the second trimester; these procedures are especially stigmatized by abortion opponents. One rarely used late-term procedure, intact dilation and extraction (D&E), developed into a focal point of controversy after it was dubbed by political opponents as a "partial-birth abortion." This term is not recognized medically,[34] but the procedure was nonetheless banned by federal law in 2003.

OTHER HEALTH-RELATED CONTROVERSIES

Several hypothesized potential side effects of abortion have been the subject of much dispute. Abortion has been postulated to increase the risk of developing breast cancer, for example. Based on extensive rigorous studies, however, the scientific consensus is that no such association exists. A comprehensive review of the available evidence does not support any hypothesis that early termination of pregnancy causes breast cancer.[35] In addition, rigorous meta-analysis of available epidemiological evidence worldwide shows no relation between induced abortion (or previous miscarriage) and the risk of subsequent breast cancer.[36]

Many claims have also been made about the potential association between abortion and long-term mental health outcomes. Research shows that these links are largely benign, at least in countries where abortion is legal and safely performed. Better-quality studies suggest few, if any, differences between women who had abortions and their respective comparison groups in terms of mental health sequelae.[37] A first abortion does not lead to any increased risk of mental health problems. Evidence for multiple terminations is more equivocal, in part due to methodological difficulties and because the same factors that predispose a woman to multiple unwanted pregnancies may also predispose her to mental health problems.[38]

PRENATAL DIAGNOSIS AND SEX-SELECTIVE ABORTION

Another bitter controversy concerns the abuse of ultrasound technology for prenatal gender selection[††] and the subsequent elimination of female fetuses, a recent trend motivated by longstanding devaluation of women. In India, China, and other East and South Asian countries with strong male gender preferences, the male-to-female live-birth sex ratio (normally about 1.04–1.06) rose quickly in the 1980s, especially for pregnancies after a first-born girl. A surplus of males has many potential ethical, social, economic and other consequences, including fewer women to marry, more mental health problems, long-term economic stresses, increased violence in young men, and a growing sex industry with coercion and trafficking of women.[39]

India banned prenatal testing in 1994 when done solely to determine the sex of the fetus, but most of its population now lives in states where selective abortion of girls is common. Jha et al.[40] assessed sex ratios by birth order from 1990 to 2005 using three nationally representative survey rounds and quantified the totals of selective abortions of girls using census cohort data. This analysis showed that the sex ratio for second-order births after a first-born girl fell from 906 girls per 1000 boys in 1990 to 836 girls per 1000 boys in 2005. The declines in the conditional sex ratio of second-order births were much greater in households with high education and wealth. Moreover, the son-biased sex ratios were more marked in states with increased availability, per person, of registered prenatal diagnostic facilities. These findings suggest that the medical community continues to meet strong son preferences through the provision of fetal sex determination with subsequent sex-selective abortion. Evidently, the financial incentives for physicians to undertake this illegal activity far exceed the penalties associated with breaking the law.

In contrast, there has been a dramatic turnaround in masculine sex ratios at birth in South Korea, which were the highest in Asia in the early 1990s and peaked in the mid-1990s. The rising child sex ratios were underpinned by the

---

[††] By the 1990s, in many Asian countries, as elsewhere, amniocentesis was also used to determine the sex of a fetus. Ultrasound in the second trimester is the safest and cheapest method, but in the 1990s would reliably work for sex identification only around 5 to 6 months' gestation (i.e., toward the end of the second trimester).

strong intensity of son preference (Figure 7–1), although this declined beginning in the mid-1980s in the course of rapid socioeconomic development and fertility decline. However, the pressure to have sons was manifested due to the increased availability of improved prenatal testing technology alongside readily accessible abortion services. Nearly three-fourths of the decline in son preference between 1991 and 2003 has been ascribed to normative changes, and the rest to increases in the proportions of urban and educated people.[41] By 2007, the overall sex ratio again fell within the biologically normal range, suggesting reduced discrimination and increased societal opportunities for women. Although the sex ratio at birth orders three and greater is still elevated, this is a very small proportion of total births in South Korea today. Notably, South Korea is the first Asian country to reverse the trend in rising sex ratios at birth.

There is some reason to believe that child sex ratios in China and India will normalize before these countries reach South Korea's level of development. Both have implemented a wide range of interventions aimed at improving public perceptions of the value of girls and bringing women into public life at an earlier stage of development than did South Korea, where public policies that upheld muscular authoritarian Confucian traditions and kept women marginalized were reversed only gradually after three decades of military rule came to an end. In contrast, gender discrimination remains highly prevalent across all strata of Indian society, and there is evidence that sex ratios in the families of physicians are even higher than the national average.[42]

Recent studies have also shown significant differences in the male-to-female sex ratio at birth for different subpopulations in the United

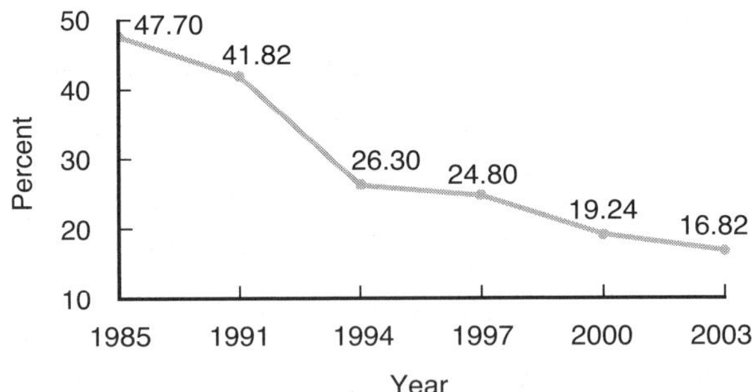

Figure 7–1 Trend in the Intensity of Son Preference (Percentage of Women Reporting "Must Have a Son"), 1985–2003

Data from Korean National Fertility and Family Health Surveys, various years. Reproduced from Chung W, Das Gupta M. Why is Son Preference Declining in South Korea?: the role of development and public policy, and the implications for China and India. The World Bank, Development Research Group, 2007.

States.[43,44] This exceeded expected biological variation for third-order and later U.S. births to Chinese, Asian Indian, and Korean parents, strongly suggesting prenatal sex selection. These findings raise serious ethical issues, although these ratios are lower than the values seen in China and other Asian countries. Sex-selective abortion has also been detected in Canada.[45] Future monitoring of child sex ratios will remain important because noninvasive, inexpensive, prenatal sex identification tests are becoming increasingly available in the first trimester.

POSTABORTION CARE

Postabortion care can save women's lives where abortion is performed unsafely. At least three types of services are needed at the facility level: emergency treatment for medical complications from incomplete or spontaneous abortions; family planning to prevent future unplanned pregnancies that would otherwise lead to repeat abortions; and other needed reproductive health counseling and services. A widely adopted PAC service delivery model developed by a consortium of family planning and reproductive health agencies includes five elements (Table 7–3).

Soon after the 1994 International Conference on Population and Development (ICPD), which recognized unsafe abortion as a major public health concern, there was a major push to develop funding and services for PAC services. Pilot programs were launched in many countries,[46] but these were often neither widely replicated nor scaled up. Turkey, several Latin

Table 7-3 Five Essential Elements of Postabortion Care

Community and service provider partnerships	• For prevention of unwanted pregnancies and unsafe abortion, mobilization of resources to help women receive appropriate and timely care for complications from abortion, and ensuring that health services reflect and meet community expectations and need
Counseling	• To identify and respond to women's emotional and physical health needs and other concerns
Treatment	• Of incomplete and unsafe abortion and complications that are potentially life threatening
Contraceptive and family planning services	• To help women prevent an unwanted pregnancy or practice birth spacing
Reproductive and other health services	• Preferably provided on-site or via referrals to other accessible facilities in providers' networks

Reproduced from the Postabortion Care Consortium Community Task Force (2002) Essential Elements of Postabortion Care: An Expanded and Updated Model. Available at: http://www.pac-consortium.org/index.php/pac-model.

American nations, and other countries successfully institutionalized the provision of the main elements of PAC through phased approaches.[47,48] In Turkey, for example, PAC family planning services were implemented and scaled up to reduce abortion through increasing contraceptive use, tilting the method mix toward more effective methods, and securing the commitment of decision makers. Elsewhere, however, PAC programs have more often not been scaled up throughout healthcare systems due to political obstacles rather than for technical reasons. Consequently, PAC is still not a readily available service for most women in many countries, impeding efforts to reduce maternal mortality further.

PAC programs are difficult to evaluate, but increases in service availability and quality are better indicators of their effectiveness than gains in the use of such services (which may be due to either improved services or increases in unsafe abortions). Overall, PAC services have been expanded in a number of countries and their quality has improved. Nevertheless, there remain numerous challenges to strengthening and scaling up PAC services where abortion is legally restricted or otherwise sensitive. In such settings, PAC services are often deficient and postabortion contraceptive counseling is poorly integrated with family planning and other reproductive health care.[49]

ABORTION LAWS AND CONSEQUENCES

Laws determine both the availability of abortion services and their safety. The most recent compilation of abortion legislation for 193 member countries of the United Nations showed that in 2011, 63% of the world's population lived in countries where early abortion is legally available on broad grounds and 42% could obtain the procedure on request—that is, without being required to specify a reason (Table 7–4). One in four (26%) women resided in a country that did not permit abortion in case of rape or incest and slightly fewer (21%–24%) could not legally obtain an abortion to preserve their physical or mental health, although nearly everyone (99%) could theoretically have an abortion if needed to save a woman's life. These numbers, however, are heavily influenced by the situation in the two demographic giants, China and India, where abortion is legal. Expressed in terms of nation states, 30% of the world's countries permitted abortion on request, 35% allowed it for economic or social reasons, just over half (51%) did so in cases of rape or incest, about two-thirds permitted it to preserve mental and physical health (65% and 68%, respectively), and 97% officially allowed abortion to save a woman's life.[50]

Nearly all Western democracies permit abortion under broad social and health grounds, enabling women to obtain a medically supervised and

Table 7–4 Grounds on Which Countries Permit Abortion, 2011

Reasons	Percentage of Countries	Percentage of Population
To save woman's life	97	99
To preserve physical health	68	79
To preserve mental health	65	76
Rape or incest	51	74
Fetal impairment	50	67
Economic or social reason	35	63
On request	30	42

Data from the United Nations. *World Abortion Policies 2013*. New York: United Nations, Department of Economic and Social Affairs, Population Division; 2013 (wallchart). http://www.un.org/en/development/desa/population/publications/policy/world-abortion-policies-2013.shtml.

extremely safe abortion early in pregnancy. England liberalized its abortion law in 1967, followed by many countries that had inherited its legal and parliamentary procedures. Six years later, the U.S. Supreme Court struck down state laws banning abortion. As elsewhere, decriminalization reduced abortion-related mortality and morbidity and led to an initial increase in abortion rates that subsequently declined as contraceptive practice improved.

Access to abortion is further determined by the actual implementation of laws as well as by societal and cultural views on sexuality and reproduction. Latin American countries have the world's most restrictive laws on abortion and generally severely stigmatize it, but safe abortion services are readily accessible for those who can pay for them.[16,30] India's experience shows that legal abortion does not guarantee safety if broad access to quality services remains impeded. India legalized abortion in 1971 but has more abortion-related deaths than any other country, with poor and rural women most likely to have clandestine procedures. Many women are unaware of the law, few primary health centers provide services despite being mandated to do so, and procedures are often performed by untrained persons in unsanitary conditions at sites other than registered government institutions.

Evidence points to a strong correlation between less-restrictive abortion laws and policies, safer abortion, and lower maternal mortality. Two well-documented case studies demonstrate how legalizing abortion increases the safety of the procedure. After South Africa legalized abortion on the request of a pregnant woman in 1997 and significantly improved PAC and family planning services, abortion-related deaths fell by 91% between 1994 and

2001, with steep declines in hospital admissions for septic and incomplete abortions, especially in younger women.[51] In Romania, the Ceaușescu regime restricted access to birth control services in a failed attempt to boost fertility, which inadvertently resulted in the highest maternal mortality rate in Europe and in thousands of unwanted children in institutions. Maternal deaths soared after abortion and contraceptives were banned in 1966. They then fell sharply (from 159 to 83 deaths per 100,000 live births within a year and continued to fall thereafter) once abortion was legalized again in 1990, with access to modern contraceptives also being significantly improved.[24,52,53]

ACCESS TO ABORTION SERVICES IN THE UNITED STATES AND CHALLENGES TO THEIR PROVISION

In the United States, the 1973 *Roe v. Wade* ruling that legalized abortion nationwide permitted a woman to have an abortion if the decision was made in consultation with a physician. Several states had already liberalized their abortion laws at the time, but most still prohibited the practice. The ruling helped improve reproductive health, but it also proved enormously controversial and reshaped national politics. The Supreme Court Justices rooted their decision in a "right to privacy" implicit in the Constitution itself, but many social conservatives mobilized to roll back what they decried as unacceptable "judicial activism." Following gains in contraceptive use and abortion-related technology, abortion is now safer and occurs earlier in pregnancy than before. Almost 9 in 10 abortions (88%) occur in the first 12 weeks of pregnancy. Use of early medication abortion is slowly growing (although it accounted for only 13% of abortions in 2005) and there has been a sustained, secular decline in the abortion rate, although it remains high relative to other developed countries.

Obtaining access to abortion is difficult for many women in the United States. The proportion of counties without an abortion provider rose from 77% to 87% over the period 1978–2000, while the proportion of women who lived in these counties increased from 27% to 34%. Abortion remains readily accessible to urban middle-class women, but poor, young, rural, and disadvantaged minority women—groups at greater risk of unintended pregnancies—find access more limited.

Access to abortion services has been stymied by opponents through multiple ways: (1) harassing abortion providers, particularly in rural areas and small towns; (2) pressuring medical schools not to train students how to perform abortions and preventing advanced practice clinicians from redressing the shortage of physician providers; (3) making women pay out-of-pocket for abortion services through insurance coverage restrictions and other means; and (4) pushing state legislatures to enact restrictions such as mandatory waiting periods, notification or consent requirements, or biased counseling, as well as imposing other bureaucratic requirements (e.g., making women undergo medically unnecessary ultrasound exams or multiple in-person visits) to create further barriers to accessing services. Thus, four decades after its legalization in the United States, abortion remains the subject of intense debate in many state capitols and among national policy makers, and many women face major impediments to accessing abortion care.

More state-level abortion restrictions were enacted in 2011 than in any prior year; 2012 brought the second highest number of restrictions ever, with 42 states and the District of Columbia enacting 122 provisions related to reproductive health and rights, more than one-third of which (43 in 19 states) sought to restrict access to abortion services. In 2013, many state legislatures continued to devote significant attention to abortion restrictions and increasingly sought to prohibit abortion even during the first trimester of pregnancy, or to declare that personhood begins at the moment of conception. Each state sets limits on when abortion can be done, up to a national maximum of 24 weeks. By mid-2013, 13 states had imposed a ban on abortions 20 weeks after fertilization (the equivalent of 22 weeks after the woman's last menstrual period—the conventional method physicians use to measure pregnancy), and still more conservative-dominated state legislatures had placed restrictions on abortion procedures and clinics, many in direct violation of U.S. Supreme Court decisions.

Abortion remains a common practice in the United States, more widespread than need be due to the continued high incidence of unintended pregnancy. It remains enshrined as the law of the land, but groups and legislators opposed to the procedure continue to employ numerous efforts in state-level strategies intended to make abortion care less accessible.

American Healthcare Reform and Abortion

Another recent development concerns the gradual implementation of the U.S. healthcare reform law known as the 2010 Patient Protection and Affordable Care Act (ACA). It will result in millions of women being able to obtain contraceptives without a copayment. The provision of free contraception is likely to lead to lower abortion and unintended pregnancy rates, further assisting women's health.

This development is also foreshadowed by data from a U.S. study that tracked 9256 young women, mostly poor or uninsured, over two years in St. Louis, Missouri. The study participants were given contraceptive counseling and their choice of a range of free contraceptives with emphasis on the superior effectiveness of long-acting reversible contraceptive (LARC) methods; 75% chose intrauterine devices (IUDs) or implants. Compared to other women in the region and nationally, the women in the study experienced statistically significantly reductions in unintended pregnancy, abortion, repeat abortion, and teenage pregnancy rates.[54]

While the provision of free contraception should decrease unintended pregnancy, it is noteworthy that ACA explicitly excluded abortion—one of the most common surgical procedures in the United States—from public funding. ACA stipulates that no federal funds can be used to purchase coverage for abortion services beyond limited conditions specified by the Hyde Amendment (saving the life of the woman or in cases of rape or incest).[55] "If an individual who receives federal assistance purchases coverage in a plan that chooses to cover abortion services beyond those for which federal funds are permitted, those federal subsidy funds (for premiums or cost-sharing) must not be used for the purchase of the abortion coverage and must be segregated from private premium payments or state funds."[56(p2)] This means that American women whose employers participate in state health exchanges mandated under ACA will be required to write a separate check for abortion coverage, intended to cover an unplanned event.

American Foreign Policy and Abortion

Internationally, *Roe v. Wade* bolstered an emerging global trend toward greater recognition of abortion rights, but this momentum was later checked. The so-called Global Gag Rule (officially the Mexico City policy, named after the location where it was first announced) is a U.S. government policy instituted by successive political administrations hostile to reproductive rights. The Mexico City policy severely restricted the ability of family planning programs receiving U.S. assistance to provide any activities that perform, counsel, or promote abortion services (even in cases of rape or incest). It also curtailed assistance for international family planning programs and reduced access to PAC, maternal and child health services, and, for some time, even to HIV-prevention and AIDS-related services,[57] before being repealed in 2009.

CONCLUSION

Abortion is a key aspect of fertility regulation and of reproductive health. Pregnancy is not always planned or welcomed—an awkward reality for public health practitioners and policy makers. Some women conceive when they do not want to, and pregnancy is not always trouble free. Programs must address such problems.

Abortion is invariably a response to unwanted pregnancy and reflects a decision that may be due to a range of circumstances. The incidence of abortion has much less to do with its legal status than with levels of unintended pregnancy, the root cause of most abortions. The best way to reduce rates of both unintended pregnancy and abortion is by giving women and couples the power to control their fertility through access to quality family planning information and services. Abortion rates will be higher where desired family size and effective contraceptive use are low, regardless of the legal status of abortion. Also, unintended pregnancy prevention efforts need to be grounded in broader antipoverty and social justice efforts, because those most disadvantaged do not fully share in the benefits of contraceptive use.

Complications arising from unsafe abortion continue to pose a serious global threat to women's health and lives. Their treatment costs impose a major economic burden on limited family and health system budgets, particularly in resource-poor settings. PAC programs are being implemented in a growing number of countries and increasingly stress a continuum of care and recognize the need to build partnerships with communities. However, there remain many obstacles to making PAC services more accessible. Much more work is needed in this vital area to prevent unsafe, incomplete, and repeat abortion, and to improve the reproductive health and well-being of women and their families.

Almost all abortion-related deaths and injuries take place in countries with highly restrictive abortion laws, and they are almost entirely preventable. New technologies allow women to obtain earlier and safer abortions, and need to be made more accessible. Laws and policies should be based on evidence that reflects health concerns, facilitating access to safe abortion services and decreasing obstacles to other reproductive health care.

DISCUSSION QUESTIONS

1. Why do nations that have legalized abortion see a drop in abortion rates? Using South Africa as an example, discuss the ways in which the legalization of abortion can be attributed to the decrease in abortion rates.
2. What are the reasons that women throughout the world seek abortion?
3. Discuss the factors that stabilize abortion rates within a society. Which factors, such as economic and female status, change the abortion rates? Do these factors vary between countries or regions? What is significant about the region that used to be under Soviet rule?

4. In 2008, the global abortion rate was estimated at 29 abortions per 1000 women aged 15–44 years, virtually unchanged since 2003, after having fallen from 35/1000 in 1995. Identify some of the factors that explain why the majority of the abortions continue to take place in developing nations.

5. Discuss the Mexico City policy and its history. What effect do you think it has had globally?

6. Abortion-related deaths are largely due to poor access to contraception, PAC, and unsafe procedures. To decrease abortion-related deaths, it would be ideal for all countries to legalize abortion and make contraception and PAC widely accessible. In the absence of those policy changes, how can women's health advocates address this problem in countries and regions where abortion remains illegal and stigmatized?

REFERENCES

1. World Health Organization (WHO). *Unsafe Abortion: Global and Regional Estimates of the Incidence of Unsafe Abortion and Associated Mortality in 2008.* 6th ed. Geneva, Switzerland: WHO; 2011. http://www.who.int/reproductivehealth/publications/unsafe_abortion/9789241501118/en/.

2. Himes NE. *Medical History of Contraception.* New York: Gamut Press; 1963.

3. Riddle JM. *Contraception and Abortion from the Ancient World to the Renaissance.* Cambridge, MA: Harvard University Press; 1992.

4. Mohr JC. *Abortion in America: The Origins and Evolution of National Policy, 1800–1900.* New York: Oxford University Press; 1978.

5. Gordon L. *The Moral Property of Women: A History of Birth Control Politics in America.* Rev. ed. Urbana, IL: University of Illinois Press; 2007.

6. Joffe CE. *Doctors of Conscience: The Struggle to Provide Abortion Before and After Roe v. Wade.* Boston, MA: Beacon Press; 1995.

7. Faúndes A, Barzelatto JS. *The Human Drama of Abortion: A Global Search for Consensus.* Nashville, TN: Vanderbilt University Press; 2006.

8. Kulczycki A. *The Abortion Debate in the World Arena.* London/New York: Macmillan/Routledge; 1999.

9. Rossier C. Estimating induced abortion rates: a review. *Stud Fam Plan.* 2003;34(2):87–102.

10. Singh S, Remez L, Tartaglione A, eds. *Methodologies for Estimating Abortion Incidence and Abortion-Related Morbidity: A Review.* New York: Guttmacher Institute; Paris: International Union for the Scientific Study of Population; 2010. http://www.guttmacher.org/pubs/compilations/IUSSP/abortion-methodologies.pdf.

11. Sedgh G, Singh S, Shah I, et al. Induced abortion: incidence and trends worldwide from 1995 to 2008. *Lancet.* 2012;379(9816):625–632.

12. Pazol K, Zane S, Parker WY, et al. Abortion surveillance: United States, 2008. *MMWR Surveill Sum.* 2011;60(15):1–41.

13. Jones RK, Kooistra K. Abortion incidence and access to services in the United States, 2008. *Persp Sex Reprod Health.* 2011;43.1:41–50.

14. Wilcox AJ. *Fertility and Pregnancy: An Epidemiologic Perspective*. New York: Oxford University Press; 2010.

15. Westoff CF, Serbanescu FI. *The Relationship Between Contraception and Abortion in the Republic of Georgia: Further Analysis of the 1999 and 2005 Reproductive Health Surveys*. Calverton, MD: Macro International; 2008.

16. Kulczycki A. Abortion in Latin America: changes in practice, growing conflict, and recent policy developments. *Stud Fam Plan*. 2011;42(3):199–220.

17. United Nations (UN). *World Contraceptive Use 2012*. (POP/DB/CP/Rev2012). New York: United Nations; 2012. http://www.un.org/esa/population/publications/WCU2012/MainFrame.html.

18. Singh S, Sedgh G, Hussain R. Unintended pregnancy: worldwide levels, trends and outcomes. *Stud Fam Plan*. 2010;41:241–250.

19. Darroch JE, Singh S. Trends in contraceptive need and use in developing countries in 2003, 2008, and 2012: an analysis of national surveys. *Lancet*. 2013;381(9879):1756–1762.

20. Bradley SEK, Croft TN, Rutstein SO. *The Impact of Contraceptive Failure on Unintended Births and Induced Abortions: Estimates and Strategies for Reduction*. DHS Analytical Studies No. 22. Calverton, MD: ICF Macro; 2011.

21. Bongaarts J, Westoff CF. The potential role of contraception in reducing abortion. *Stud Fam Plan*. 2000;31(3):193–202.

22. Kulczycki A, Potts M, Rosenfield A. Abortion and fertility regulation. *Lancet*. 1996;347(9016):1663–1668.

23. Marston C, Cleland J. Relationships between contraception and abortion: a review of the evidence. *Int Fam Plan Persp*. 2003;29(1):6–13.

24. David H, ed. *From Abortion to Contraception: A Resource to Public Policies and Reproductive Behavior in Central and Eastern Europe from 1917 to the Present*. Westport, CT: Greenwood Press; 1999.

25. Finer LB, Frohwirth LF, Dauphinee LA, et al. Reasons U.S. women have abortions: quantitative and qualitative perspectives. *Persp Sex Reprod Health*. 2005;37(3):110–118.

26. Jones RK, Kavanaugh ML. Changes in abortion rates between 2000 and 2008 and lifetime incidence of abortion. *Obstet Gynecol*. 2011;117(6):1358–1366.

27. Christiansen LR, Collins KA. Pregnancy-associated deaths: a 15-year retrospective study and overall review of maternal pathophysiology. *Am J Forensic Med Pathol*. 2006;27(1):11–19.

28. Raymond EG, Grimes DA. The comparative safety of legal induced abortion and childbirth in the United States. *Obstet Gynecol*. 2012;119(2 pt 1):215–219.

29. World Health Organization (WHO). *Safe Abortion: Technical and Policy Guidance for Health Systems*. 2nd ed. Geneva, Switzerland: WHO; 2012. http://www.who.int/reproductivehealth/publications/unsafe_abortion/9789241548434/en/.

30. Kulczycki A. A comparative study of abortion policymaking in Brazil and South America: the salience of issue networks and policy windows. *J Compar Policy Anal*. 2013;15(5):1–17.

31. Paul M, Lichtenberg ES, Borgatta L, et al., eds. *Management of Unintended and Abnormal Pregnancy: Comprehensive Abortion Care*. Oxford, UK: Wiley-Blackwell; 2009.

32. Johnston HB, Oliveras E, Akhter S, Walker DG. Health system costs of menstrual regulation and care for abortion complications in Bangladesh. *Int Persp Sex Reprod Health.* 2010;36(4):197–204.

33. Winikoff B, Sheldon W. Use of medicines changing the face of abortion. *Int Persp Sex Reprod Health.* 2012;38(3):164–166.

34. Johnson TRB, Harris LH, Dalton VK, Howell JD. Language matters: legislation, medical practice, and the classification of abortion procedures. *Obstet Gynecol.* 2005;105(1):201–204.

35. National Cancer Institute (NCI). *Summary Report: Early Reproductive Events and Breast Cancer Workshop.* Washington, DC: National Institutes of Health; 2003 (updated 2010). http://www.cancer.gov/cancertopics/causes/ere/workshop -report.

36. Collaborative Group on Hormonal Factors in Breast Cancer. Breast cancer and abortion: collaborative reanalysis of data from 53 epidemiological studies, including 83,000 women with breast cancer from 16 countries. *Lancet.* 2004;363(9414):1007–1016.

37. Charles VE, Polis CB, Sridhara SK, Blum RW. Abortion and long term mental health outcomes: a systematic review of the evidence. *Contraception.* 2008;78(6):436–450.

38. Major B, Appelbaum M, Beckman L, et al. *Report of the APA Task Force on Mental Health and Abortion.* Washington, DC: American Psychological Association; 2008.

39. Hesketh T, Xing ZW. Abnormal sex ratios in human populations: causes and consequences. *Proc Natl Acad Sci USA.* 2006;103:13271–13275.

40. Jha P, Kesler MA Kumar R, et al. Trends in selective abortions of girls in India: analysis of nationally representative birth histories from 1990 to 2005 and census data from 1991 to 2011. *Lancet.* 2011;377(9781):1921–1928.

41. Chung W, Das Gupta M. The decline of son preference in South Korea: the roles of development and public policy. *Pop Develop Rev.* 2007;33(4):757–783.

42. Patel AB, Badhoniya N, Mamtani M, Kulkarni H. Skewed sex ratios in India: "physician, heal thyself." *Demography.* 2013;50(3):129–134.

43. Almond D, Edlund L. Son-biased sex ratios in the 2000 United States census. *Proc Natl Acad Sci USA.* 2008;105(15):5681–5682.

44. Egan JF, Campbell WA, Chapman A, et al. Distortions of sex ratios at birth in the United States; evidence for prenatal gender selection. *Prenatal Diagn.* 2011;31(6):560–565.

45. Almond D, Edlund L, Mulligan K. Son preference and the persistence of culture: evidence from South and East Asian immigrants to Canada. *Pop Develop Rev.* 2013;39(1):75–95.

46. Curtis C. Meeting health care needs of women experiencing complications of miscarriage and unsafe abortion: USAID's Postabortion Care Program. *J Midwifery Women's Health.* 2007;52(4):368–375.

47. Billings DL, Crane BB, Benson J, et al. Scaling-up a public health innovation: a comparative study of post-abortion care in Bolivia and Mexico. *Soc Sci Med.* 2007;64:2210–2222.

48. Senlet P, Cagatay L, Ergin J, Mathis J. Bridging the gap: integrating family planning with abortion services in Turkey. *Int Fam Plan Persp.* 2001;27(2):90–95.

49. RamaRao S, Townsend JW, Diop N, Raifman S. Postabortion care: going to scale. *Int Persp Sex Reprod Health.* 2011;37(1):40–44.

50. United Nations (UN). *World Abortion Policies 2013.* New York: United Nations, Department of Economic and Social Affairs, Population Division; 2013 (wall-chart). http://www.un.org/en/development/desa/population/publications/policy/world-abortion-policies-2013.shtml.

51. Jewkes R, Rees H, Dickson K, et al. The impact of age on the epidemiology of incomplete abortions in South Africa after legislative change. *Br J Obstet Gynaecol.* 2005;112(3):355–359.

52. Kligman G. *The Politics of Duplicity: Controlling Reproduction.* Berkeley: University of California Press; 1998.

53. Stephenson P, Wagner M, Badea M, Serbanescu F. Commentary: the public health consequences of restricted induced abortion: lessons from Romania. *Am J Public Health.* 1992;82(10):1328–1331.

54. Peipert JF, Madden T, Allsworth JE, Secura GM. Preventing unintended pregnancies by providing no-cost contraception. *Obstet Gynecol.* 2012;120(6): 1291–1297.

55. McFarlane DR. Reproductive health politics and policies. In: Morone JA, Ehlke, eds. *Health Politics and Policy.* 6th ed. Stamford, CT: Cengage Learning; 2013: 292–305.

56. Kaiser Family Foundation. *Focus on Health Reform: Summary of New Health Reform Law.* June 18, 2010.

57. Kulczycki A. Ethics, ideology, and reproductive health policy in the United States. *Stud Fam Plan.* 2007;38(4):333–351.

Chapter 8

Benefits of Family Planning

E. Hazel Denton

INTRODUCTION

Two major forces have propelled the *contraceptive revolution* that began in the early 1970s. First, the increased number of effective contraceptive methods, the dissemination of contraceptive knowledge, and expanded access to contraceptives have made the long-established inclination for planning the arrival of a child much easier to accomplish. Second, the expanded education of women along with their growing participation in the labor force, together with global socioeconomic development and increased understanding of the health benefits of family planning, have generated a greater preference for a smaller family size.

Throughout the world, fertility rates have dropped. The total fertility rate (TFR) in less-developed countries[1] has fallen from 5.2 children in 1970–1975 to 2.7 in 2005–2010, and in least-developed countries from 6.7 to 4.4.[2(p26)] (See Box 8–1 for definitions of country classifications.) In 1960, only 10% of women of reproductive age (15–49) used any contraceptive method (modern or traditional). This rate has risen to 59% in less-developed countries and 33% in least-developed nations, while the percentages using modern methods are 54% and 27%, respectively.[3]

Improved access to reliable contraception has had a major impact on global society. Not only have women and their children become healthier,

---

Box 8–1   Classifying Countries

Following the UN classification,[1] "more developed" countries comprise all of Europe and North America, plus Australia, Japan, and New Zealand. All other regions and countries are classified as "less developed" or "developing." The "least developed" countries consist of 49 countries with especially low incomes, high economic vulnerability, and poor human development indicators; 33 of these countries are in sub-Saharan Africa, 14 are in Asia, and 1 is in the Caribbean.

Data from United Nations. *World Population Prospects: The 2010 Revision.* New York: United Nations Department of Economic and Social Affairs, Population Division, Classification of Countries; 2010.

Box 8–2   Reduced Fertility, Population Growth, and Poverty

"Reduced fertility and population growth, by themselves, will not automatically achieve aspirations for a better world, such as those enshrined in the Millennium Development Goals but they make achievement much more feasible. Their cross-cutting contributions to poverty reduction, better health, enhanced education, gender equality, and the environment make continued investment in family planning compelling."

Reproduced from Cleland J, Bernstein S, Ezeh A, et al. Family planning: the unfinished agenda. *Lancet.* 2006;368:1815, with permission from Elsevier.

---

but in many cases their families have been transformed economically. Family planning has not just improved lives but it has also been promoted as a tool for economic development[4(p1815)] and poverty reduction (Box 8–2).

This chapter examines options and opportunities for family planning, emphasizing conditions in developing countries. The chapter first lays out the historical preference for planned families. Next, it discusses the means used to attain preferred family size, including modern and traditional contraceptive methods as well as induced abortion. The third section reviews the health benefits of family planning for women and their families, by avoiding births that are "too early, too late, too many or too frequent."[5] The final section discusses how these benefits may be reinforced and expanded by educating girls.

CONTRACEPTION, UNMET NEED, AND UNINTENDED PREGNANCY

The practice of contraception involves taking actions that indicate a preference to *not* conceive at a particular point in time. This preference has been exhibited around the world for centuries. For example, long before modern contraceptives were available, English data reveal that family size by socioeconomic group fell steadily for all classes (Table 8-1). The main methods used as birth control during this time were withdrawal (i.e., coitus interruptus) and abortion.[6]

Brief History of Contraception

Throughout history, many practices have been used to delay, space, or stop pregnancies. For example, prolonged breastfeeding reduces fecundity. Taboos against intercourse for a lengthy period following birth (i.e., postpartum abstinence) also limit conception rates. Periodic abstinence from sexual intercourse has been widely practiced as well.

As medical knowledge grew in the 19th and 20th centuries, efforts expanded to improve both the understanding of conception and contraceptive efficacy. A first step, realized in the mid-19th century, was appreciating that the man generated the sperm and fertilized the woman's egg, rather than the man creating a life, which the woman then nurtured for nine months until its birth.[7] Identification of a "safe period" when ovulation was less likely to occur was another crucial step, albeit one that was initially misunderstood.[8]

During the 19th century, various chemical and barrier methods of contraception became available, including douching solutions, vaginal sponges and diaphragms, condoms, and spermicides. Condoms, made from animal skin, had been used in Europe as prophylactics against sexually transmitted infections (STIs) since at least the 17th century. Until the development of vulcanized rubber in the 1850s, condoms were too expensive for widespread use.[9] Moreover, they were tainted by their association as being for use outside of marriage to prevent sexually transmitted infections—a message reinforced by their widespread distribution to troops during wartime.

Today, contraceptives are grouped into two categories: modern methods and traditional methods. Modern methods include female sterilization, male sterilization, the pill, the intrauterine device (IUD), injectables, implants (such as Norplant), the female condom, the male condom, lactational amenorrhea method (LAM), emergency contraception, the diaphragm, and foam/jelly. Traditional methods include periodic abstinence, withdrawal, and any country-specific traditional methods.[10]

Table 8-1 The Fall in Family Size in Great Britain by Socioeconomic Groups

	Professional	Employers	Own Account	Salaried Employees	Non-Manual Wage Earners	Manual Wage Earners	Farmers and Farm Managers	Agricultural Workers	Labourers	All Groups
1890–1899	2.80	3.28	3.70	3.04	3.53	4.85	4.30	4.71	5.11	4.34
1900–1999	2.33	2.64	2.96	2.37	2.89	3.96	3.50	3.88	4.45	3.53
1915	2.02	2.07	2.13	1.88	2.20	2.91	2.69	2.74	3.54	2.61
1925	1.69	1.71	1.82	1.48	1.77	2.48	2.22	2.62	3.05	2.24

Note: Live births to all completed marriages in which the wife was under 45 at marriage. The dates refer to the years in which the marriages were contracted.

Data from Wrigley EA. *Population and History.* New York: World University Library (McGraw-Hill); 1969.

Use of Contraception

Both awareness and use of contraception in developing countries have increased rapidly in recent decades (Figure 8–1).[11] The mix of birth control methods varies throughout the world because of availability and cultural differences. The IUD is the method most widely used in Asia, the pill and condoms in Europe, and female sterilization in North America and Latin America.[12]

What leads a couple to make decisions about family size? Three conditions must be present for a substantial decline in fertility preferences. First, childbearing must be an option that couples can consciously consider rather than leaving it to fate or to divine will. Second, it must be advantageous for them to consider family planning. Third, methods to control fertility must be readily available.[13] In every country, rich or poor, those who are better educated—those with more life options—will consider and use contraception more regularly than others. As a result, the more educated have smaller families. Figure 8–2 illustrates how differences in fertility are correlated with the education of women.[11]

If most people do want to plan their families, why is contraception not more widely used? There are multiple answers to this question. First, methods must be culturally acceptable. Preferences may vary by culture and region, and they may change over time. Second, supplies must be consistently available, a condition often determined by donor countries or organizations. Third, the quality of service delivery is important. A study of

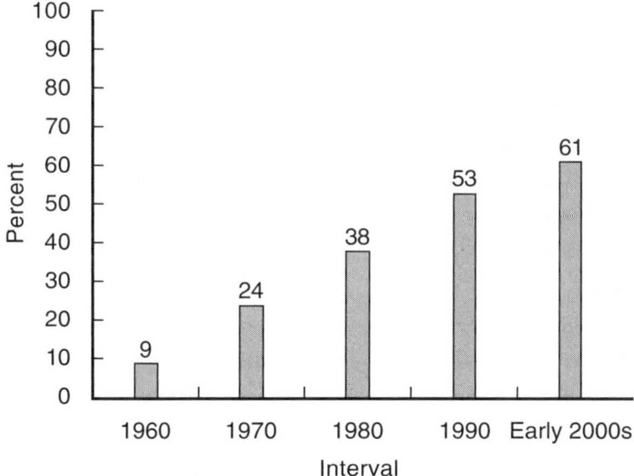

Figure 8–1 Rising Family Planning Use, Developing Countries, 1960 to Early 2000s

Reproduced from Population Reference Bureau. Graphics bank: family planning use worldwide. http://www.prb.org/Multimedia/Graphics-Bank/FamilyPlanning.aspx. Accessed January 8, 2014.

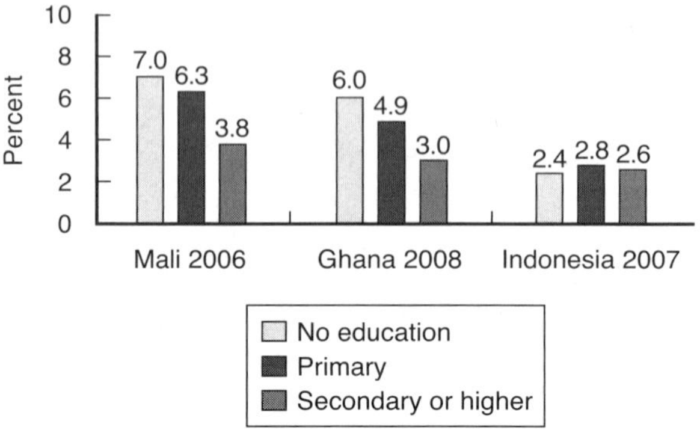

Figure 8–2 Total Fertility Rate by Mother's Education

Reproduced from Population Reference Bureau. Graphics bank: disparities in childbearing and contraceptive use. http://www.prb.org/Multimedia/Graphics-Bank/FamilyPlanning.aspx. Accessed January 8, 2014.

contraceptive use in 15 developing countries revealed that within the first year of starting a method, 7% to 27% of women stopped using contraception because of low quality of services rather than a change in desire to stop or delay childbearing.[14]

Contraceptive Prevalence

The contraceptive prevalence rate (CPR) "is the number of women of reproductive age who are using contraception per 100 women of reproductive age. This measure provides an indication of the number of women who have a lower risk of conception at a given time."[15(p15)] It may be calculated for all women or subpopulations such as married women, unmarried women, or women who are sexually active. The CPR is usually published for all contraceptive methods, including modern methods and traditional methods. It ranges from less than 20% in many African countries to 75% or more in many European countries, Australia, Brazil, and a few countries in East and Southeast Asia.[15]

The CPR is closely related to a decline in family size. Indeed, the CPR and total fertility rate (TFR) have a close linear relationship. The TFR gives the best picture at any point in time of the number of children women are currently having.[15] It is estimated that each increase of 10 points in the CPR is associated with a decline in the TFR of 0.7 children. Thus, to reach the replacement level of fertility (2.1), a CPR of between 60% and 70% is needed.[16]

Unmet Need

Unmet need for family planning is defined as the percentage of women who do not want to become pregnant but are not using contraception. Tracking unmet need can provide useful information for policy makers and program managers, but the trends need to be carefully understood (Figure 8–3). For example, if contraceptive use increases and desired family size stays constant, unmet need will generally decrease. In contrast, if contraceptive use increases at the same time that desired family size decreases, the level of unmet need may increase.[17]

Although contraceptive use has been steadily rising and the unmet need falling, the estimated level of unmet need for 2000–2009 was still moderate to high, especially in the least-developed countries.[18] In 2012, the total number of women with an unmet need for modern contraception was estimated at 222 million.[19] An analysis of unmet need in developing countries has shown that as women's education increases, unmet need usually declines.[17]

Unintended Pregnancies

Unmet need is closely linked to unintended pregnancies. As smaller families become the norm, the risk of an unwanted pregnancy may increase. In some countries, there may be an increase in abortion before effective methods

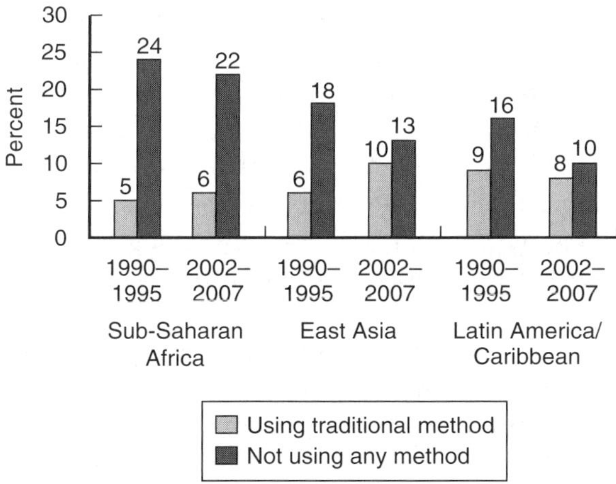

Figure 8–3 Trend in Unmet Need for Modern Contraception

Reproduced from Population Reference Bureau. Graphics bank: family planning costs and demand. http://www.prb.org/Multimedia/Graphics-Bank/FamilyPlanning.aspx. Accessed January 8, 2014.

of contraception are widely available. For example, Vietnam experienced a large number of induced abortions toward the end of the 20th century, as there was a rapid fall in desired family size, but the family planning method mix available in the country was limited.[2(p139)]

Unplanned births are either births that are mistimed (occurring two or more years sooner than desired) or not wanted at all. Of the 208 million births worldwide in 2008, an estimated 41% were unintended. Of these 86 million unplanned pregnancies, almost half resulted in abortions.[20]

Even in developed countries with relatively easy access to modern methods and high levels of education, the extent of unintended births can be high. Approximately one in two pregnancies in the United States and one in three in the United Kingdom and France are estimated to be unintended, primarily because of inconsistent, inefficient, or lack of use of contraception.[21]

ABORTION

The root cause of induced abortion is unintended pregnancy, which varies by region. More than half of all pregnancies in Latin America and the Caribbean region were unintended, of which an estimated one-third resulted in abortion. By contrast, only 39% of pregnancies in Africa were unintended. As larger families are more desired on the African continent, just one-third resulted in an abortion.[22]

Abortion incidence is also related to a country's socioeconomic development and demographic transition. Figure 8-4 highlights the relationship between socioeconomic development, use of effective contraception, and the role of induced abortion.[23] Ironically, the initial use of abortion is positively correlated to contraceptive use. As noted, this relationship reflects the fact that socioeconomic development lowers the desired size of the family before use of effective contraception is well established and readily accessible. As access is improved, the use of contraception increases concomitantly and abortion levels fall.[24]

During the period 1995–2008, both the number and the rate of unintended pregnancies throughout the world declined as contraceptive use increased. Had the unmet demand for modern contraception in the developing world been fully met, an estimated 54 million unintended pregnancies would have been averted annually, including 22 million unplanned births, 25 million induced abortions, and 7 million miscarriages.[20]

Abortions are carried out in all parts of the world—either legally or illegally. The abortion rate in developed and developing countries is similar, at slightly less than 30. Legal restrictions do not decrease the incidence of abortion, but they do affect its safety.[23]

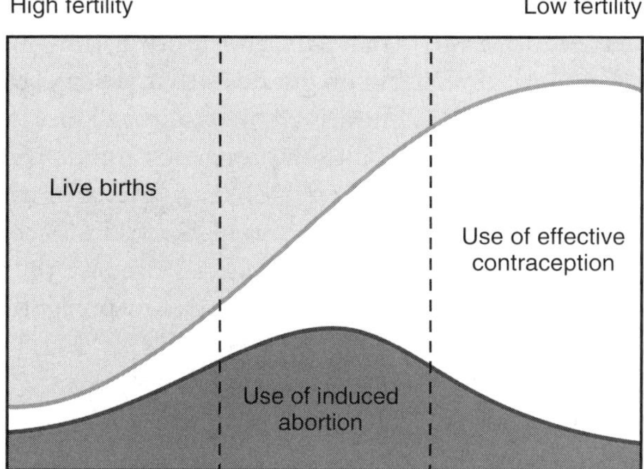

Figure 8–4 Transition Trajectory from High to Low Fertility and the Relative Levels of Induced Abortion, Effective Contraception, and Live Births

Reproduced from World Health Organization. *Unsafe Abortion: Global and Regional Estimates of the Incidence of Unsafe Abortion and Associated Mortality in 2008.* 6th ed. Geneva, Switzerland: WHO; 2011:10.

Globally, the proportion of abortions that are unsafe is believed to have increased. In 2008 nearly half of all abortions were unsafe, compared to 44% in 1995. Most abortions in developed countries are legal and considered safe. Where abortion is illegal, however, it is much more likely to be unsafe; therefore, the risk to the mother's health can be significant, including high rates of morbidity and mortality.[24-26]

Relationship of Abortion to Contraception

At the societal level, increased interest in using contraception generally accompanies emerging preferences for smaller families. However, in some countries, there may be a period where rates of both induced abortion and contraception rise. During this time, the demand for smaller families exceeds the availability of family planning services or contraceptive supplies. South Korea is a notable example of a country in which the preference for smaller families emerged before effective contraception was readily available. As effective modern methods became more widespread, the country's abortion rate fell.[27,28]

BENEFITS OF CONTRACEPTION TO WOMEN

Women who have ready access to effective contraception can protect their short-term and long-term health as well as the health of their children. Contraception can play a key role in helping women to manage the timing and spacing of their children, as well as in defining the period of motherhood. Using contraception lowers the maternal mortality risk by modifying the hazards of birth—namely, avoiding pregnancies that are "too early, too late, too many, or, too frequent."[5] While a small mortality risk is associated with contraception, all methods are safer than pregnancy and delivery.[29]

As fertility has declined, so has the risk of pregnancy. For example, from 1990 to 2005, an estimated 1 million maternal deaths in 68 countries were directly averted by the reduction in fertility rates. Moreover, the reduction in high-risk births, especially high-parity births, certainly prevented additional maternal deaths.[30]

Birth Spacing and Limiting

A woman can suffer from various health complications during pregnancy, especially when births are close together. Maternal mortality, morbidity, and nutrition are all affected by the length of time between pregnancies. Spacing births enables a woman to recover her strength, provide adequate nutrition to the new infant, and give due care and attention to her existing family.

Utilizing contraception for birth spacing can be an effective intervention to improve the health of both women and children.[31] Thinking about spacing children also encourages parents to consider the *number* of children they want, as well as when they prefer to have them. Spacing can be a generally acceptable way of approaching the topic of family planning. Indeed, at one rural health clinic in Eastern Nigeria, a sign labeled one room's purpose in just this fashion: "Birth Spacing Room."

Research on birth spacing in developing countries has generated evidence that for infants and children younger than 5 years of age, births spaced at least 36 months apart are associated with the lowest mortality risk. Birth to conception intervals of less than 6 months, as well as abortion–pregnancy intervals of less than 6 months are associated with increased risk of preterm births, low birth weight, and small-for-gestational-age infants. In addition, birth to conception intervals of less than 6 months are associated with increased risk of maternal mortality and morbidity.[31] World Health Organization (WHO) guidelines recommend a minimum of 2 years after a live birth before attempting the next pregnancy, so as to reduce the risk of adverse maternal, perinatal, and infant outcomes.[32]

Breastfeeding and Birth Spacing

For generations, women have experienced the spacing of children through breastfeeding. The reverse is also true: contrary to the generally observed inverse relationship to income and family size, some upper-income women in the premodern era in Britain had higher fertility than the poor due to their use of wet nurses (which meant the mother did not breastfeed her own child) as well as being better fed and housed.[33]

During breastfeeding, the fecundity of a woman is reduced through postpartum amenorrhea. When women fully or nearly fully breastfeed their children and they maintain appropriate breastfeeding practices to prolong lactational infertility, such behavior can have 98% contraceptive effectiveness.[34] The length of the period of breastfeeding, plus the extent to which breastmilk is supplemented, affects the contraceptive effects.[35]

However, breastfeeding is not a fully reliable method for preventing conception given the flexibility with which breastfeeding may occur. Research in developing countries has shown that longer length of each breastfeed, greater number of breastfeeds, and greater proportion of total feeds at the breast will all lower the risk of ovulation.[36] Analysis of data from 38 countries in Africa, Asia, and the Americas showed that given the range of breastfeeding durations, as breastfeeding rates or durations decline, there is an increased need for contraceptive use for birth spacing and birth limiting purposes.[37] Therefore, providing advice to a mother on the eventual use of contraception after birth, even while breastfeeding, is relevant to her health and that of her child.

Sexually Transmitted Infection Prevention

Women risk contracting infection during unprotected sexual intercourse. Some barrier methods of contraception—namely, male and female condoms—offer *dual protection* against both unintended pregnancy and STIs, including HIV. Ironically, the extensive publicity surrounding the HIV/AIDS epidemic has increased opportunities to discuss the role of condoms in reducing the risk of unintended pregnancy.[38]

Male condoms, which are typically the most accessible contraceptive, have been widely used to prevent STIs. They are relatively cheap, can be purchased over the counter, require no special storage, and are easy to explain to new users. However, their acceptance and use vary throughout the world.

HIV/AIDS and Pregnancy

The Convention on the Elimination of All Forms of Discrimination Against Women (CEDAW) states that all women, including those infected with HIV, have the right "to decide freely and responsibly on the number and spacing of their children, and to have access to the information, education and means to enable them to exercise these rights."[39] Throughout the world, there are HIV-positive women who wish to conceive.

In this situation, contraception can be useful to optimize maternal health prior to pregnancy and to prevent transmission to an HIV-uninfected sexual partner.[40] Nevertheless, the probability of maternal death increases when the woman is HIV positive, especially in East and Southern Africa. Unfortunately, maternal health interventions seldom include access to antiretroviral drugs.[41] Also important is ensuring that birth occurs where adequate services are available to assist the pregnant woman to avoid mother-to-child transmission of HIV.[39,42]

Mother's Age

Births to very young women or to women at older ages generate higher risks for both mothers and children. If only women aged 18–35 years were to deliver offspring, it is estimated that global maternal mortality could be reduced by 20% to 25%. The role of contraception in reducing obstetric mortality and morbidity is well established.[30,43]

Despite the risks, teenage pregnancy rates remain high in many countries. Adolescent girls (aged 10–19 years) account for 11% of all births worldwide, ranging from a low of 29 per 1000 live births in Europe, to 58 per 1000 live births in Asia, to a high of 130 per 1000 live births in Africa.[44] Indeed, complications in pregnancy and childbirth are the leading causes of death among adolescent girls ages 15–19 in low- and middle-income countries. Adolescents are more likely than women in their 20s to deliver preterm and low-birth-weight babies.[45] In addition to the health impacts, very young mothers often have to forfeit their own opportunities to obtain an education, thus limiting their future career and work choices, as well as the opportunities for their children.

Using family planning to delay first births could improve lives as well as save them. In many resource-poor countries, high maternal mortality and morbidity have multiple causes. Very young women are particularly at risk of obstructed labor, a problem that can result in obstetric fistula.[46]

Births to older women also generate higher risks than births to women 18–35 years of age. After age 35, a woman is at increased risk of hypertension, preeclampsia, and diabetes. Her offspring are more likely to be stillborn or low birth weight.[47]

Maternal Mortality

Maternal mortality is a multiple tragedy—to the child who may be born without a mother, to the family who may lose a mother, to the spouse who loses a partner. How to prevent death from motherhood is well understood, and WHO has identified family planning as one of the core strategies for addressing maternal mortality. In many parts of the world, however, resources to address maternal deaths are inadequate. As a result, more than 99% of maternal deaths occur in developing countries. As previously discussed, global maternal mortality is concentrated in sub-Saharan Africa and South Asia. Table 8-2 shows maternal mortality ratios for different regions in sub-Saharan Africa and within those regions, the range of national maternal mortality ratios for 1980 and 2008.

Maternal deaths are uncommon in developed countries. However, the risk of dying from maternal causes was as high in Europe and North America in the nineteenth century as it is today in some parts of the world, such as Afghanistan, Angola, Malawi, Niger, Rwanda, Sierra Leone, and Tanzania.[43] A comparison of current maternal mortality death rates in sub-Saharan Africa with the clear progress in England (close to the Western European average) highlights the potential for reducing the rate (see Table 8-3).[41] Over the past century, about half the reduction in mortality in Europe was due to family planning, which enabled women to space and to limit their pregnancies. The other half was from improved obstetric care.[48]

Table 8–2 Maternal Mortality Ratios (MMR) in Sub-Saharan African Regions (with National MMR Ranges)

Year	Central	East	Southern	West
1980	711 (487–1072)	707 (586–854)	242 (184–319)	683 (577–818)
2008	586 (392–839)	508 (430–610)	381 (188–496)	629 (508–787)

Reproduced from Hogan MC, Foreman KJ, Naghavi M, Ahn SY, Wang M, Makela SM, Lopez AD, Loranzo R, and Murray CJ. Maternal Mortality for 181 Countries, 1980–2008: a systematic analysis of progress toward Millennium Development Goal 5. *Lancet.* May 8, 2010;375.

Table 8–3 Maternal Mortality Ratios in England, 1650–2008 (Maternal Deaths Per 100,000 Births)

Year	Ratio
1650–1700	1,600 per 100,000 live births
1700–1750	1,050 per 100,000 live births
1750–1800	750 per 100,000 live births
1800–1870	550 per 100,000 live births
2013 (United Kingdom)	8 per 100,000 live births

Data for 1650–1870 from Riley, JC. Rising Life Expectancy. Cambridge, UK: Cambridge University Press; 2001. Data for 2013 from the World Bank. Maternal mortality ratio (modeled estimate, per 100,000 live births). http://data.worldbank.org/indicator/SH.STA.MMRT.

The number of unwanted pregnancies and the unmet need for contraceptives remain high in many developing countries. A review of maternal deaths in 172 countries estimated that 342,203 women died of maternal causes in 2008, but that contraceptive use averted 272,040 maternal deaths. Thus, without contraceptive use, the number of maternal deaths would have been significantly higher. Using the WHO estimate of maternal deaths (roughly 358,000) for the same year, Figure 8-5 shows that satisfying the unmet need for contraception could have prevented an additional 104,000 maternal deaths per year. Clearly, the use of contraception is a major and effective primary prevention strategy to lower maternal mortality in developing countries.[49]

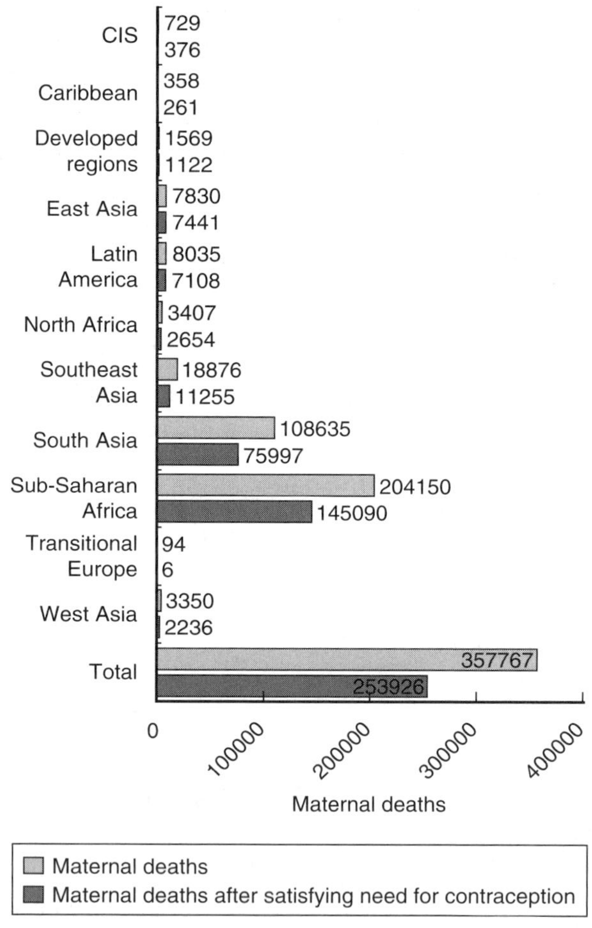

Figure 8–5 Expected Reduction of Maternal Deaths If Unmet Needs for Contraception Are Fulfilled

Reproduced from Ahmed S, Li Q, Liu L, Tsui AO. Maternal deaths averted by contraceptive use: an analysis of 72 countries. *Lancet.* July 14, 2012;380, with permission from Elsevier.

The lack of education in developing countries has routinely been flagged in various studies as an indicator for obstetric morbidity and mortality, probably because less-educated women, for a multitude of reasons (including early marriage and poverty), are less likely to gain access to the healthcare system when obstetric complications arise.[46(p315)] Increased education of women is seen as a critical lever in reducing the maternal mortality rate.

BENEFITS OF FAMILY PLANNING TO CHILDREN AND FAMILIES

Well-Being of Children

A child who is wanted and born to a healthy mother has a head start in life. While acknowledging the measurement challenges inherent in its definition, the *wantedness* of a pregnancy is a useful concept. Multiple analyses have shown that unwanted pregnancies pose greater risks to both child and maternal health because less time and money are invested in antenatal and postnatal care.[32]

Young children and infants are especially vulnerable. Protection from disease, malnutrition, and polluted air are critical to ensure their good health in the early years of life. Provision of hygienic conditions, access to adult care and attention, adequate nutrition, and eventually education enable a healthy child to develop to his or her potential. The infant mortality rate (IMR)— defined as the probability of dying between birth and 1 year of age expressed per 1000 live births—reflects the ability of a woman and her family to ensure the survival of the newborn child. As such, the IMR has been referred to as a *flash indicator of development.*

The variability of the IMR among countries and within a country can indicate the quality of life and general standard of living. The majority of women in developed countries enjoy good prenatal and postpartum care, access to skilled birth attendants, and access to emergency obstetric care, if necessary. Most women living in developed countries are also well-nourished mothers and are more likely to live in safe and clean environments than women in developing countries. Consequently, the IMR in developed countries is currently at a record low of 5 infant deaths per 1000 births.[50]

Without the supports described previously, the IMR can be far higher. For developing countries, the infant death rate is 45/1000; for the least-developed countries, it is 72/1000. Regional differences are also significant. The IMR in sub-Saharan Africa is 72/1000; in Latin America and the Caribbean, 20/1000; in Asia, 37/1000; and in Europe 5/1000.[50] To give a view of progress parallel to that of maternal mortality, the IMR in the United States of 47 deaths per 1000 live births in 1940 is comparable to the rate in developing countries today.[51] Not only are infant mortality rates higher in developing countries, but child survival rates vary similarly. Figure 8–6 shows that inequities *within* a country as well as *between* countries are wide.[52]

By lengthening the gap between pregnancies in developing countries, contraception can improve perinatal outcomes and child survival. The risk of prematurity and low birth weight was found to double when conception occurs within six months of a previous birth, and children born within two years of an older sibling are 60% more likely to die in infancy than those born more than two years after the sibling.[53]

Other studies have elucidated similar findings. A recent 17-country study of the impact of birth intervals on child mortality and nutritional status confirmed that the longer the birth interval, the lower the risks. Findings included an increased risk of mortality and undernutrition if a subsequent

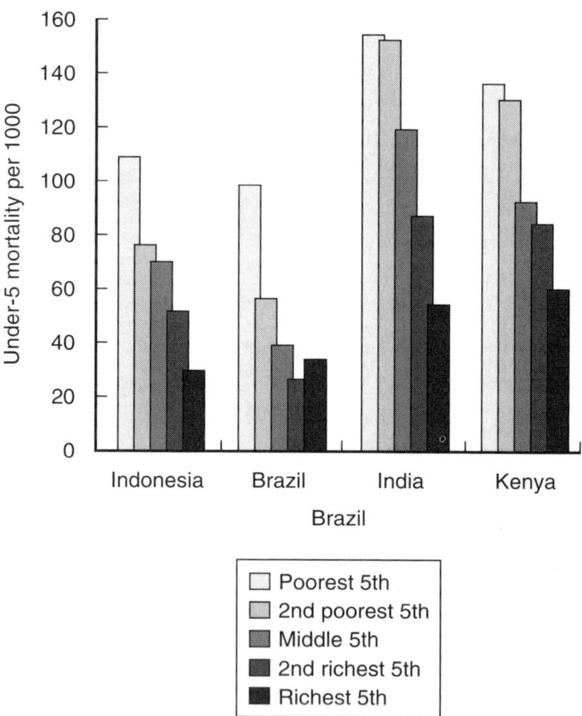

Figure 8–6 Under-5 Mortality Rates by Socioeconomic Quintile of the Household for Selected Countries

Reproduced from Cesar VG, Wagstaff A, Schellenberg JA, et al. Applying an equity lens to child health and mortality: more of the same is not enough. *Lancet.* 2003;362:233–241.

birth occurred earlier than 36 months after the previous birth. The study concluded that the optimal birth interval is between 36 and 59 months.[54] Another study in Bangladesh confirmed the relationship between malnutrition, birth spacing, and number of siblings. Children are at higher risk of malnutrition if their mothers receive less education and they have several siblings.[55]

In designing initiatives to address the high level of child mortality in developing countries, shockingly it is often the case that public subsidies benefit the rich more than the poor. Effective interventions need to be specifically targeted at those most in need.[52]

Benefits to the Family

A family faces constraints on resources when a new member is added. These constraints may be looser in a richer country. However, for a family with multiple children in a poorer country, the constraints can challenge the family budget so that limited food, health care, space, and education must be shared by even more people. If the child is born close to an older sibling and has to compete with several other siblings for resources, the child's start in life can be particularly challenging, especially in a family that struggles to find sustainable income for the existing family.

When children are spaced out, there is the opportunity for the mother's health to recover, for the family circumstances to improve, and for the child to become more self-sufficient before the possible arrival of a sibling. The child also benefits from increased parental attention and training. If a mother has multiple children, she cannot provide as much supervision and support to each. If her children are close in age and her own support systems are limited, each child has a reduced share of the mother's attention. Improved management of fertility also allows women more opportunity to learn skills that can raise their lifetime earnings, to accumulate capital assets, and to make greater investments in their children's health and education.[56]

Benefits of Family Planning at the National Level

Initiatives to provide supplies of culturally acceptable family planning methods where the desire to use them exists have now demonstrated an impact on reducing the total fertility rate throughout the developing world. For example, the TFR in Africa fell from 6.60 to 4.64 over the period 1950–2010; it fell from 5.82 to 2.28 in Asia, and from 5.86 to 2.30 in Latin America and the Caribbean region over the same period.

At the core of effective programs is ensuring that the three conditions mentioned earlier are met—namely, that couples appreciate that they have a choice of when and how many children to have, recognize that it is to their advantage to plan their family size, and see that the methods to do so are readily available.[13] Making contraceptive methods available is one component of programs that seek to lower the growth rate of population. This initiative must be balanced by providing knowledge about and understanding of the options. To achieve this goal, expanded access to education, especially for girls, plays a major role.

ROLE OF GIRLS' EDUCATION

In 1948, the United Nations (UN) issued a Universal Declaration of Human Rights, which included this statement: "Everyone has the right to education. Education shall be free, at least in the elementary and fundamental stages. Elementary education shall be compulsory."[57] At the time the UN issued its declaration, access to education in developing countries was severely limited, especially for girls. Access to education for both boys and girls has now expanded; the gender gap has narrowed considerably at all levels of income; but children not in school tend to come from the poorest households.[58] There does not appear to be evidence of gender discrimination in promoting education in various forms of governance (i.e., political institutions), but such influence is often evident through culture and religion.[59]

Seven decades after the promulgation of the Universal Declaration of Human Rights, girls' education is recognized as a critical factor in efforts to reduce high levels of fertility in developing countries. Identifying and quantifying the linkage between education and women's reproductive choices remains challenging because myriad factors influence childbearing decisions. Nevertheless, numerous studies confirm that women with more education can make more informed decisions about reproduction and gain more secure positions in the labor force, thereby providing more adequately for their children.[60,61]

Studies on this topic have been conducted throughout the world and in different time periods. For example, a review of 157 countries over the period 1991–2006 concluded that education provides many tools to women that help them to make pivotal judgments in their lives, coalescing around preferences for smaller families.[62] A survey of fertility decline across India (1981–1991) found that women's education was the most important factor explaining fertility differences across the country and over time.[63] A World Bank study of 14 countries in sub-Saharan Africa found a strong inverse relationship between female schooling and cumulative fertility in all countries in both urban and rural areas.[64]

Figure 8–7 shows a young school girl in India. Even just a few years of education can impact fertility, but secondary school has a more consistent and stronger effect on delay of childbearing, being linked with increased use of contraception, a preference for fewer children, and reduced fertility. The more educated the girl, the more opportunities she will have to be self-supporting and, therefore, the more likely she is to consider delayed marriage. Such a delay will help to ensure that she is physically more able to handle pregnancy and psychologically better equipped to understand the use of contraception for spacing her children. With more education, she gains more options to negotiate for safe sex through condom use and to make decisions on the age at which to marry.

When women have less education, marriage (or union) is likely to occur at a younger age, to produce more children, and to risk more obstetric difficulties. A study of DHS data from 31 countries showed that the likelihood of using modern contraception and attending at least four antenatal care visits was significantly higher for women with complete primary education than for those with less education.[65]

Figure 8–7 Indian School Girl
© Jayakumar/Shutterstock.com

The evidence is clear that more educated women marry later, are better informed about and have greater access to contraception, use such methods more effectively, and have greater autonomy in decision making about reproductive issues. They are also more motivated to implement a preference for a smaller family because of the higher opportunity costs of unintended childbearing.[66] The woman in this situation can make more informed decisions as to when to become pregnant, when to add to her family, and when the family is complete. She will also be empowered to use her education to improve her health and that of her children through lifestyle choices.

CONCLUSION

Couples have always attempted to plan their families. However, until relatively recently, the knowledge and means to do so effectively and easily were limited. The introduction of modern contraceptive methods has reinforced and overtaken age-old practices. Couples can now achieve their desired family size with far less difficulty. Avoiding pregnancies that are "too early, too late, too many, or too frequent" makes a significant contribution to ensuring that a healthy mother will give birth to a healthy child.

In recent decades, worldwide social and economic development has increased the opportunities for both men and women to live a family life that enriches both them and their children. However, ensuring that they have ready access to the means to plan their families has not always been straightforward. The growing knowledge and understanding of family planning options has expanded recognition of unmet need and increased the demands for help. Facilitating access to and availability of family planning services is linked to the growing understanding of the role that family planning can play in contributing to the good health of the mother, the child, and the family as a whole.

To achieve these benefits, a critical step is providing young women with the tools to understand both their bodies and their options. A more educated girl is more likely to practice behavior that leads to a healthier lifestyle, to form a union at a later age, and to produce fewer and healthier children. To achieve this outcome, she needs ready access to culturally acceptable contraception in a supportive environment.

Economic development can be overwhelmed by rapid population growth that outstrips available resources. If such population growth comes from unintended births, then making it easier for members of the community to shape their families will encourage growth that can be fully supported. Offering opportunities that enable individuals to plan their families benefits all members of the community. Where the policies of an individual country seek to encourage a healthier population growing at a slower rate, the essential components are support for education (especially of girls) and ensuring access to a range of contraceptive methods to plan and space the family.

DISCUSSION QUESTIONS

1. Which factors precipitated the global contraceptive revolution that began in the 1970s?
2. Discuss the health benefits of using contraception.
3. What is the relationship between the desire for small families, contraceptive practice, and induced abortion?
4. Which family planning methods did couples use before the advent of modern contraception?
5. Why is girls' education an important component of family planning?

REFERENCES

1. United Nations. Classification of countries. In: *World Population Prospects: The 2010 Revision*. New York: United Nations Department of Economic and Social Affairs, Population Division; 2010: vii.
2. May JF. *World Population Policies: Their Origin, Evolution, and Impact*. Dordrecht, Netherlands; New York: Springer; 2012.
3. Population Reference Bureau. World population data sheet. 2012. http://www.prb.org/Publications/Datasheets/2012/world-population-data-sheet.aspx.
4. Cleland J, Bernstein S, Ezeh A, et al. Family planning: the unfinished agenda. *Lancet*. 2006;368:1810–1827.
5. UNICEF. *Plan of Action for Implementing the World Declaration on the Survival, Protection and Development of Children in the 1990s*. New York: UNICEF, September 20, 1990, pp. 1–23. http://www.cf-hst.net/unicef-temp/Doc-Repository/child_protection.html. Accessed March 19, 2014.
6. Wrigley EA. *Population and History*. New York: World University Library, McGraw-Hill; 1969.
7. Public Broadcasting Corporation. Timeline: the pill (Genesis–1950). *American Experience*. 1999–2000. http://www.pbs.org/wgbh/amex/pill/timeline/index.html.
8. Potts M, Campbell M. History of contraception. *Gynecol Obstet*. 2002;6(8):18–22.
9. McFarlane DR, Meier KJ. *The Politics of Fertility Control: Family Planning and Abortion Policies in the American States*. New York and London: Chatham House Publishers/Seven Bridges Press; 2001.
10. MEASURE Demographic and Health Surveys (DHS). Family planning. n.d. http://www.measuredhs.com/topics/Family-Planning.cfm.

11. Population Reference Bureau. Graphics bank: family planning. n.d. http://www.prb.org/Multimedia/Graphics-Bank/FamilyPlanning.aspx.

12. United Nations Population Division, Department of Economic and Social Affairs. Contraceptive prevalence by region wallchart. 2011. http://www.un.org/esa/population/publications/contraceptive2011/wallchart_graphs.pdf.

13. Coale A. The demographic transition. In: *Proceedings of the International Population Conference, Liege.* 1973;1:53–72. Liege, Belgium: IUSSP.

14. Blanc AK, Curtis SL, Croft TN. Monitoring contraceptive continuation: links to fertility outcomes and quality of care. *Stud Fam Plan.* 2002;33(2):127–140.

15. Haupt A, Kane TT, Haub C. *Population Handbook.* 6th ed. Washington, DC: Population Reference Bureau; 2011.

16. Futures Group. The policy circle: FamPlan. Computer model funded by USAID. n.d. http://www.policyproject.com/policycircle/content.cfm?a0=5b.

17. Bradley SEK, Croft TN, Fishel JD, Westoff CF. *Revising Unmet Need for Family Planning.* DHS Analytical Studies No. 25. Calverton, MD: ICF International; 2012.

18. United Nations, Department of Economic and Social Affairs, Population Division. *World Fertility Report, 2009.* ST/ESA/SER.A/304. New York: United Nations; 2011.

19. Greene M, Joshi S, Robles O. *UNFPA State of the World Population 2012: By Choice, Not by Chance: Family Planning, Human Rights, and Development.* New York: United Nations Population Fund; 2012.

20. Singh S, Sedgh G, Hussain R. Unintended pregnancy: worldwide levels, trends, and outcomes. *Stud Fam Plan.* 2010;41(4):245.

21. Black KI, Gupta S, Rassi A, Kubba A. Why do women experience untimed pregnancies? A review of contraceptive failure rates. *Best Pract Res Clin Obstet Gynaecol.* August 2010;24(4):441–455.

22. Sedgh G, Henshaw S, Singh S, et al. Induced abortion: estimated rates and trends worldwide. *Lancet.* 2007;370(9595):1338–1345.

23. World Health Organization. *Unsafe Abortion: Global and Regional Estimates of the incidence of Unsafe Abortion and Associated Mortality in 2008.* Geneva, Switzerland: WHO; 2011.

24. Marston C, Cleland J. Relationships between contraception and abortion: a review of the evidence. *Int Fam Plan Persp.* 2003;29(1):6–13.

25. Sedgh G, Singh S, Shah IH, et al. Induced abortion: incidence and trends worldwide from 1995–2008. *www.thelancet.com.* January 19, 2012. doi: 10.1016/S0140-6736(11)61786-8.

26. Myers JE, Self MW. Global perspective of legal abortion: trends analysis and accessibility. *Best Pract Res Clin Obstet Gynaecol.* 2010;24:457–466.

27. Bongaarts J, Westoff C. The potential role of contraception in reducing abortion. *Stud Fam Plan.* 2000;31(3):193–202.

28. Marston C, Cleland J. *The Effects of Contraception on Obstetric Outcomes.* Geneva, Switzerland: World Health Organization, Department of Reproductive Health and Research; 2004.

29. Campbell OMR, Graham WJ. Strategies for reducing maternal mortality: getting on with what works. *Lancet.* September 28, 2006;368:1284–1299.

30. Stover J, Ross J. How increased contraceptive use has reduced maternal mortality. *Matern Child Health J.* 2010;14:687–695. http://link.springer.com .libproxy.unm.edu/article/10.1007/s10995-009-0505-y#page-2. Accessed March 19, 2014.

31. Norton M. New evidence on birth spacing: promising findings for improving newborn, infant, child, and maternal health. *J Gynecol Obstet.* 2005;89:S1–S6.

32. World Health Organization. *Report of a WHO Consultation on Birth Spacing.* Geneva, Switzerland: Department of Making Pregnancy Safer, Department of Reproductive Health and Research; June 2005.

33. Stone L. *The Family, Sex and Marriage in England 1500–1800.* Abridged ed. New York: Harper Torchbooks; 1979.

34. Labbok MH, Perez A, Valdes V, et al. The lactational amenorrhea method (LAM): a postpartum introductory family planning method with policy and program implications. *Adv Contraception.* 1994;10(2):93–109.

35. Zohoori N, Popkin BN. Longitudinal analysis of the effects of infant-feeding practices on postpartum amenorrhea. *Demography.* 1996;33(2):167–180.

36. Winikoff B, Mensch B. Rethinking postpartum family planning. *Stud Fam Plan.* 1991;22(5):294–307.

37. Smith DP. Breastfeeding, contraception, and birth intervals in developing countries. *Stud Fam Plan.* 1985:16(3):154–163.

38. Brady M. Preventing sexually transmitted infections and unintended pregnancy, and safeguarding fertility: triple protection needs of young women. *Reprod Health Matt.* 2003;11(22):134–141.

39. Wilcher R, Cates W. Reproductive choices for women with HIV. Bull WHO. 2009;87:833–839. doi: 10.2471/BLT.08.059360.

40. Hoyt MJ, Storm DS, Aaron E, Anderson J. Preconception and contraceptive care for women living with HIV. *Infect Dis Obstet Gynecol.* 2012. doi: 10.1155/20/6041832012.

41. Hogan MC, Foreman KJ, Naghavi M, et al. Maternal mortality for 181 countries, 1980–2008: a systematic analysis of progress toward Millennium Development Goal 5. *Lancet.* May 8, 2010;375(9726):1609–1623.

42. De Cock KM, Fowler MG, Mercier E, et al. Prevention of mother-to-child HIV transmission in resource-poor countries: translating research into policy and practice. *JAMA.* 2000;283(9):1175–1182. doi: 10.1001/jama.283.9.1175.

43. Diamond-Smith N, Potts M. A woman cannot die from a pregnancy she does not have. *Int Persp Sex Reprod Health.* 2011;37(3):155–157.

44. Cherry AL, Byers L, Dillon M. A global perspective on teen pregnancy. In: Ehiri JE, ed. *Maternal and Child Health: Global Challenges, Programs and Policies.* Boston, MA: Springer Science+Business Media; 2009:Chapter 25;375–397.

45. World Health Organization. Interventions for preventing unintended pregnancies among adolescents. Cited in UNFPA. *Fact Sheet on Adolescent Girls' Sexual and*

Reproductive Health Needs. Health Statistics and Health Information Systems; July 5, 2012.

46. Wall LL. The global burden of obstetric fistulas. In: Ehiri JE, ed. *Maternal and Child Health: Global Challenges, Programs and Policies.* Boston, MA: Springer Science+Business Media; 2009;Chapter 17; 311–321.

47. Hansen JP. Older maternal age and pregnancy outcome: a review of the literature. *Obstet Gynecol Surv.* 1986;41(11):726–742.

48. Potts M, Henderson CE. Global warming and reproductive health. *Int J Gynecol Obstet.* 2012;119:S64–S67.

49. Ahmed S, Li O, Li L, Tsui AO. Maternal deaths averted by contraceptive use: an analysis of 72 countries. *Lancet.* July 14, 2012;380(9837):111–125.

50. Population Reference Bureau. *Data Sheet 2012.*

51. Centers for Disease Control and Prevention. Deaths: final data for 2006. *National Vital Statistics Reports.* 2009;57(14):13. http://www.cdc.gov/nchs/data/nvsr/nvsr57/nvsr57_14.pdf. Accessed May 2013.

52. Cesar VG, Wagstaff A, Schellenberg JA, et al. Applying an equity lens to child health and mortality: more of the same is not enough. *Lancet.* 2003;362:233–241.

53. Cleland J, Augustin C-A, Peterson H, et al. Contraception and health. *Lancet.* July 14, 2012;380(9837):149–156.

54. Rutstein SO. Effects of preceding birth intervals on neonatal, infant and under-five years mortality and nutritional status in developing countries: evidence from the Demographic and Health Surveys. *Int J Gynecol Obstet.* 2005;89:S7–S24.

55. Khorshed Alam Mozumder ABM, Barkat-E-Khuda, Kane TT, et al. The effect of birth interval on malnutrition in Bangladesh infants and young children. *J Biosocial Sci.* 2000;32:289–300.

56. Canning D, Schultz TP. The economic consequences of reproductive health and family planning. *Lancet.* July 14, 2012;380(9837):149–156.

57. United Nations. Universal Declaration of Human Rights. December 10, 1948.

58. Lewis MA, Lockheed ME. *Inexcusable Absence: Why 60 Million Girls Still Aren't in School and What to Do About It.* Washington, DC: Centre for Global Development; 2006:33.

59. Cooray A, Potrafke N. Gender inequality in education: political institutions or culture and religion. *Eur J Politic Econ.* 2011;27:268–280.

60. Vavrus F, Larsen U. Girls' education and fertility transitions: an analysis of recent trends in Tanzania and Uganda. *Econ Develop Cultural Change.* 2003;51(4):945–975.

61. Vavrus F, Larsen U. Economic development and cultural change. University of Chicago Press. doi: 10.1086/377461. http://www.jstor.org/stable/10.1086/377461.

62. Basu AM. Why does education lead to lower fertility? A critical review of some of the possibilities. *World Develop.* 2002;30(10):1779–1790.

63. Dreze J, Murthi M. *Fertility, Education and Development: Further Evidence from India.* Discussion Paper No. DEDPS 20. Suntory Centre, Suntory and Toyota International Centres for Economics and Related Disciplines, London School of Economics and Political Science; January 2000.

64. Ainsworth M, Beegle K, Nyamete A. *The Impact of Women's Schooling on Fertility and Contraceptive Use: A Study of Fourteen Sub-Saharan African Countries.* World Bank; 1996.

65. Ahmed S, Creanga AA, Gillespie DG, Tsui AO. Economic status, education and empowerment: implications for maternal health service utilization in developing countries. *PLoS One.* 2010;5(6):e11190. doi: 10.1371/journal .pone.0011190.

66. Bongaarts J. *Vienna Yearbook of Population Research.* 2010;8;31–50.

Part IV

RELATED ISSUES

Chapter 9

Women's Status and Reproductive Rights

Deborah R. McFarlane

INTRODUCTION

Gender inequalities remain "deeply entrenched in every society."[1] Throughout the world, "women lack access to decent work and face occupational segregation and gender wage gaps."[1] Too often, they lack access to basic education and health care. Many women and girls are not free to marry whom they wish or have the number of children that they desire. In every country, women suffer violence and discrimination. Moreover, they are underrepresented in political and economic decision-making processes.[1]

The process by which unequal power relations are transformed and women gain greater equality vis-à-vis men is called *women's empowerment*. At the global level, the empowerment of women has been recognized as a basic human right. It was a central policy goal for the International Conference on Population and Development (ICPD) in Cairo in 1994 and the Fourth World Conference on Women (FWCW) in Beijing in 1995. Moreover, both conferences reaffirmed that access to reproductive health services is necessary for advancing the status of or empowering women.

Transforming unequal power relations so that women gain greater equality relative to men is a multifaceted undertaking. At the government or macro level, social, economic, and political rights can be extended to women. For example, some countries require that a specific number or percentage of legislative seats be reserved for women. On the micro or individual level, women and girls can gain the ability to control their own lives in a variety of ways, including education, later marriage, and access to economic opportunities. An example of this process is the powerful impact of extending micro-credit to women in Bangladesh, where formal banking institutions are unlikely to loan them money (Box 9–1).[2]

Changing gender roles is often controversial. In many cultures, strong social norms exist for how girls are educated, who chooses their husbands, how many children they bear, and with whom they socialize. Outside the home, expectations remain concerning which types of vocations are acceptable for young women, with whom they should socialize, what is appropriate dress, and how much they can participate in the political system. Once women are married, husbands may decide how household resources are spent and whether women can seek health care outside the home (Figure 9–1).

Challenging gender roles or social norms can even result in violence. Throughout the world, domestic violence is the most common. Some cultures even condone honor-based violence, which disproportionately affects women and girls.[3] The most extreme form of this violence is honor killings, estimated by the United Nations to number 5000 deaths per year.[4] Many authorities believe that this estimate is greatly undercounted. Moreover, this statistic does not reflect other forms of honor-based violence, such as forced abortion and hymen repair, abduction and imprisonment, forced marriage, or honor suicide.[5]

Box 9–1 Financial Credit Eludes Women

Throughout the world, formal banking institutions neglect women's needs for credit.

- In Bangladesh, women contribute 27% of total formal-sector deposits, but their share in formal credit is less than 2%.
- Women entrepreneurs in South Africa also face major barriers in accessing finance. For the black economic empowerment equity fund of a major bank in the country, only 5% of clients were women.
- In Uganda, women have just 9% of the available credit, and that credit amount declines to 1% in rural areas.

Data from United Nations. World survey on the role of women in development: at a glance: women's control over economic resources and access to financial resources, fact sheet. 2009. http://www.un.org/womenwatch/daw/ws2009/documents/DESA_Survey_FactSheet.pdf.

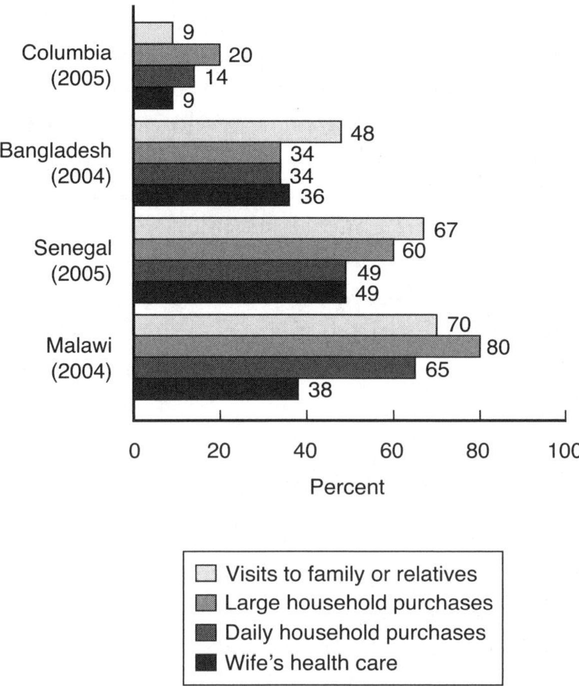

Figure 9–1 Household Decisions Made by Husbands Alone

Reproduced from Population Reference Bureau. *The World's Women and Girls: 2011 Data Presentation.* Washington, DC: Population Reference Bureau; 2011. http://www.prb.org/Publications/ Datasheets/2011/worlds-women-and-girls.aspx. Accessed January 8, 2014.

The advancement of women, however, need not be a "battle of the sexes." Indeed, the education and economic advancement of women can have tangible health and economic benefits for families, including their husbands. Conversely, men's participation in and acceptance of changed gender roles has been important for the women's empowerment. An example of this process has been the influence of fathers in curtailing female genital mutilation in Senegal and the Gambia.[6]

Reproductive and sexual health rights are central issues for women's empowerment, and they serve as a cornerstone of economic development.[7] Controlling fertility can give women options and a degree of freedom not previously available.[8] The United Nations Population Fund (UNPFA) has posited that to achieve sustainable and equitable economic development, people must be able to exercise control over their sexual and reproductive lives.[7(p2)]

Nevertheless, many women and their partners still lack access to reproductive health services. In developing countries, an estimated 222 million women do not have the means to delay pregnancies and childbearing. Moreover, millions of women in developed countries are unable to plan their families because they lack access to contraceptive information and methods or because they face social and economic barriers.[9]

The first section of this chapter summarizes the status of women throughout the world. Next, it reviews global indicators of sexual and reproductive health. It then explores the human rights justification for sexual and reproductive health. Finally, it juxtaposes the sexual and reproductive rights agenda with the demographic imperative to limit human population growth.

THE GLOBAL STATUS OF WOMEN

Identifying the position of women within any society has many dimensions. At the micro or individual level, the women of the world are disproportionately poor. Women and girls, who represent the majority of the global population, account for 70% of the world's poor living on $1 or less per day. They represent two-thirds of the world's illiterate. Although women produce more than half of the world's food supply, they own less than 1% of the world's property. Globally, 10 million more girls than boys do not attend school, a fact that portends the continuation of female poverty.[4]

Throughout the world, women and girls suffer discrimination while working for others or self-employed. They are often paid less than men for the same work. Many women face discrimination when they apply for jobs or credit. Female workers are concentrated in insecure, unsafe, and low-wage work. Eight out of 10 women workers are considered to be in vulnerable employment, with global economic changes taking a huge toll on their livelihoods.[10]

Violence Against Women

Violence against women is pervasive. Domestic violence is the most common type of violence against women; its incidence ranges from 8% in Albania to 49% in Ethiopia and Zambia. Domestic violence and rape account for 5% of the disease burden for women aged 15 to 44 years in developing countries and 19% of women in that age group in developed countries.[4] (Note that these data do not indicate that domestic violence and rape have fewer consequences in developing countries; rather, they reflect substantially higher health risks, including maternal mortality, faced by women in poorer countries.) Outside the home, rape is increasingly being used as a genocidal tool. For example, hundreds of thousands of women have been raped and mutilated in the ongoing conflict in Eastern Congo.

Femicide—the murder of women—often involves sexual violence (Figure 9-2). An estimated 40% to 70% of women murdered in Australia, Canada, Israel, South Africa, and the United States are killed by their husbands or boyfriends. Over the past two decades, hundreds of women have been abducted, murdered, and raped in Juarez, Mexico, yet these crimes have not been solved.[4] Honor killings persist in Pakistan, India, and elsewhere.[5]

In Africa alone, an estimated 3 million girls per year are subjected to female genital mutilation (FGM). The World Health Organization (WHO) estimates that between 100 million and 140 million girls and women world-wide have been subjected to one of the first three types of female genital mutilation (Box 9-2). The most recent prevalence data estimate that 91.5 million girls and women older than age 9 years in Africa are currently living with the consequences of FGM. Although FGM is most common in Africa and Yemen, immigrant women and girls in the West are also at risk (Figure 9-3).[11-13] Worldwide, about 30 million girls are at risk of being circumcised in the next decade.[14]

Human Trafficking

"Human trafficking is a crime that ruthlessly exploits women, children, and men for numerous purposes, including forced labour and sex."[15(p1)] With this underground industry generating billions of dollars in profits for traffickers, an estimated 20.0 million people are victims each year.

This form of modern slavery is a global crime affecting nearly all countries in every region of the world. Recent data show that at least 136 nationalities have been detected in 118 different countries. However, victims are disproportionately from poorer countries. "Trafficking from East Asia remains the most conspicuous globally."[15(p1)]

Across the world, women and girls are more likely than males to be victims of such violence, accounting for more than three-fourths of persons trafficked. Sexual exploitation is more common than forced labor. Figure 9-4 shows that women represent 59% of victims while girls represent 17%. A disturbing trend is the increase in child victims. From 2003 to 2006, 20% of all detected victims were children. That percentage rose to 27% during the period 2007 to 2010. Two out of three child victims are girls, and the majority of them are trafficked for sexual exploitation.[15(pp1-2)]

Signs of Progress

Yet there is progress. The number of girls receiving formal education is increasing. Girls now represent only 57% of the children globally not

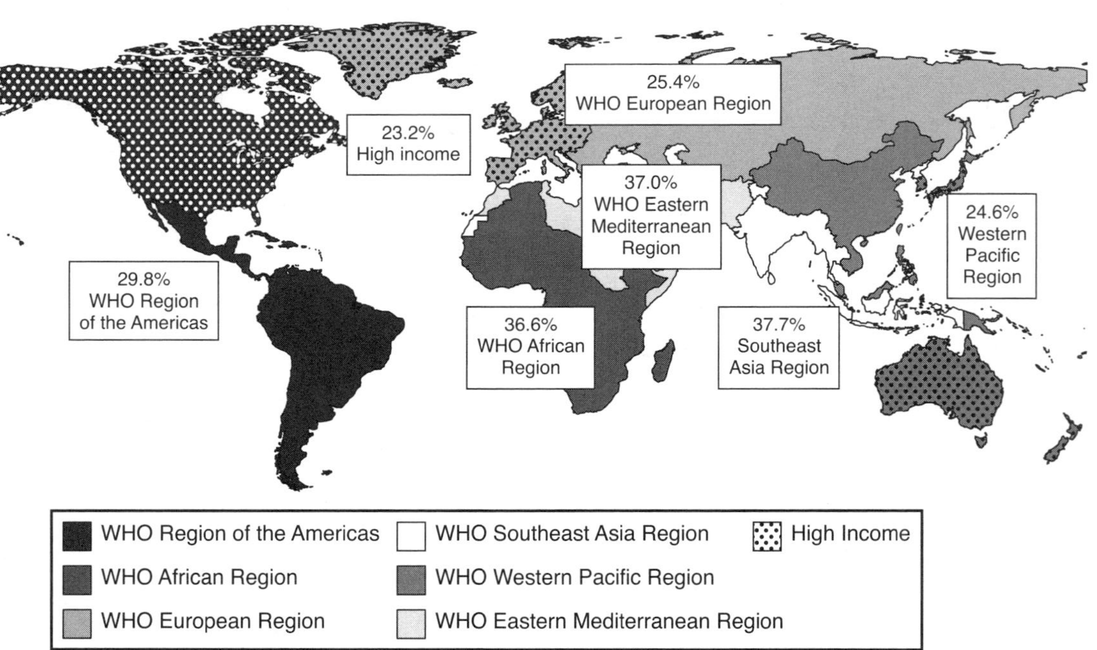

Figure 9–2 Prevalence of Intimate Partner Violence and Non-Partner Sexual Violence (World Health Organization Poster)

Reproduced from World Health Organization. *Global and Regional Estimates of Violence Against Women: Prevalence and Health Effects of Intimate Partner Violence and Non-Partner Sexual Violence.* Geneva, Switzerland: WHO; 2013. http://www.who.int/reproductivehealth/publications/violence/VAW_Prevelance.jpeg.

> **Box 9–2**   Classifications for Female Genital Mutilation or Cutting
>
> Type I: Partial or total removal of the clitoris and/or the prepuce (clitoridectomy).
>
> Type II: Partial or total removal of the clitoris and the labia minora, with or without excision of the labia majora (excision).
>
> Type III: Narrowing of the vaginal orifice with creation of a covering seal by cutting and appositioning the labia minora and/or the labia majora, with or without excision of the clitoris (infibulation).
>
> Type IV: All other harmful procedures to the female genitalia for nonmedical purposes—for example, pricking, piercing, incising, scraping, and cauterization.
>
> Modified from the World Health Organization. *Female Genital Mutilation.* Factsheet 241. Geneva, Switzerland: WHO; 2013. http://www.who.int/mediacentre/factsheets/fs241/en/.

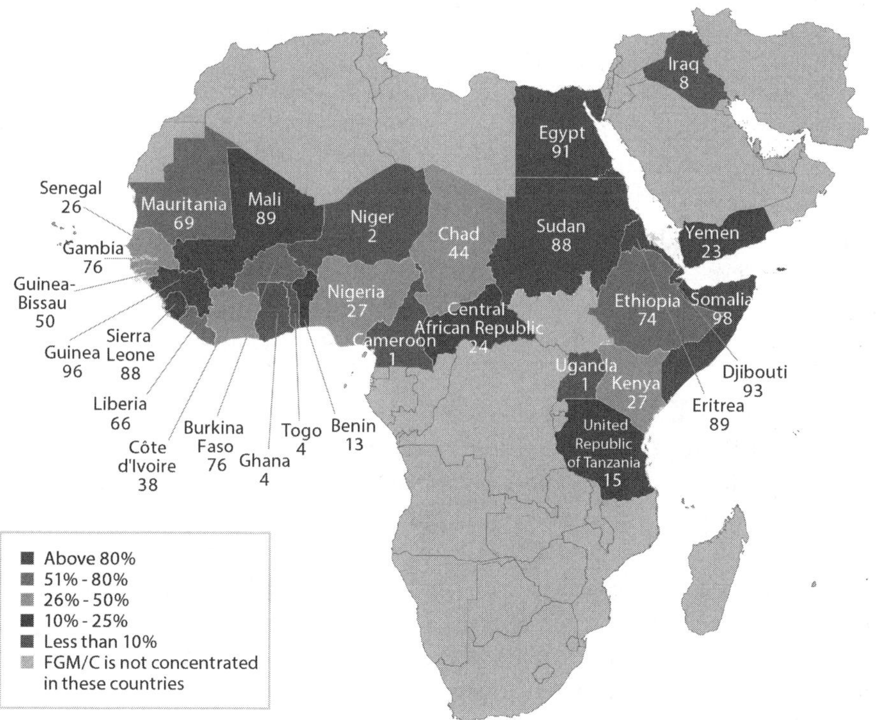

Figure 9–3 Prevalence of Female Genital Mutilation in Africa and Yemen (Women Ages 15–49), 1997–2006

Courtesy of UNICEF Division of Policy and Strategy

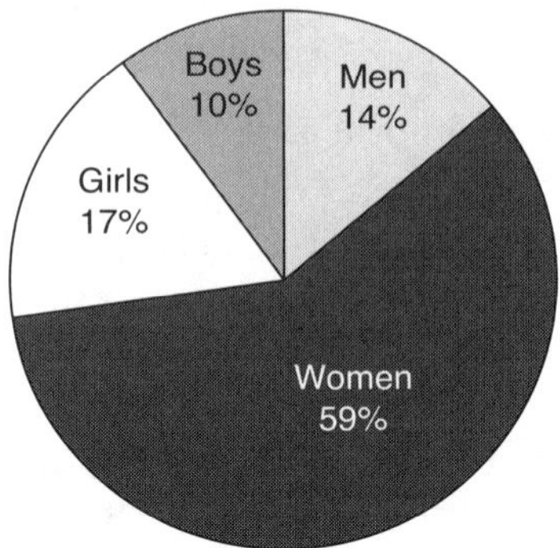

Figure 9–4 Gender and Age Profile of Trafficking Victims Detected Globally, 2009

Reproduced from United Nations, Office on Drugs and Crime. *Global Report on Trafficking in Persons 2012*. Vienna/New York: United Nations; 2012:25. http://www.unodc.org/unodc/data-and-analysis/glotip.html. Accessed January 8, 2014.

Box 9–3   Millennium Development Goals

Goal 1: Eradicate extreme hunger and poverty.

Goal 2: Achieve universal primary education.

Goal 3: Promote gender equality and empower women.

Goal 4: Reduce child mortality.

Goal 5: Improve maternal health.

Goal 6: Combat HIV, malaria, and other diseases.

Goal 7: Ensure environmental sustainability.

Goal 8: Develop a global partnership for development.

Data from the United Nations. *We Can End Poverty by 2015: Millennium Development Goals*. New York: United Nations; 2013. http://www.un.org/millenniumgoals/.

---

attending school; that figure was 67% in the 1990s. A recent United Nations Children's Fund (UNICEF) study[14] indicates that female genital mutilation or cutting is becoming less prevalent in half of the 29 countries where it is practiced. Women have made gains in the political arena as well. In 2013, 19 women served as heads of state and women now hold nearly 22% of the world's parliamentary seats.[16,17] Although the fifth Millennium Development Goal's target of reducing the maternal mortality ratio by three-fourths from 2000 to 2015 certainly has not be achieved, maternal mortality has dropped significantly since 1990 (Box 9–3).[18]

INDICATORS OF SEXUAL AND REPRODUCTIVE HEALTH

The status of women is a powerful predictor of both fertility levels and reproductive health. In countries where women experience parity in educational opportunities, fertility rates are lower. When women have economic security and other opportunities, they experience better reproductive health outcomes.

These are complex, not just unilateral, relationships. While the status of women affects fertility rates, levels of fertility also influence women's opportunities, including their life expectancies. Bearing and raising larger numbers of children often thwarts women's progress toward education, economic independence, and political equality.[19] Economic deprivation means that women are less likely to have access to reproductive health services, including contraception and safe maternity care. Poverty increases the likelihood that young women will marry young and have more and less-healthy children than more affluent women.

Child Marriage and Early Childbearing

Child marriage (before age 18) is a major concern for women's education, fertility, and reproductive health. In some parts of Asia and Africa, most girls are married by age 18—a fact that poses serious consequences for their health and economic prospects (Figure 9-5).[20] In populous India, nearly half of women are married by age 18. Child marriage is far more common for girls than boys, and young girls are often married to significantly older husbands. "Risks of HIV/AIDS infection are higher among young girls as their negotiation skills and experience to ensure a healthy sexual life are less developed."[21(p1)] Child brides frequently drop out of school and begin childbearing at an early age, "when the body is not fully prepared for child bearing."[21(p1)]

Health risks for both mothers and infants accompany very early childbearing. These risks are exacerbated when the mother is malnourished or stunted. Girls aged 15–19 years are more likely to experience delivery complications than older women are. Figure 9–6 shows that neonatal, infant, and child mortality rates are much higher for the offspring of younger girls. Moreover, the risk of malnutrition (stunting and underweight) is higher in children younger than 5 years of age born to young mothers (younger than age 20) than in children born to older women.[21(p2)]

Obstetric Fistula

One of the consequences of early childbearing is vesicovaginal fistula (VVF).[22] VVF is a debilitating, humiliating, and life-threatening result of obstructed labor, which WHO calls "the most dramatic aftermath of neglected childbirth."[23] Fistulas are injuries to the bladder, urethra, and lower end of the bowel causing constant leakage of urine and, sometimes, the vaginal excretion of feces. Fistula is prevalent where maternal mortality is highest, especially where emergency obstetric care, referral systems, and

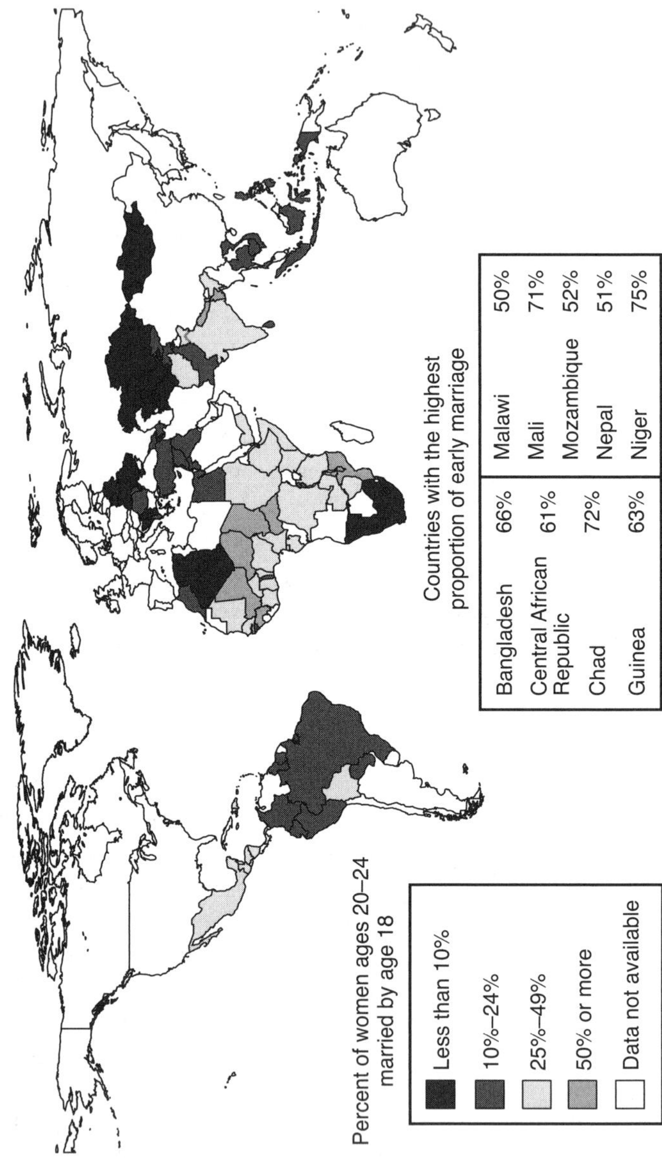

Figure 9–5 Prevalence of Early Marriage Around the World

Reproduced from Population Reference Bureau. *The World's Women and Girls: 2011 Data Presentation.* Washington, DC: Population Reference Bureau; 2011. http://www.prb.org/Publications/Datasheets/2011/worlds-women-and-girls.aspx. Accessed January 8, 2014.

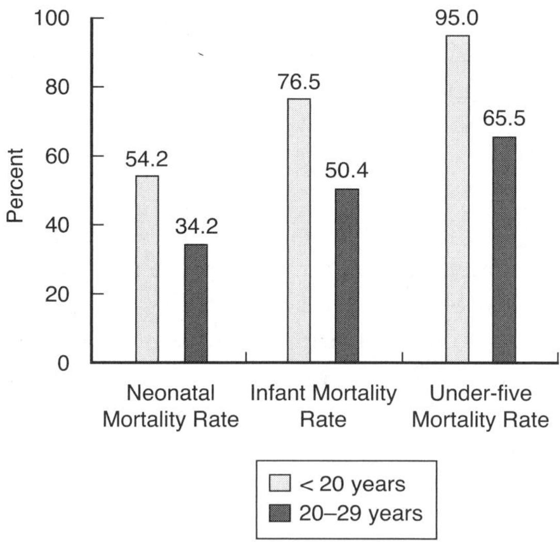

Figure 9–6 Child Mortality Rates by Age of Mother: India, 2005–2006

Data from International Institute for Population Sciences (IIPS) and Macro International. National Family Health Survey (NFHS-3), Volume I. Mumbai, India: IIPS; 2005–2006.

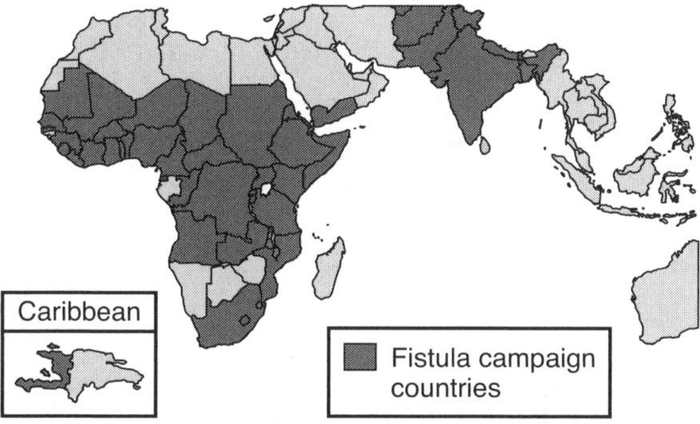

Figure 9–7 Countries with High Prevalence of Vesicovaginal Fistula

Reproduced from United Nations Population Fund. *Campaign to End Fistula: 2008, The Year in Review.* New York: UNFPA; 2009. http://www.unfpa.org/webdav/site/global/shared/documents/publications/2009/fistula_annual_report_2008.pdf. Accessed January 23, 2014.

infrastructure are poor (Figure 9–7).[24] Adolescent VVF victims suffer infection and may be made infertile, becoming social outcasts as a result of divorce or of being forced into prostitution.

The exact prevalence rates for VVF are unknown, but it is estimated that worldwide at least 2 million women live with untreated obstetric fistula. There are at least 75,000 new cases every year. More than 90% of babies do not survive the traumatic labor that causes this condition. Obstetric fistula is preventable; it can generally be avoided by delaying the age of first pregnancy, stopping harmful traditional practices (e.g., FMG), and ensuring timely access to obstetric care.[23]

Other Indicators

Around the world, improvements in women's status have been uneven. Literacy among women has improved, because more girls attend school. Globally, however, boys still lead in literacy, primary school completion, and secondary school enrollment. Educating girls is important for reproductive health because education influences how many children women want (ideal number) as well as how many children they will actually bear (Figure 9–8).

Fertility rates have been falling in most countries. The average number of lifetime births per woman worldwide is 2.5. In more-developed countries, women have an average of just one or two children (1.7); the lowest fertility rates are in East Asia and Europe. In less-developed regions (excluding China), the average is three children per woman (3.1), but fertility levels in some regions are much higher. In sub-Saharan Africa, each woman has an average of more than five children; in nine countries, the average is six or more. Outside of Africa, the highest fertility levels are found in Afghanistan (5.7 children per woman) and Yemen (5.5).[25]

Women and men often hold very different views about the ideal number of children a family should have. Couples who live in urban areas,

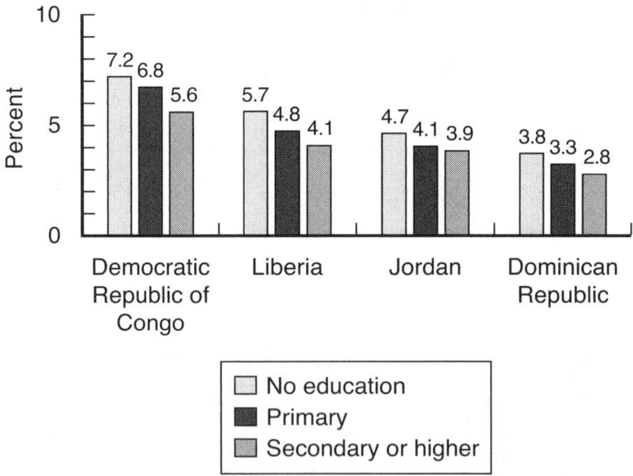

Figure 9–8 More Educated Women Prefer Fewer Children

Reproduced from Population Reference Bureau. Graphics bank: family size preferences. http://www.prb.org/Publications/GraphicsBank/FamilyPlanning.aspx. Accessed January 8, 2014.

have a secondary school education, and make up the wealthiest 20% of the population frequently prefer fewer children and smaller families. In Kenya, however, men consistently want more children than women. The diverse range of views about ideal family size suggests that programs need to reach out to men as well as to women, and to focus on improved couple communication regarding desired family size.[25]

Many cultures place a premium upon male infants as opposed to girl babies. In this context, parents have a preference for sons because boys are thought to bring more prestige or wealth to a family than girls do. As a result, the practice of sex-selective abortion has become widespread in a number of countries where parents can afford such technology. In the absence of sex-selective abortion, the natural sex ratio is 1.05 (about 105 boys born for every 100 girls; Figure 9–9). However, in countries where sex-selective abortion takes place, the birth ratios are much higher than 1.05, meaning a disproportionate number of boys are born. The United Nations projects that highly skewed sex ratios in most countries will decline in the coming decades. India, however, is projected to remain steady at 1.08 through the mid-21st century.[26]

Around the world, much work remains to ensure that gender equality is achieved. Fertility rates remain high in many sub-Saharan African

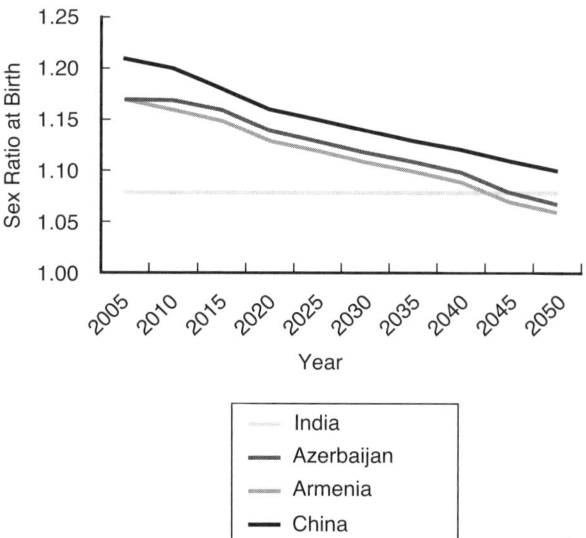

Figure 9–9 Estimated and Projected Sex Ratios at Birth, Selected Countries

Reproduced from Population Reference Bureau. *PowerPoint Presentation: The World's Women and Girls.* Washington, DC: Population Reference Bureau; 2011. http://www.prb.org/Reports/2011/international-womens-day-ppt.aspx. Accessed January 8, 2014.

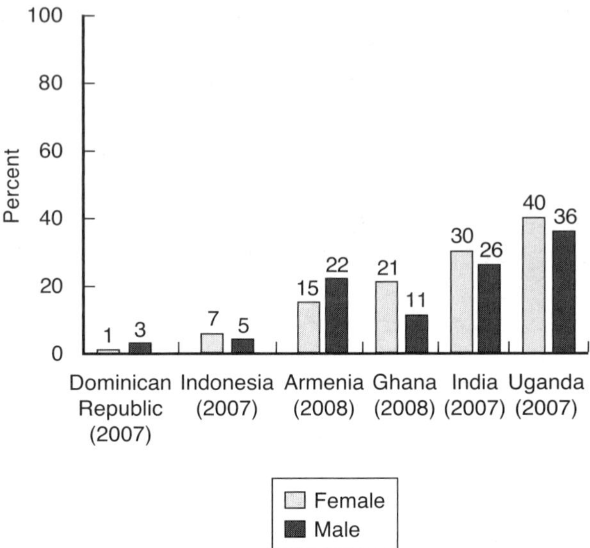

Figure 9–10 Percentage of Women and Men Who Agree That Wife Beating Is Acceptable if a Wife Argues with Her Husband

Reproduced from Population Reference Bureau. *PowerPoint Presentation: The World's Women and Girls.* Washington, DC: Population Reference Bureau; 2011. http://www.prb.org/Reports/2011/international-womens-day-ppt.aspx. Accessed January 8, 2014.

countries where contraceptive prevalence is low. Maternal mortality and morbidity remain high in sub-Saharan Africa as well as in Afghanistan and India. Figure 9–10 shows that in some societies, both women and men still consider wife-beating acceptable.[26]

SEXUAL AND REPRODUCTIVE RIGHTS

Recognition of gender inequities has led to increased global support for sexual and reproductive rights, which include the following:

1. The freedom to decide how many children to have and when to have them
2. The right to have the information and necessary means to regulate one's fertility
3. The right to control one's own body

The United Nations established the first and second rights in a series of treaties, conventions, and agreements.[9,27] The third right has "emerged from long-standing feminist discourse"[27(p12)] and is inferred in some UN documents.

Historical Evolution of Sexual and Reproductive Rights

The notion of sexual and reproductive rights has evolved from human rights precepts. Enlightenment theories in the 17th and 18th centuries posited that men had *natural* or *unalienable rights*. These ideas about individual liberties influenced both the American and French Revolutions and provided the foundation for later ideas about economic and social rights or *entitlements*. In the 18th and 19th centuries, a few liberal reformers, radical utopians, and feminists even suggested that "the rights of man might extend to women, too." [27(p3)] In 1792, Mary Wollstonecraft published "A Vindication of the Rights of Women," later hailed as "the feminist declaration of independence." [4(p131)]

The concept of *women's rights* was powerful, propelling 19th-century feminist reformers to push for women's suffrage and other opportunities. This notion also spawned the women's liberation movement in the second half of the 20th century. Because reproductive freedom is so important to women's exercise of other freedoms, women's rights advocates made the issue of bodily integrity or the freedom to control one's own body a cornerstone of the women's liberation movement.[27(p12),28]

The notion that family planning is a human right was first formalized at the United Nations International Conference on Human Rights held in Teheran, Iran, in 1968.[9,27] Table 9–1 shows that this idea was been expanded and affirmed in subsequent decades. Of particular interest is the International Conference on Population and Development (ICPD) held in Cairo in 1994, because so many subsequent documents and policies refer to its 20-year Programme of Action.

Table 9–1 Selected International Agreements and Documents Relevant to Sexual and Reproductive Rights

Year	Document	Key Points
1948	Universal Declaration of Human Rights	Pivotal for human rights discourse, many laws, and constitutions.
1968	Teheran Conference on Human Rights	Declares right of individuals and couples to family planning information and services.
1969	United Nations General Assembly Declaration on Social Progress and Development	Resolves that parents have the exclusive right to plan their families and this right requires necessary information and services.

(Continues)

Table 9–1 Selected International Agreements and Documents Relevant to Sexual and Reproductive Rights (*Continued*)

Year	Document	Key Points
1974	World Population Plan of Action	Reiterates couples' rights to choose the size of their families and governments' responsibility to provide necessary means to exercise rights.
1976	International Covenant on Economic, Social, and Cultural Rights	Universal right to highest attainable standard of physical and mental health.
1979	Convention on the Elimination of All Forms of Discrimination Against Women	Specifies family planning as key for health and well-being of families. Promotes gender equality.
1986	Declaration on the Right to Development	Calls for economic development that benefits entire population.
1989	Convention on the Rights of the Child	Ensures the right of children to have access to certain services, such as health care, including information on sexuality and reproduction.
1993	United Nations World Conference on Human Rights	Women's reproductive rights are indivisible from basic human rights.
1994	International Conference on Population and Development (ICPD)	Links population and development. Women's empowerment, including reproductive health needs, required for "balanced development." 179 governments adopted the 20-year Programme of Action.
1995	Beijing Declaration and Platform for Action	Reiterates broad definition of right to family planning from ICPD Programme of Action.

Data from Greene M, Joshi S, Robles O. *UNFPA State of the World Population 2012: By Choice, Not by Chance: Family Planning, Human Rights, and Development.* New York: United Nations Population Fund;2012; Kantner JF, Kantner A. *The Struggle for International Consensus on Population and Development.* New York: Palgrave MacMillan; 2006; May J. *World Population Policies: Their Origin, Evolution, and Impact.* Washington, DC: Springer; 2013.

Social Entitlements and Binding Covenants

The Universal Declaration of Human Rights adopted by the United Nations in 1948 incorporated two conceptions of human rights. The first was that *human rights are unalienable or natural.* This notion is embodied in the English Bill of Rights of 1689, the American Bill of Rights of 1789, and the Declaration of the Rights of Man and of the Citizen adopted by the French National Assembly in 1789. These and subsequent documents set forth a concept of individual liberty in which the primary role of the state is to ensure freedom of the citizenry from abuses of power (e.g., from governments not freely elected, from arbitrary arrest and detention).

The second concept involved social entitlement—that is, the responsibility of the state to *achieve results,* not just guarantee freedom of opportunity or personal rights. Emanating from 19th- and 20th-century socialist theory, social entitlement means that all persons should have tangible and material rights. The Universal Declaration included "the right to an adequate standard of living, to education, to work, to just and favorable conditions of work, and to protection from unemployment."[27(p6)]

The extension of the concept of human rights to social entitlement also implies obligations for citizens. The right to education becomes a moral, and often material, imperative to send children to school, at least up to a certain age. Similarly, the right to health becomes an obligation to vaccinate one's children against infectious diseases or seek health care for them when warranted. The right to decide the number and spacing of one's children may imply collective reproductive responsibility. Until very recently, the world's extreme example of this concept was found in China, where family planning policy was more collective duty than individual right.

Proclamations of human rights by the United Nations have multiplied since 1948. These include *declarations and resolutions,* which are not binding, and *covenants and conventions,* which, at least theoretically, bind ratifying member states to translate principles into action. In practice, it is often difficult to ensure that meaningful implementation is accomplished, whether the proclamation is binding or not.

For example, the Convention on the Elimination of All Forms of Discrimination Against Women (CEDAW) was adopted by the General Assembly in 1979 and ratified by 187 countries by 2013. Often described as an international bill of rights for women, the Convention defines discrimination against women as "any distinction, exclusion or restriction made on the basis of sex which has the effect or purpose of impairing or nullifying the recognition, enjoyment or exercise by women, irrespective of their marital status, on a basis of equality of men and women, of human rights and fundamental freedoms in the political, economic, social, cultural, civil or any other field." In principle, states that ratified CEDAW committed to undertake a series of measures to end discrimination against women in all forms, including opportunities in political and public life as well as education, health, and employment.[29]

CEDAW "is the only human rights treaty which affirms the reproductive rights of women and targets culture and tradition as influential forces shaping gender and family relations."[29]Among the disparate countries that have ratified the Convention are Afghanistan, Germany, India, Iraq, Saudi Arabia, Uganda, and the United Kingdom. Noteworthy is the fact that although the United States signed the document in 1980, it has never been ratified by the U.S. Congress. Regardless of their CEDAW ratification status, individual countries differ greatly in the importance they ascribe to political rights and social rights, particularly those related to gender.[27]

Notwithstanding the difficulty of implementing even binding covenants and conventions, many public health advocates and scholars maintain that these formalized rights can be useful. Human rights provide tools for advocates to hold governments legally accountable for their failure to address specific issues in binding resolutions. For example, because maternal mortality is recognized as a human rights violation, this approach is being used to compel national governments to make pharmaceuticals from WHO's Essential Medicines List (EML) available for free or at minimal cost in health facilities or in communities. The EML includes the full range of contraceptives and emergency contraception, along with other medicines required for maternal and newborn health.[30]

This application of human rights to maternal health care has produced mixed results, particularly regarding the availability of emergency contraception (EC) in Latin America. While not failsafe, the human rights approach is at least "part of the solution." Human rights mechanisms have transformed global understanding of maternal death from a perception of mere misfortune to an injustice that countries are obligated to remedy.[30(p6)] Many believe that the human rights approach has been useful for promoting reproductive health more generally.

The real global challenge is to ensure that "the abstract principle of sexual and reproductive rights becomes a reality in both women and men's lives."[27(p27)] To produce tangible results, policy implementers and program planners need to consider the gender inequities faced by many of the world's women—that is, the circumstances of their lives. For example, effective contraceptive information programs must take into account how and with whom both women and men communicate about matters related to sexuality. Similarly, reproductive health services that address the unmet need for family planning services will have to consider actual social mores (as opposed to ideology), transportation issues, childcare provision, and the hours when women, in particular, are able to access services.

SEXUAL AND REPRODUCTIVE RIGHTS AND ICPD

Widespread support exists for the promises embodied within the concept of sexual and reproductive rights. The United Nations Population Fund (UNPFA) states

> The right of the individual to freely and responsibly decide how many children to have and when to have them has been a guiding principle for decades, but especially since 1994, when 179 governments came together and adopted the groundbreaking Programme of Action of the International Conference on Population and Development, the ICPD.[9]

ICPD

The International Conference on Population and Development held in Cairo in 1994 was the third decennial on population and development. "Prior to the ICPD, the United Nations organized six expert meetings, eight round tables on technical subjects, and three preparatory meetings that were open to all countries."[31(p108)] At the conference, delegates from 189 member states and representatives from 1200 accredited nongovernmental agencies agreed to the set of 15 principles. Each principle was associated with a specific set of actions. Together, they constituted the Programme of Action to be monitored by UNFPA until 2014. Box 9–4 shows Principles 4, 8, 9, 10, and 11, which are especially relevant to women's rights.

Principle 4 ties gender equality, equity, and empowerment to population and development. It calls on the international community to ensure the equal and full participation of women. It also calls for curtailing violence against women. It advocates for the right of women to control their fertility, but remains silent on abortion.

Principle 8 reiterates the right of individuals to decide freely the number and spacing of their children. Principle 9 states that women must be free to choose their spouses and that within marriage they should be equal partners. Principle 10 calls for girls' education. Principle 11 gives priority to children, specifically mentioning two risks to girls—sexual abuse and trafficking.

The Programme of Action that emerged from the ICPD "explicitly called for dropping demographic and family planning program targets in favor of a broader policy agenda that included a range of reproductive health measures, including family planning, that would cater to women's overall reproductive health needs as well as a series of social and economic policy measures designed to empower women and to strengthen

Box 9–4   ICPD Principles Most Pertinent to Women's Issues

Principle 4. Advancing gender equality and equity and the empowerment of women, and the elimination of all kinds of violence against women, and ensuring women's ability to control their own fertility, are cornerstones of population and development- related programmes. The human rights of women and the girl child are an inalienable, integral and indivisible part of universal human rights. The full and equal participation of women in civil, cultural, economic, political and social life, at the national, regional and international levels, and the eradication of all forms of discrimination on grounds of sex, are priority objectives of the international community.

Principle 8. Everyone has the right to the enjoyment of the highest attainable standard of physical and mental health. States should take all appropriate measures to ensure, on a basis of equality of men and women, universal access to health-care services, including those related to reproductive health care, which includes family planning and sexual health. Reproductive health-care programmes should provide the widest range of services without any form of coercion. All couples and individuals have the basic right to decide freely and responsibly the number and spacing of their children and to have the information, education and means to do so.

Principle 9. The family is the basic unit of society and as such should be strengthened. It is entitled to receive comprehensive protection and support. In different cultural, political and social systems, various forms of the family exist. Marriage must be entered into with the free consent of the intending spouses, and husband and wife should be equal partners.

Principle 10. Everyone has the right to education, which shall be directed to the full development of human resources, and human dignity and potential, with particular attention to women and the girl child. Education should be designed to strengthen respect for human rights and fundamental freedoms, including those relating to population and development. The best interests of the child shall be the guiding principle of those responsible for his or her education and guidance; that responsibility lies in the first place with the parents.

Principle 11. All States and families should give the highest possible priority to children. The child has the right to standards of living adequate for its well-being and the right to the highest attainable standards of health, and the right to education. The child has the right to be cared for, guided and supported by parents, families and society and to be protected by appropriate legislative, administrative, social and educational measures from all forms of physical or mental violence, injury or abuse, neglect or negligent treatment, maltreatment or exploitation, including sale, trafficking, sexual abuse, and trafficking in its organs.

Reproduced from the United Nations Population Fund. Chapter 2: principles. In: *Programme of Action of the International Conference on Population and Development.* http://www.unfpa.org/public/home/sitemap/icpd/International-Conference-on-Population-and-Development/ICPD-Programme#ch2.

their rights."[32](p10) While the ICPD principles were silent on abortion, paragraph 8.25 in the Programme of Action did address unsafe abortion. Governments and relevant nongovernmental organizations were urged "to strengthen their commitment to women's health, to deal with the health impact of unsafe abortion as a major public health concern and to reduce the recourse to abortion through expanded and improved family planning services." The document stopped short of advocating the legalization of abortion in countries where the procedure was illegal, but stated that where "abortion is not against the law, such abortion should be safe. In all cases women should have access to quality services for the management of complications arising from abortion. Post-abortion counselling, education and family planning services should be offered promptly which will also help to avoid repeat abortions."[33]

UNFPA calls the ICPD "a great paradigm shift in the field of population and development, replacing a demographically driven approach with one that is based on human rights and the needs, aspirations, and circumstances of each woman." This sentiment was echoed by DeJong,[34] who reiterated the term paradigm shift, declaring that "progress has been significant, particularly in the recognition of gender inequities, of the importance of previously unrecognized reproductive health problems and for the attention given to ethical abuses and the need for a broader approach to services beyond the provision of family planning alone."[34](p951)

Critics of ICPD

Not everyone agrees that this paradigm shift—that is, moving from family planning services to reproductive health and more—was a wise turn for the world's women. Critics include many demographers and the "population establishment," the Vatican and other conservative parties, as well as some feminists.

Within the demographic community, pejoratively labeled the "population establishment" by some feminist groups,[34] serious concerns emerged about the ICPD Programme of Action. One issue is the "extraordinary lack of concern about the long term implications of rapid population growth, which may have eroded the will of donor governments to strain themselves to support family planning programs or the will of Third World governments to give them high priority."[35] Related to this concern was the prohibition of demographic targets, which it was argued, made monitoring programmatic progress more difficult. Moreover, while broadening family planning services is a laudable goal, many critics worried that it constituted program overreach. They argued that ICPD was visionary, but not attainable given resource constraints and ultimately it would not benefit the developing world.[35(p5)] Related to that concern was the use of a nebulous term *reproductive health*, which lacked precision particularly in different cultures and languages, thereby making monitoring its progress very difficult. Finally, many demographers and population advocates worried that pressing for the inclusion of reproductive health in the already stretched primary healthcare systems of the developing world would diminish successful community-based distribution of contraceptives.

The Vatican and other conservative allies have also criticized the ICPD Programme of Action, particularly its call for safe abortion. Because of its unique status at the United Nations,[36] the Vatican was an official observer and very much a political player at the ICPD.[37(p503)] In 1994, representatives of the Holy See, conservative Islamic states, and several Latin American countries were unsuccessful in their attempts to resist paragraph 8.25, but the clout of this coalition thwarted calls within the ICPD forum for legalizing abortion in countries where it was illegal. At ICPD follow-ups, these conservative voices continued to worry that reproductive health services imply abortion services, and many did not support the ICPD goal of making family planning services universally available by 2015. Some analysts have argued that "the success of the progressive women's movement in securing the new framework provoked an ongoing campaign, led by the Vatican and its allies, to reverse the cumulative gains made by women." [38(p98)]

Although many analysts view ICPD as a feminist victory, not all feminists were satisfied with the Programme of Action. Betsy Hartmann, for example, questioned the alliance of feminists and population advocates or neo-Malthusians, interpreting it "as a co-optation of the interests of women to the greater force of population control." She believes that use of the term *reproductive health* merely allowed population community "to cloak their narrow objectives within a more palatable guise."[34(p947)] Some radical feminists abroad echoed her concerns, arguing that any population policy is "eugenic, racist, and sexist."[37(p505)]

Implementation of ICPD

While there are many opinions about ICPD, the Cairo conference was a truly a watershed moment in women's health. "Some have already seen it as the end of the family planning movement, an event celebrated by many feminists and women's rights and human activists as a paradigm shift and equally regretted by traditional population advocates, including many demographers and others concerned about high fertility rates as an abandonment of a decades-long commitment to population stabilization."[32(p10)]

More agreement exists that ICPD implementation has been less than optimal. The 1994 ICPD Programme of Action advocated making family planning universally available by 2015 or sooner, as part of a broader approach to reproductive rights. It also provided estimates of levels of national resources and international assistance needed to meet this goal. Over time, however, funding delivered by ICPD participants did not match the rhetoric.

Several reasons explain this shortfall. Many argue that HIV/AIDS programs supplanted reproductive health funding. Others point to conservative politics, particularly in the United States. Still others note that the broad reproductive rights agenda was parceled out to many implementers (e.g., nongovernmental organizations) so the reproductive rights agenda remained fragmented and unfocused.[31(pp145–146)]

The need to bridge current ideological divides for the benefit of the world's women and girls is clear. Over two decades ago, Ruth Dixon-Mueller noted, "There is an urgent need for a gathering of feminist forces that will (1) recognize that population growth at the global level is a problem that must be solved; (2) articulate a set of feminist principles on which a responsive population (reproductive) policy could be based; (3) build alliances with human rights advocates, development specialists, government planners, legislators, and the population/family planning community in the policy-making process; and (4) work to extend to all women the full range of sexual and reproductive rights to which they are, in principle, entitled."[27(px)]

CONCLUSION

Gender inequalities are globally pervasive. Women and girls are poorer, less educated, more prone to being victims of violence, and less likely to be politically active than their male counterparts. Throughout the world, female workers are concentrated in insecure, low-wage, and unsafe work. More often than males, they are trafficked for the purpose of sexual exploitation or forced labor.

Women's status is directly related to both reproductive health status and fertility outcomes. Poor and uneducated women are more likely than to marry early and experience negative health outcomes, such as obstetric fistula or the mortality of their offspring. Girls who are lucky enough to receive formal education are less likely to have high fertility than uneducated girls.

The notion of sexual and reproductive rights emanated from human rights discourse. Recognized as a human right in 1968, the individual right to voluntary family planning has evolved into sexual and reproductive rights, articulated at the third International Conference on Population and Development held in Cairo in 1994. Although ICPD has been hailed as a paradigm shift in population policy, criticism of its implementation persists. There remains a need to bridge ideological divides for the benefit of women and girls throughout the world.

DISCUSSION QUESTIONS

1. Discuss the status of women throughout the world. Which indicators are especially important?
2. How does the status of women affect their health and that of their children?
3. What are sexual and reproductive rights, and how have they developed?
4. What should be the relationship between sexual and reproductive rights and human population growth?
5. Discuss the ICPD from the perspectives of feminists, demographers, and conservative parties.

REFERENCES

1. United Nations Entity for Gender Equality and the Empowerment of Women. About us. 2013. http://www.un.org/womenwatch/daw/daw/index.html. Accessed March 19, 2014.
2. United Nations. World survey on the role of women in development: at a glance: women's control over economic resources and access to financial resources, fact sheet. 2009. http://www.un.org/womenwatch/daw/ws2009/documents/DESA_Survey_FactSheet.pdf.
3. Metropolitan Police Services. *Honour Based Violence: Get the Facts.* London, UK: Metropolitan Police Services; 2013. http://hbv-awareness.com/forms-of-hbv/. Accessed July 24, 2013.

4. Foerstel K. Women's rights: are violence and discrimination against women declining? *CQ Global Researcher: Exploring International Perspectives.* 2008;2(5):115–147.

5. Honour Based Violence Awareness Network. International Resource Centre. 2013. http://hbv-awareness.com. Accessed July 22, 2013.

6. Garcia-Moreno C, Guedes A, Knerr W. *Violence Against Women: Female Genital Mutilation.* Geneva, Switzerland: World Health Organization; 2012.

7. Pillsbury B, Maynard-Tucker G, Nyguen F. *Women's Empowerment and Reproductive Health: Links Throughout the Life Cycle.* New York, NY/Los Angeles, CA: United Nations Population Fund and Pacific Institute for Women's Health; 2000.

8. Bulatao R. *The Value of Family Planning Programs in Developing Countries.* Population Matters series. Santa Monica, CA: Rand; 1998. http://rand.org/.

9. Greene M, Joshi S, Robles O. *UNFPA State of the World Population 2012: By Choice, Not by Chance: Family Planning, Human Rights, and Development.* New York: United Nations Population Fund; 2012.

10. UN Women. Women, poverty, & economics. 2013. http://www.unifem.org/gender_issues/women_poverty_economics/.

11. Garcia-Moreno C, Pallitto C, Devries K, et al. *Global and Regional Estimates of Violence Against Women of Violence Against Women: Prevalence and Health Effects of Intimate Partner Violence and Non-partner Sexual Violence.* Geneva, Switzerland: World Health Organization; 2013. http://apps.who.int/iris/bitstream/10665/85239/1/9789241564625_eng.pdf.

12. World Health Organization. *Female Genital Mutilation.* Factsheet 241. Geneva, Switzerland: WHO; 2013. http://www.who.int/mediacentre/factsheets/fs241/en/.

13. World Health Organization. *Female Genital Mutilation and Other Harmful Practices.* Geneva, Switzerland: WHO; 2013. http://whqlibdoc.who.int/publications/2008/9789241596442_eng.pdf.

14. United Nations Children's Fund. 2013. *Female Genital Mutilation/Cutting: A Statistical Overview and Exploration of the Dynamics of Change.* New York: UNICEF; July 2013.

15. United Nations, Office on Drugs and Crime. *Global Report on Trafficking in Persons 2012.* Vienna, Austria/New York, NY: United Nations; 2012. http://www.unodc.org/unodc/data-and-analysis/glotip.html.

16. McCullough JJ. Female world leaders currently in power. *Fillibuster Cartoons.* 2014. http://www.filibustercartoons.com/charts_rest_female-leaders.php. Accessed March 26, 2014.

17. Inter-Parliamentary Union (IPU), Women in National Parliaments: Situation as of 1st January 2014. Geneva, Switzerland: IPU; 2014. http://www.ipu.org/wmn-e/arc/world010114.htm. Accessed March 26, 2014.

18. United Nations. *We Can End Poverty by 2015: Millennium Development Goals.* New York, NY: United Nations; 2013. http://www.un.org/millenniumgoals/.

19. Conway-Turner K, Cherrin S. *Women, Families, and Feminist Politics: A Global Exploration.* New York: Haworth Press;1998.

20. Population Reference Bureau. *The World's Women and Girls: 2011 Data Presentation.* Washington, DC: Population Reference Bureau; 2011. http://www.prb.org/Publications/Datasheets/2011/worlds-women-and-girls.aspx.

21. United Nations Children's Fund. *Child Marriage*. New York: UNICEF; November 2011. http://www.unicef.org/india/Child_Marriage_Fact_Sheet_Nov2011_final.pdf.

22. Cook RJ. International human rights and women's reproductive health. *Stud Family Plan*. 1993;24(2):73–86.

23. World Health Organization. *10 Facts on Obstetric Fistula*. Geneva: WHO; 2010. http://www.who.int/features/factfiles/obstetric_fistula/en/.

24. United Nations Population Fund. *Campaign to End Fistula: 2008. The Year in Review*. New York: UNFPA; 2009. http://www.unfpa.org/webdav/site/global/shared/documents/publications/2009/fistula_annual_report_2008.pdf.

25. Population Reference Bureau. *The World's Women and Girls: 2011 Data Sheet*. Washington, DC: Population Reference Bureau; 2011. http://www.prb.org/pdf11/worlds-women-and-girls-factsheet.pdf.

26. Population Reference Bureau. *PowerPoint Presentation: The World's Women and Girls*. Washington, DC: Population Reference Bureau; 2011. http://www.prb.org/Reports/2011/international-womens-day-ppt.aspx.

27. Dixon-Mueller R. *Population Policy and Women's Rights: Transforming Reproductive Choice*. Westport, CT: Praeger; 1993.

28. Steinem G. Q and A: interview by Marianne Schnall. *The Official Website of Author and Activist Gloria Steinem*. 2010. http://www.gloriasteinem.com/qa.

29. United Nations. *Treaty Collection: Chapter IV. Human Rights, 8. Convention on the Elimination of All Form of Discrimination Against Women*. New York: United Nations; August 4, 2013. http://treaties.un.org/Pages/ViewDetails.aspx?src=TREATY&mtdsg_no=IV-8&chapter=4&lang=en#EndDec.

30. Shaw D, Cook RJ. Applying Human Rights to Improve Access to Reproductive Health Services. *Int J of Gyn and Obst*. October 2012;119(1):S55–S59. http://www.sciencedirect.com/science/article/pii/S0020729212001592. Accessed March 26, 2014.

31. May J. *World Population Policies: Their Origin, Evolution, and Impact*. Washington, DC: Springer; 2013.

32. Sinding SW. Overview and perspective. In: *The Global Family Planning Revolution*. Washington, DC: World Bank; 2007: 1–12.

33. United Nations Population Fund. 1994. Chapter 8: health, morbidity, and mortality. In: *Programme of Action of the International Conference on Population and Development*. 1994. http://www.unfpa.org/public/home/sitemap/icpd/International-Conference-on-Population-and-Development/ICPD-Programme#ch8. Accessed November 17, 2013.

34. DeJong J. The role and limitations of the Cairo International Conference on Population and Development. *Soc Sci Med*. 2000;51:941–953.

35. Kantner JF, Kantner A. *The Struggle for International Consensus on Population and Development*. New York: Palgrave MacMillan; 2006.

36. McIntosh CA, Finkle JL. The Cairo Conference on Population and Development: a new paradigm? *Pop and Dev Rev*. June 1995;21(2):223–260.

37. Hodgson D, Watkins SC. Feminists and neo-Malthusians: past and present alliances. *Pop Develop Rev*. 1997;23(3):469–523.

38. Antrobus P. *The Global Women's Movement*. London, UK: Zed Books; 2004.

Sustainability, Population, and Environmental Degradation

John E. Becker
Deborah R. McFarlane

INTRODUCTION

Sustainability is based on a simple principle: everything that humans need for survival and well-being depends, either directly or indirectly, on the natural environment. Sustainability creates and maintains the conditions under which humans and nature can exist in productive harmony, and that permit fulfilling the social, economic, and other requirements of present and future generations.[1] *Sustainable practices* are important for ensuring the

---

Box 10–1   Lester Brown, Environmental Leader, Comments on the 20th Century
for Humanity

The twentieth century has been extraordinarily successful for the human species perhaps too successful. As our population has grown and the economy exploded since 1900, we have overwhelmed the natural systems from which we emerged and created the dangerous illusion that we no longer depend on a healthy environment.

Reproduced from Brown LR, Worldwatch Institute. *State of The World, 1998: A Worldwatch Institute Report on Progress Toward a Sustainable Society.* New York: Norton; 1998.

---

continuity of water, materials, and resources necessary to protect human health and the environment.

As Box 10-1 explains, the current relationship between humanity and the natural environment is far from sustainable.[2] Figure 10-1 shows that demographic growth greatly increased during the 20th century, a trend that continues in the 21st century.[3] The Population Division at the United Nations projects that global population will reach 8 billion by 2025, 9 billion before 2045, and 10 billion by 2085. Nearly all of that growth will occur in developing countries—that is, nations that are trying to raise their standard of living and, therefore, their resource consumption (Figure 10-2).

The increase in their numbers, along with the expanding world economy, particularly since 1950, altered the connection between people and their environment. This skewed relationship continues, accompanied by increasingly negative environmental repercussions. The main drivers of this unsustainable situation are rapid human population growth and extreme resource consumption.

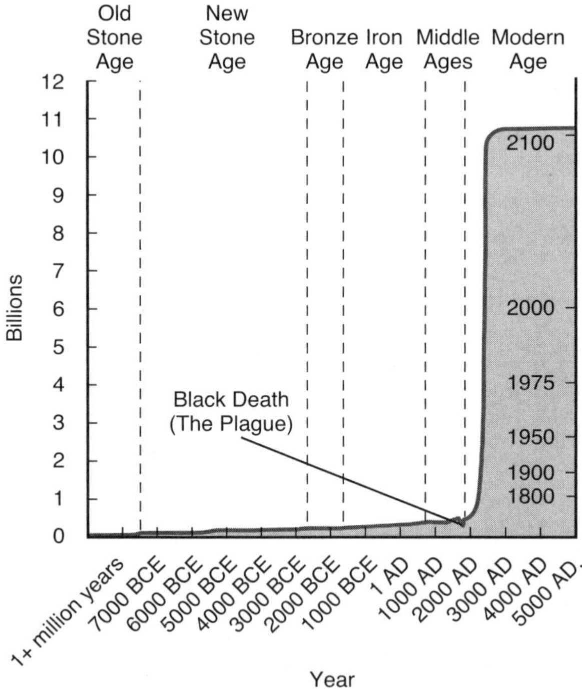

Figure 10-1 World Population Growth Through History

Reproduced from Population Reference Bureau. Graphics bank: population dynamics. http://www.prb.org/Multimedia/Graphics-Bank/PopulationTrends.aspx. Accessed January 8, 2014.

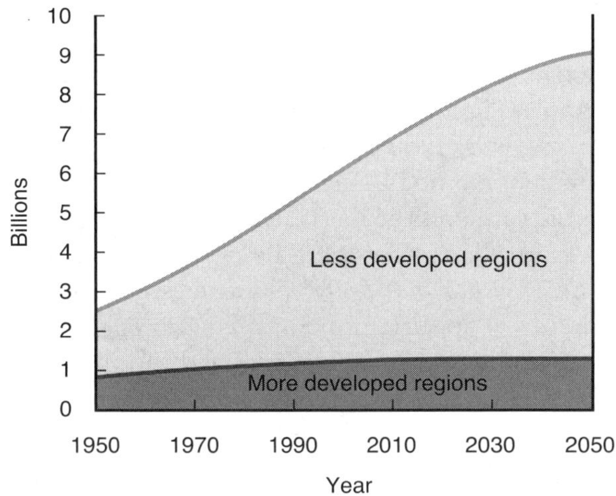

Figure 10–2 Growth in More- and Less-Developed Countries

Reproduced from Population Reference Bureau. Graphics bank: population dynamics. http://
www.prb.org/Multimedia/Graphics-Bank/PopulationTrends.aspx. Accessed January 8, 2014.

Obviously, each human being requires food, water, and air for survival. However, not every person uses resources equally. An estimated 26% of children worldwide are stunted because of malnutrition; 2 billion people suffer from one or more micronutrient deficiencies; and yet, "1.4 billion people are overweight, of whom 500 million are obese."[4(p1)]

Fossil fuel provides another example of disparities in resource use. In terms of annual per capita carbon emissions from fuel combustion, the United States leads the world with 20 tons per person, followed by Japan (10.3 tons per person), the European Union (8.3 tons per person), and China (6.3 tons per person). The corresponding rate for the average African is 1.1 tons per person.[5]

Both population growth and consumption patterns continue to degrade the earth's ecosystems, including fresh water resources, oceans, wetlands, forests, fisheries, biodiversity, the atmosphere, and even the climate. Environmental depletion may be as old as human civilization, but its current scale and velocity are unprecedented.[6,7]

In this chapter, we examine effects of population growth and human consumption on the natural environment. In doing so, we present ways of measuring human impact as well discussing global trends. We also consider approaches to change this paradigm, including sustainable development and green growth. Finally, we examine global efforts to address both population and environmental degradation and conservation, including the role of women.

MEASURING ENVIRONMENTAL IMPACT

Measuring the overall impact of human activity on the environment is anything but simple (Figure 10–3). Several approaches have been developed, and more promising work is on the horizon. In this section, we examine five of these measurement methods. *Environmental resource accounting* attempts to place an economic value on the use of environmental goods and services. *IPAT* provides a formula to "describe the overall impact of humanity on the environment."[7(p4)] *Ecological footprint analysis* estimates the resource and waste requirements of specific populations. *Carrying capacity* is a related notion, referring to the number of people earth or one of its regions can support.[7(p4)] *Population situation analysis* is a more recent method that considers specific population and environment interactions.[8]

Environmental Resource Accounting

Environmental resource accounting places a monetary value on environmental goods and services—natural resources that, conventionally, economists have regarded as free and used by all. These include unpolluted

Figure 10–3 Human Impact on Planet Earth

© mini.fini/Shutterstock.com

fresh water, clean air, ocean life, forest resources, and wetland ecosystems. Using this methodology, the total value of ecosystems and products has been estimated to exceed the total value of the global economy as conventionally measured.[7]

Some economists argue that the value of environmental goods and services should be calculated through estimates of the gross domestic product (GDP), similar to manufactured assets. This recommendation deviates from conventional practice. "Unlike manufactured capital, which depreciates in value over time, environmental capital (such as forests, fisheries, and clean air and water) is not considered to depreciate, and no charge is made against current income as it is used."[7(p4)] "A country can cut down its forests, erode its soils, pollute its aquifers and hunt its wildlife and fisheries to extinction, but its measured income is not affected as these assets disappear."[9(p94)] Therefore, environmental degradation is likely to be measured as progress.

The policy implication that accompanies environmental resource accounting is that countries and subnational units (e.g., states, provinces, regions) might use their natural resources more efficiently as well as conserve them, if they were aware of their true value.

IPAT

IPAT is an equation that attempts to describe the overall impact of humanity on the environment, emphasizing population growth. It was first unveiled in 1971 when Stanford University scientists Paul Ehrlich and John Holdren presented the idea to the President's Commission on Population Growth and the American Future (Box 10–2).[10] IPAT is useful for conceptualizing the increased demands on natural populations from growing populations (Box 10–3). When applied to the absolute decline of global fisheries, for example, conditions such as growing water pollution from terrestrial agriculture production and the overexploitation of fisheries from developed nations are partly to blame. In turn, both of these conditions are exacerbated by the disproportionate impact of population growth in developing countries.[11]

The equation is $I = P \times A \times T$, where

- I is environmental impact;
- P is population (including size, growth, and distribution);
- A is the level of affluence (consumption per capita); and
- T is the level of technology.

The IPAT model postulates that environmental impact (I) is the product of population (P), per capita affluence (A), and technology (T). In some

Box 10–2   President Richard Nixon Appoints the Commission on Population Growth and the American Future

One of the most serious challenges to human destiny in this last third of this century will be the growth of the population. Whether man's response to that challenge will be a cause for pride or despair will depend very much on what we do today. If we now begin our work in an appropriate manner, and if we continue to devote a considerable amount of attention and energy to this problem, then mankind will be able to surmount this challenge as it has surmounted so many during the long march of civilization.

—Richard M. Nixon, July 18, 1969

Reproduced from Nixon R. *Special Message to the Congress on Problems of Population Growth, July 18, 1969.* Public Papers of the Presidents, No. 271, p. 521, Office of the Federal Register, National Archives, Washington, DC, 1971.

applications, affluence has been expressed as gross national product (GNP) per capita, and technology has been used as a general variable to consider everything not previously included in population and affluence. Because of its conceptual clarity, the IPAT formula is useful for discussing population and environmental degradation.[12]

When IPAT is applied to real-world situations, complexities arise. Concern is growing that assigning equal weights to each multiplier may over simplify its relative contribution to environmental degradation. Indeed, analyses of the dynamic and complex role of growing populations suggest that environmental degradation may occur faster than population increase. One reason for the disproportional effect of population size is the exhaustion of natural processes that aid in managing pollution. For example, when forests and oceans are overburdened, they become compromised in their ability to absorb continually growing quantities of carbon dioxide (CO_2) emitted by human processes. Eventually a point may be reached where each additional contribution of CO_2 emissions causes a disproportional effect because of lack of natural resources to process the pollution. Consequently, a doubling of population size may more than double CO_2 pollution in the atmosphere given the compromised pollution sinks.[13]

Ecological Footprint Analysis

Ecological footprint analysis (EFA) was developed in 1990 by sustainability advocates Mathis Wackernagel and William Rees. EFA is an accounting tool for estimating the resource consumption and waste assimilation requirements of a defined population or economy in terms of a corresponding productive land area. Wackernagel and Rees used the analogy of a city enclosed in a glass or plastic hemisphere, which allowed light to enter, but prevented tangible materials from entering or leaving. In this scenario, the health and integrity of human systems depend on whatever was trapped in the initial hemisphere. Predictably, this imaginary human terrarium has insufficient "carrying capacity" to support the ecological load imposed by the contained population. Such a model serves as a reminder of human-kind's continuing ecological vulnerability.[14]

For Wackernagel and Rees, *ecological footprints* represent spatial indicators used to measure human demand on the earth's ecosystems often designated in "global hectares" or "planets." The footprint of a nation, for example, is constrained by the amount of land and water regionally available from nature for food, materials, and built-up areas, and for absorbing waste emissions. These indicators serve as measure of the extent to which the planet, a region, or nation is practicing sustainability. Footprints vary between countries at different stages of economic development and with varying geographic characteristics. When used for resource accounting, ecological footprints draw attention to the inequalities between nations in terms of the environmental burdens they place on the earth (Figure 10–4).

A stark divide exists between the resource consumption of developed and developing countries (Box 10–3). Acknowledging the lack of a globally established convention for the designation of "developed" versus "developing" nations, the United Nations Millennium Development Report does suggest regional development groupings (Figure 10–5).[15] "Developed" regions with slowing fertility rates tend to be located in the Northern Hemisphere, while "developing regions" with higher fertility rates are further subdivided mainly in the Southern Hemisphere. Per capita ecological footprints in the industrialized (prosperous) Northern Hemisphere, where most developed economies are located, greatly exceed those in the economically disadvantaged Southern Hemisphere. While it is expected that in the future "the South" will

---

Box 10–3   Different Consumption Patterns Across the Globe

Most *developed* economies currently consume resources much faster than they can regenerate. Most *developing* countries with rapid population growth face the urgent need to improve living standards.

Reproduced from Hinrichsen D, Robey B. Population and the environment: the global challenge. *Population Reports*. Fall 2000;XXVIII(3, Series M): 15, 1. http://www.k4health.org/toolkits/info-publications/population-and-environment-global-challenge. Accessed September 28, 2013.

Figure 10–4 Illustration of the Ecological Footprint of Mankind on the Planet

© Fernando Jose V.Soares/Shutterstock.com

grow at a faster pace as it attempts to reach the levels of prosperity seen in "the North," it is apparent that the industrialized nations have contributed greatly to existing ecological degradation due to greenhouse gas emissions lingering in the earth's atmosphere.[16]

"In 1997, as part of the five year review of environmental conditions following the Rio Earth Summit, the Earth Council of Costa Rica sponsored a major Ecological Footprints of Nations study,"[7(p4)] commissioning Mathis Wackernagel, co-founder of EFA, to be the principal investigator. This study focused on 52 nations containing 80% of the world's population and 95% of the world domestic product. The research team analyzed the amount of biologically productive areas needed to provide the resources consumed by the population and absorb their wastes, under the prevailing levels of technology. It is alarming to note that almost two decades ago, Wackernagel and his colleagues concluded that "the world's people are using about one-third more of the earth's biological productivity than can be regenerated."[7(p4)]

Carrying Capacity

Carrying capacity refers to the number of persons that the earth can support. This concept has been debated since at least 1798, when Thomas

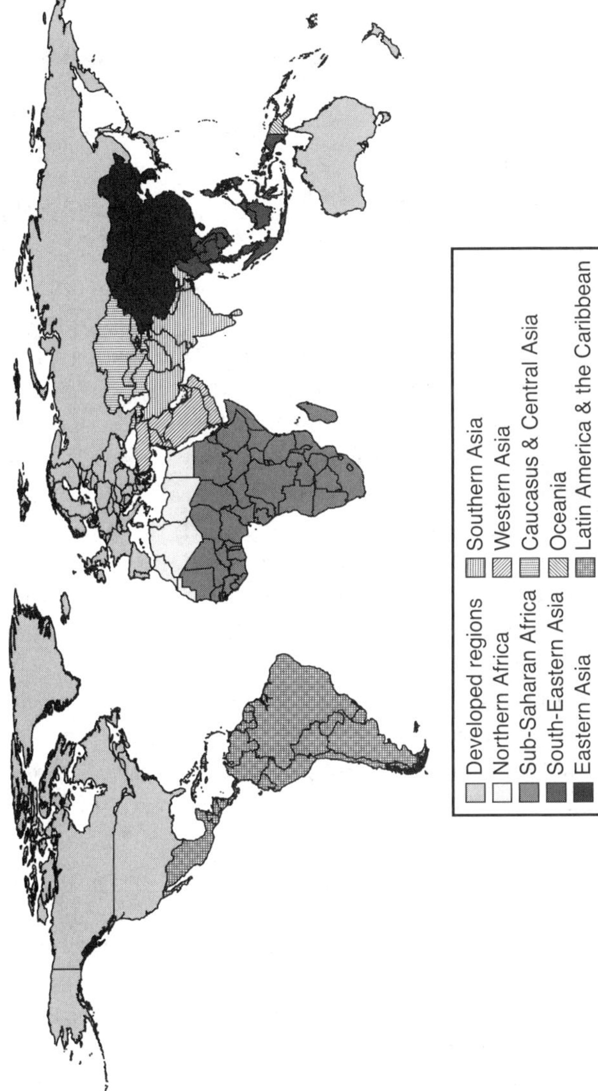

Figure 10-5 Developed Countries and Other Regional Areas

Reproduced from United Nations. *The Millennium Development Goals Report 2012.* New York: United Nations; 2012. http://www.un.org/millenniumgoals/pdf/MDG%20Report%202012.pdf. Accessed January 8, 2014.

Malthus predicted that population growth would outstrip the food supply. The logic is that population growth must stop at some point, or the earth will become overcrowded and its resources depleted. This line of reasoning leads to the question of what is the maximum possible population.[7]

Besides being a useful concept, carrying capacity can refer to the methodology of calculating how much population an area can sustain. In this respect, it is akin to EFA.[17,18]

Population Situation Analysis

An important aspect of the environment–population interaction is how the specific population inhabiting that space can rationally manage a specific area, given that there are wide variations in natural resources and topographies. Governments and local authorities can address population dynamics in a manner tailored to country-specific challenges and priorities by using population situation analysis (PSA). As endorsed by the United Nations Population Fund, PSA optimizes processing of available population data to analyze how current and projected population trends affect development. At the country level, PSA guidance provides the basis for an integrated appraisal of the population and reproductive health dynamics and their linkages to and impacts on poverty, inequality, and development.

While the links between local population dynamics and environmental degradation are not yet formalized in a way that lends itself to a simple assessment of their relationship, analyzing population dynamics can clarify who is most vulnerable to environmental changes, why they are affected, and how interventions can most effectively reach them. The PSA methodology suggests that the factors influencing population vulnerability include location, poverty, demographic characteristics, and the extent of protection provided to people by their housing, infrastructure, and social and economic support structures. Each of these factors must be considered to customize interventions.

In the case of climate change, the PSA methodology suggests an evaluation of at-risk populations by size and composition in relation to projected urbanization rates as well as the extent to which urban areas are located in places where climate change impacts are likely. Better urban planning, including slum improvements, could help mitigate greenhouse gas emissions, while also providing resilient and adaptive environments to reduce vulnerability, particularly for impoverished urban dwellers.

One PSA application considered the problem of deforestation due to increased demand for cooking fuel in urban slums within the Democratic Republic of the Congo (DRC), Tanzania, and Kenya. Two major population interventions that could potentially break this vicious cycle emerged from the PSA analysis. The first was to promote family planning in rural areas, particularly rural areas that experience environmental stress. The second was to accept a certain amount of rural–urban migration as inevitable and then to plan for it.[8]

EXPLORING THE ENVIRONMENTAL CONSEQUENCES

This section explores selected natural resources and processes that are particularly at risk of degradation due to population growth and unbridled consumption.

Water Resources

Roughly 2.5% of all of the earth's water is fresh water, the majority of which is relatively inaccessible because it is underground or part of large ice and snow deposits. In fact, the amount of surface freshwater in lakes and rivers, and therefore readily obtainable to serve most of humanity's water needs, is less than 0.0002% of the total water reserves on earth.[19] Given current energy, economic, and technological constraints, desalination of water remains a relatively inaccessible option. As human populations grow, less renewable water is available for each person.

Current estimates and projections indicate that water stress and scarcity issues are already prevalent in much of Northern and Eastern Africa, as well as in Western Asia (Box 10–4 and Figure 10–6).[20] Alarmingly, it is estimated that by 2025, between 2.4 billion and 3.2 billion people may have difficulty obtaining sources of freshwater because of depleting resources. These projections do not account for water quality concerns or irregularities in availability due to climactic change—conditions that may further complicate water supplies.[21] The magnitude of climate change will vary from one region to another, but because of variability in precipitation, semi-arid regions, such as those in East Africa, are expected to have more frequent periods of drought.

Box 10–4   Water Scarcity

"Water scarcity is among the main problems to be faced by many societies in the XXIst century. Water use has been growing at more than twice the rate of population increase in the last century," ...and "as an increasing number of regions are chronically short of water."

United Nations Department of Economic and Social Affairs, International Decade for Action, 'Water for Life' 2005–2015, Water Scarcity. http://www.un.org/waterforlifedecade/scarcity.shtml. Accessed January 23, 2014.

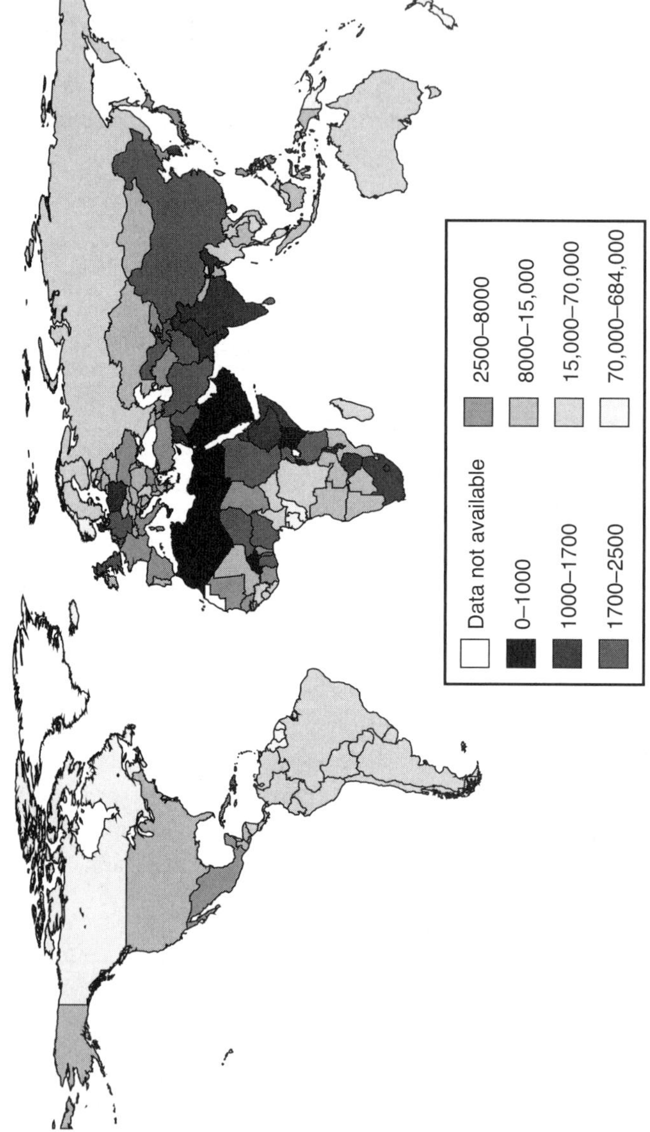

Figure 10-6 Fresh Water Availability per Capita by Country, 2007

Reproduced from United Nations Environment Programme (UNEP). *Vital Water Graphics*. New York: UNEP; 2008. http://www.unep.org/dewa/vitalwater/article192.html. Accessed January 8, 2014.

Central to water scarcity issues is how populations use water. While population growth itself is a leading driver of fresh water scarcity, urbanization is another important factor. Growing cities concentrate the demand for water within smaller areas. Asian cities alone are expected to grow by 1 billion people in the next 20 years. Exacerbating matters, per capita domestic water use is expected to rise as the world becomes more developed.

Not only is water in short supply, but much of what is available is not safe to drink—that is, it is not *potable*. In 2011, approximately 768 million people lacked access to an improved source of water.[21] Eighty-three percent of the population without access to an improved drinking water source (636 million people) live in rural areas; more than 40%of all people without improved drinking water live in sub-Saharan Africa.[22] (An *improved drinking water source* is one that is constructed to adequately protect the water source from outside contamination, in particular from fecal matter.[23]) Moreover, concerns about the quality and safety of many improved drinking water sources persist. As a result, the number of people without access to safe drinking water may be two to three times higher than official estimates.[21]

Most people around the world aspire to have piped drinking water supplies on their premises. Yet 38% of the 6.2 billion people globally using an improved drinking water source do not enjoy the convenience and associated health and economic benefits of piped drinking water at home. Instead, they spend valuable time and energy queuing up at streams or public water supplies and carrying heavy loads of water home, with those supplies often sufficient only to meet minimal drinking water needs.

Women and girls in rural areas are most affected by the inaccessibility of drinking water and lack of sanitation (Figure 10–7). For the most part, they bear the responsibility for traveling long distances to water sources to fetch the water necessary for their families' drinking, bathing, cooking, and other household needs. Additionally, women face the challenge of maintaining basic household hygiene and caring for sick children suffering from diarrheal disease borne out of contact with unsanitary water. A lack of clean water and sanitation facilities aggravates an already vicious poverty cycle by robbing girls of educational opportunities, causing disease and malnutrition, and ultimately reducing lifelong productivity.[24,25]

Notably, the greatest demand for water comes from relatively inefficient uses associated with crop production. Globally, of all the water that is extracted for human purposes, an estimated 70% is used in agriculture. This figure actually rises to 90% in some fast-growing economies.[26] Indeed, to produce enough food to satisfy a single person's daily diet requires about 2000 to 5000 liters of water.[27] In contrast, only 2 to 4 liters is required for

Figure 10-7 Women Carrying Water in Ethiopia

The daily effort to secure sanitary water disproportionately affects women in developing nations.

© Stockbyte/Thinkstock

drinking purposes, and 20 to 300 liters is sufficient for domestic needs. Farmers, in particular, face the challenge of accessing this increasingly scarce resource to meet their food production needs.

Related to increased demands for water for agriculture is the diminished availability of water for natural systems. Water that supports ecosystems is threatened by reductions in water flows and water quality standards. Fully half of the world's wetlands disappeared during the 20th century, many rivers no longer reach the sea, and fish species are endangered.[20,7]

Further exacerbating the impact of population growth and economic development are the global changes in dietary preferences. With rising incomes in many regions, food habits are changing toward richer and more varied diets. This change in consumption patterns shifts more agricultural production away from cereal crops and toward livestock, fish, and high-value crops that consume even more water.[28]

Land Use and Food Production

Poverty, food production, and environmental degradation are inextricably intertwined. From 1975 to 2000, the number of people living in countries scarce in per capita cropland grew from 175 million to 420 million, a statistic that will continue to increase as national populations grow. Here the dividing line between sufficient and scarce cropland is the amount of land necessary to feed one person a vegetarian diet (without relying on chemical inputs such as pesticides). The Food and Agriculture Organization of the United Nations (FAO) has documented that much of sub-Saharan Africa already lacks adequate nourishment for more than 35% of the population, creating countries that are dependent on food inputs from outside their borders.[24]

Traditionally, increased food production has been made possible by converting more land to agricultural production; however, most of the world's prime agricultural land is already in production. Furthermore, each year additional prime agricultural land is lost to urban uses or degraded through imprudent agricultural methods leading to erosion, salinization, leaching of nutrients, and increased toxicity from the use of chemical inputs. Fulfilling nutritional requirements for growing human populations requires changes in land use. These changes can be achieved through forest clearing and intensifying production on already cultivated land.

During the past three centuries, the amount of earth's cultivated land has grown by more than 450%, which has caused dramatic deforestation throughout the globe.[29] This phenomenon has only intensified in recent decades. "Globally, about 16 million hectares of forest , an area roughly the size of Nepal, are cut, bulldozed, or burned each year"[7(p3)] Beef and soya production, in particular, threaten the Amazon rainforest (Figure 10–8). "Between 1995 and 2006, cropland in the Brazilian Amazon more than tripled, and cattle herds increased by nearly 80%."[30]

Forests

As the global population grows and forests retreat, the ratio of forest land to human beings has declined significantly. This per capita decline is

Figure 10–8 Agricultural Conversion of Amazon Rainforest from the Air
© Phototreat/iStockphoto.com

important because forests play an essential role not only in providing products, but also in providing ecosystem services, including natural carbon capture necessary to mitigate climate change. Remarkably, forests harbor more than half of the world's plant and animal species, and they provide many with renewable sources of energy. [7,31]

The FAO's Global Forest Resources Assessment[32] indicates that although 130 million hectares of forest lands were lost between 2000 and 2010, approximately 78 million hectares were recovered by active resource management in the form of planted forests and natural forest expansion, including in many of the world's most forested regions such as in the Russian Federation, the United States, China, India, and Europe. Nevertheless, losses in Latin America and in Africa outweighed the gains over the past two decades, with about 10% of all forest land in Africa being converted to other uses in the 20-year period between 1990 and 2010, mainly for agriculture, animal grazing, fuel, and production of forest products.

While the need to feed and provide for a growing population is a key driver in forest losses, it is widely believed that active management can make a difference in compensating for the failure of markets to capture all of the important ecosystem services provided by forests. Market-based schemes can bridge the gap between industrial consumption of timber and consideration of the public good generated by forests in climate regulation and water resource conservation.[33] One example of such a plan is a national program in Costa Rica that compensates farmers to protect biodiverse forests.[34] As result of its efforts, Costa Rica has managed to increase its forest cover in recent decades (Box 10–5).[33]

Box 10–5   Costa Rica: A Leader in Sustainable Practice and Policy

Reforestation and sustainable timber production has long been a focus in Costa Rica, and governmental incentives for investing in such projects has been provided. For example, a forest preservation project may receive up to $300 per year per hectare; a reforestation project is eligible for $150 per year per hectare. Due to this political prioritization, the forest cover in Costa Rica has grown from a 21% low in 1987 to an impressive 52% in 2005.

Reproduced from the United Nations Environmental Programme (UNEP). *Forests: Costa Rica.* New York: UNEP. http://www.unep.org/forests/Portals/142/docs/Costa_Rica.pdf. Accessed October 6, 2013.

Fisheries

Fish are among the world's most important renewable resources and traditionally account for a relatively inexpensive and vital portion of the growing coastal population's animal protein intake and health. However, while the production of fish continues to grow roughly at the rate of human population, these increases are coming almost entirely from the rapid expansion of aquaculture—that is, the business of farming aquatic plants and animals. In 2010, total production of fish, crustaceans, mollusks, and other aquatic animals reached 148.5 million tons and continues to increase, while capture production has remained steady around 90 million tons since 2001. This trend away from capture fisheries is especially pronounced in the production of freshwater fish: aquaculture production in 2012 accounted for 56.4% of the total freshwater fishes consumed by humans. Not surprisingly, the world's fishing fleet, which comprises approximately 4.4 million vessels, has remained relatively stable since 1998,[35] with fishing fleets having to travel farther and farther to yield decreasing catches in both quantity and size of the fish.

To preserve fisheries, policy makers must acknowledge the limits of what the oceans can provide as well as the fact that rebuilding depleted fish populations is necessary to maximize current and future yields. To this end, essential marine habitats must be protected and fishing activities arranged to minimize pollution and overall impacts on the environment. Many fish stocks are now classified as overexploited or depleted, and the FAO warns that if action is not taken to address current fishing trends, only half the amount of fish available in 1970 will be available by 2015 and one-third in 2050. Aquaculture also has limits as a source of cost-efficient protein, with production growth rates peaking in the 1980s and slowly decreasing through the 2000s.[35,36]

Global marine fisheries are failing not only at the social level, but also economically. Existing global and national subsidies for fishing and fuel greatly contribute to overfishing and depletion of fish stocks. In the short term, greening efforts may require reorienting public spending while reducing the excess capacity of fishing fleets and temporary relocation of employment that relies on fishing until at-risk marine ecosystems can recover. Such efforts must take place not only in developing nations that rely economically on small-scale fishing, but also in nations that operate disproportionately larger fleets. Measured by fleet capacity, Russia, Japan, China, Taiwan, and the United States are the top five fishing countries.[35]

Compounding the effects of shortfalls in wild catch are the increasing costs of land, fresh water, and antibiotics used to raise healthy stocks of farm-raised fish. Indeed, farm-raised fish are too costly to be counted on as a major food source for many in low-income countries. The predictable shortfall in fish supplies in the future will likely affect developing nations disproportionally, as higher process costs are likely to encourage exports to developed nations at the expense of local consumption by lower-income populations.[36] As is the case with other environmental consequences mentioned in this chapter, long-term solutions require innovative international conservation efforts.

Energy

Transforming the way that energy is supplied, transformed, delivered, and used by growing populations is central to the issues of environmental protection. Existing energy systems are key contributors to climate change, representing approximately 60% of total greenhouse gas emissions. Emissions from the combustion of fossil fuels are major contributors to the unpredictable effects of climate change as well as to urban air pollution

Figure 10–9 Wind Turbine Generators: A Sustainable Energy Alternative to Fossil Fuel

© Brejeq/iStock/Thinkstock

and acidification of land and water. As the increasing global population prompts the global economy to double in size over the next two decades, there will be a commensurate increase in energy supply given current inefficient conversion and use of energy. Incentivizing modern energy supply systems with reduced greenhouse gas emissions, as well as increased efficiency by end users, is critical for reducing the risk of irreversible, catastrophic climate change[37] (Figure 10–9).

Developing countries in particular must modernize their existing energy systems to address the economic needs of the world's poor, which threaten to jeopardize the achievement of global development goals, including those related to environmental sustainability. The United Nations Secretary-General's Advisory Group on Energy and Climate Change (AGECC) estimates that 3 billion people rely on relatively inefficient traditional biomass for cooking and heating, with about 1.5 billion lacking any access to electricity. In the coming decades, lower-income countries will need to expand their energy services substantially to meet the needs of the several billion people whose current reliance on traditional biomass is dictated by inadequate and unreliable access to energy services.[37]

Climate Change

Given that climate has a profound influence on life on earth, the prospect of human-induced climate change is a matter of great concern. The components of the earth's climate system are complex and inherently chaotic; in turn, predicting changes in climate systems is particularly challenging. Notably, the ability of climate scientists to predict future impacts of human activity on climate change is constrained by their ability to predict population change, economic change, technological development, and other relevant characteristics of future human activity.

However, increasingly sophisticated climate models are enhancing the ability of climate scientists to account for numerous variables in the global ecosystem. The contribution of large-scale human activities to the emission of greenhouse gases, along with widespread changes in land use, indicate that human activities have impacted and are highly likely to continue to affect global climate.[38]

The Intergovernmental Panel on Climate Change (IPCC) reports that prior to the beginning of the Industrial Revolution in the mid-18th century, the amount of greenhouse gases in the atmosphere remained relatively constant. During the unprecedented economic and population growth that followed, the concentration of various greenhouse gases—most notably carbon dioxide (CO_2)—increased significantly. In large part because of the combustion of fossil fuels and deforestation, CO_2 concentrations in the atmosphere have increased 30% since pre-industrial times, and they continue to increase at an unprecedented rate of, on average, 0.4% per year. The increase in greenhouse gas and aerosol concentrations in the atmosphere, along with land-use changes, affects processes and feedbacks in the earth's climate system.[39]

BIOLOGICAL DIVERSITY

Among the most troubling of the ecological consequences related to population growth is loss of biological diversity. Human-imposed alterations of the earth's ecosystems are causing accelerated losses of both plant and animal species. The growth of the human population, combined with the associated increased demand in resources required for human enterprises, plays a significant role in the transformation of land surfaces, changes in major biogeochemical cycles, and the addition or removal of species from earth's ecosystems. These pervasive human influences lead to major alterations in the functioning of the earth system. Among other consequences, human activity is effecting irreversible losses of biological diversity through the extinction of particular species as well as the overall loss of ecosystems.[40]

At the beginning of this millennium, it was estimated that plant and animal species were disappearing at rates 1000 times higher than occurred in the prehuman past. Conservative estimates project that roughly 10 million species on the earth today may become extinct during the next few centuries, largely due to the unprecedented expansion of human activities and the human population. In 1995, about 1.1 billion people were concentrated in 25 "biodiversity hotspots," rich in concentrations of unique species. Notably, these human populations grew at a pace almost 40% faster than the world's population as a whole. Additionally, three major tropical wilderness areas were home to 75 million people, a population that grew at more than twice the rate for the world as a whole. The high threat of human enterprises concentrated in areas of high potential impact accounts for the detrimental impact on biological diversity.[7]

The young mountain gorilla shown in Figure 10–10 is a member of a critically endangered species. "Since the discovery of the mountain gorilla subspecies in 1902, its population has endured years of war, hunting, habitat destruction, and disease—threats so severe that it was once thought the species might be extinct by the end of the twentieth century." Today, there are only about 880 mountain gorillas remaining on earth.[41]

The issues discussed in this chapter do not comprise a comprehensive inventory of the environmental degradation that is occurring globally. Other

Figure 10–10 Young Mountain Gorilla (Critically Endangered Species)

© ajber/iStock/Thinkstock

serious issues exist as well, such as air pollution, hazardous wastes, water pollution, and food insecurity. All of these issues are affected by population growth, unsustainable resource consumption, and economic development. Addressing these complex problems in meaningful ways will require collective action and international cooperation.

TOWARD MEETING THE NEEDS OF THE FUTURE: SUSTAINABLE DEVELOPMENT

Development experts, national leaders, and others increasingly recognize that the destruction of natural resources "to meet current needs or to make a quick profit is short-sighted and potentially disastrous for future generations."[7(p4)] The alternative, "meeting people's current needs while preserving nature's capacity for the future," is *sustainable development*. The concept of sustainable development rejects the "view that economic development is a necessity, but environmental protection is a luxury."[7(p4)] Advocates of sustainable development argue that economic development and environmental protection are inextricably linked.[7(p4)]

Sustainable development is not a new concept. The idea of living within the earth's natural limits was discussed even before Malthus's famous *Essay on Population* was published in 1798. In recent decades, the trends of increasing population, industrialization, and environmental degradation have given sustainable development new traction. For example, the 1969 National Environmental Policy Act introduced sustainable development into the U.S. regulatory code.[42]

At the global level, the 1987 United Nations World Commission on the Environment and Development, chaired by Norwegian Prime Minister Gro Harlem Brundtland, adopted a strategy of integrating conservation and development globally. Its seminal report, *Our Common Future*, defined sustainable development as "development that meets the needs of the present without compromising the ability of future generations to meet their own needs." This definition implies a concern for social equity both in the present (intragenerational needs) and in the future (intergenerational needs).[43] The Brundtland Report reinforced the principle that environmental, economic, and societal needs must be considered in economic development decisions. This triad of needs is often called the "triple bottom line" and describes the balance necessary to achieve sustainable outcomes.

United Nations Environmental Conferences

The 1992 United Nations Conference on Environment and Development (known as the "Earth Summit"; Figure 10–11) held in Rio de Janeiro, Brazil, provided a forum for national leaders to endorse an action plan for sustainability. Known as Agenda 21, this plan presented 27 principles for sustain-

Figure 10–11 Earth Summit Illustration by Terrance Cummings
Courtesy of United Nations

able development called the Rio Declaration. The strong linkages that exist between population dynamics and sustainable development are reflected in Principle 8 of the Rio Declaration: "To achieve sustainable development and a higher quality of life for all people, States should reduce and eliminate unsustainable patterns of production and consumption and promote appropriate demographic policies."[44]

Other United Nations conferences addressing sustainable development followed the Earth Summit in Rio de Janeiro. In June 1997, Agenda 21 from the Earth Summit was reaffirmed as the fundamental program of action for achieving sustainable development, and signatories recommitted to working together to meet equitably the needs of present and future generations.[45] In 2000, world leaders adopted the Millennium Declaration, committing themselves to combat poverty, hunger, disease, illiteracy, environmental degradation, and discrimination against women. The 2002 United Nations Summit on Sustainable Development, held in Johannesburg, South Africa, reaffirmed commitment to the Rio principles, the implementation of Agenda 21, and internationally agreed-upon development goals such as the Millennium Declaration. At the 2012 United Nations Conference on Sustainable Development (Rio+20) held in Rio de Janeiro, leaders from 193 member states approved a nonbinding document, *The Future We Want*, which provides general guidance to shape policies promoting global prosperity, reducing poverty, and advancing social equity and environmental protection.

Green Economy

The term *green economy* has recently entered the mainstream policy discourse as a new economic paradigm in the context of sustainable development and poverty eradication. A green economy is one where material wealth is not delivered at the expense of growing risks, ecological scarcities, and social disparities. Rather, it encourages incentives and policies that level the playing field for products that inflict reduced harm upon ecosystems. Similarly, focus is placed on eliminating harmful subsidy systems that may foster short-term economic welfare at the expense of achieving sustainable objectives. A 2011 United Nations Environment Program (UNEP) report, *Towards a Green Economy*, laid the foundation for policy makers to transition toward a green economy that can enable economic growth and investment while increasing environmental quality and social inclusiveness.[46]

As noted earlier, the world population is expected to exceed 9 billion by midcentury, thus raising the stakes for addressing the dual concerns of poverty and the depletion of increasingly scarce natural resources. Given the disproportionate effects of environmental degradation on the world's least-developed countries, these nations actually have the most to gain from a transition to a green economy. As a group, the least-developed countries contribute less than 1% of the world's total greenhouse gas emissions, yet their economic reliance on natural resources and climate-sensitive economic sectors render them disproportionately vulnerable to the effects of changing climate conditions.[47]

Women and Sustainable Development

"Despite their key role as suppliers of food, fuel, and water for household subsistence," economic development programs have often ignored women.[7(p4)] This exclusion is a serious omission, one that compromises the overall success of sustainable development.[7(p4)] "Women are the main users of natural resources from their immediate environment and therefore have a direct stake in preserving it."[7(p4)] Moreover, in developing countries, women are heads of more than one-third of all households.[7(p4)]

Women have led indigenous environmental movements in Brazil, Costa Rica, Kenya, and the Philippines. The best known is the late environmental and political activist, Wangari Maathai of Kenya (Figure 10–12). Founder of

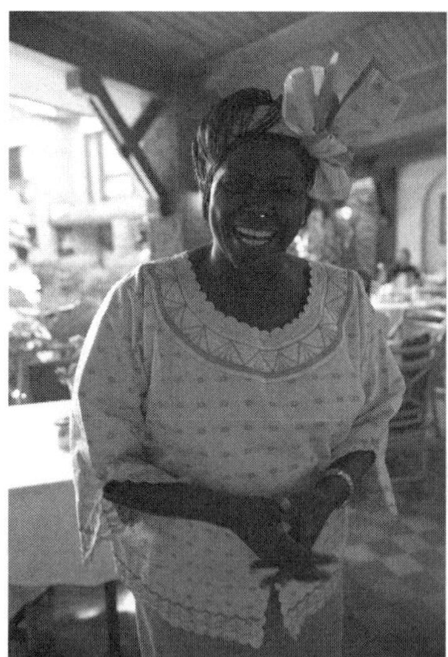

Figure 10–12 Wangari Maathai (1940–2011), Environmental Activist
© spirit of america/Shutterstock.com

the Green Belt Movement, she organized women to plant 30 million trees in a country that had lost 90% of its forests from 1950 to 2000.[48] Awarded the Nobel Peace Prize in 2004, Maathai was the first African woman to receive this distinction. The Norwegian Nobel Committee stated unequivocally, "Peace on earth depends on our ability to secure our living environment."[49] Box 10–6 discusses Maathai's achievements.

International agencies concerned with environmental protection are increasingly recognizing women and taking their economic needs into account. To maximize women's contributions to sustainability, women need to be equal partners in planning and decision making—not just passive beneficiaries. It is worth noting that reproductive health and family planning programs throughout the world have many female village health workers and managers, with large reservoirs of social change knowledge that could be applied to environmental issues.[7(p4)]

Box 10–6   Wangari Maathai: Nobel Prize for Green Belt Movement

The Nobel Peace Prize was awarded to Wangari Maathai "for her contribution to sustainable development, democracy and peace." The announcement of the prize proclaimed that "Maathai stands at the front of the fight to promote ecologically viable social, economic and cultural development in Kenya and Africa. She has taken a holistic approach to sustainable development that embraces democracy, human rights and women's rights in particular. She thinks globally and acts locally."[48]

Maathai's actions demonstrated recognition of the inextricable connection between environmental degradation and poverty. As a biologist, Maathai recognized the problems that deforestation and soil erosion were causing in rural areas, especially for the women who do most of the physical work. Women were required to go farther and farther in search of resources such as wood for cooking. In 1977, Maathai founded the Green Belt Movement with the aim of restoring Africa's forests and thereby putting an end to the poverty that deforestation was causing.[49]

Overcoming the political obstacles presented by a previously oppressive political regime in Kenya, Professor Maathai's Green Belt Movement mobilized thousands of poor women in an effort to plant more than 30 million trees over the period of nearly 30 years. Through education, family planning, nutrition, and the fight against corruption, the Green Belt Movement paved the way for development at the grass-roots level, in turn revitalizing exploited forest lands while encouraging women to better their social and economic standing.[49]

CONCLUSION

Sustainable practices are important for ensuring the continuity of water, forests, and other resources necessary to protect human health and the environment. Dramatic population growth and economic development, particularly in the 20th century, has skewed the balance between humanity and natural resources. The negative impact of this imbalance can be measured in several ways, including ecological footprints and natural resource accounting. Human populations are still growing, and environmental degradation on many fronts (e.g., water scarcity, climate change, loss of species) continues. In some cases, these processes are irreversible. Nevertheless, significant global resources and policies are trying to alter these trends and promote sustainable development. Increasing opportunities for women in promoting sustainable practices have shown positive results.

DISCUSSION QUESTIONS

1. How do developed and developing countries impact the environment in different ways?
2. What is sustainable development, and how can it be measured?
3. What is the role of population growth in environmental degradation? Is it more or less important than economic development?
4. What is required to reverse global environmental degradation?
5. How can women in the developing world contribute to sustainable development?

REFERENCES

1. United States Environmental Protection Agency. What is sustainability? What is EPA doing? How can I help? http://www.epa.gov/sustainability/basicinfo .htm#sustainability.
2. Brown LR, Worldwatch Institute. *State of the World, 1998: A Worldwatch Institute Report on Progress Toward a Sustainable Society.* New York: Norton; 1998.
3. Population Reference Bureau. Graphics bank: population dynamics. http:// www.prb.org/Multimedia/Graphics-Bank/PopulationTrends.aspx. Accessed September 28, 2013.
4. Food and Agriculture Organization of the United Nations (FAO). *The State of Food and Agriculture 2013: Executive Summary.* Rome, Italy: FAO; 2013. http://www .fao.org/docrep/018/i3301e/i3301e.pdf. Accessed September 28, 2013.
5. Fu-Bertaux X. *Carbon Dioxide Emissions and Concentrations on the Rise as Kyoto Era Fades. Vital Signs: Global Trends That Shape Our Future.* Washington, DC: Worldwatch Institute; April 26, 2012. http://vitalsigns.worldwatch.org/vs-trend/carbon -dioxide-emissions-and-concentrations-rise-kyoto-era-fades. Accessed September 28, 2013.

6. Ehrlich PR, Erlich AH. Symposium on population law: the population explosion: why we should care and what we should do about it. *Environ Law.* 1997;27:1187–1373.

7. Hinrichsen D, Robey B. Population and the environment: the global challenge. Population Reports Fall 2000;XXVIII(3, Series M):15. http://www.k4health.org/toolkits/info-publications/population-and-environment-global-challenge. Accessed September 28, 2013.

8. United Nations Population Fund (UNFPA). Population situation analysis (PSA): a conceptual and methodological guide. 2010. http://www.unfpa.org/webdav/site/global/shared/documents/publications/2011/PSA_Guide.pdf. Accessed September 29, 2013.

9. Repetto R. Accounting for environmental assets. *Sci Am.* 1992;266(6):94–100.

10. United States Commission on Population Growth and the American Future. *The Report of the Commission on Population Growth and the American Future* [Advance copy]. New York: Signet Special from New American Library; 1972.

11. Ehrlich PR, Holdren JP. Impact on population growth. *Science.* 1971;171(3977):1212–1217.

12. Fischer-Kowalski M, Amann C. Beyond IPAT and Kuznets curves: globalization as a vital factor in analyzing the environmental impact of socio-economic metabolism. *Pop Environ.* 2001;23(1):7–47.

13. Harte J. Human population as a dynamic factor in environmental degradation. *Pop Environ.* 2007;28(4–5):223–236.

14. Wackernagel M, Rees WE. *Our Ecological Footprint: Reducing Human Impact on the Earth.* 6th ed. Gabriola Island, BC: New Society Publisher; 1998.

15. United Nations. *The Millennium Development Goals Report 2012.* New York: United Nations; 2012:69. http://www.un.org/millenniumgoals/pdf/MDG%20Report%202012.pdf. Accessed September 28, 2013.

16. Cranston GR, Hammond GP. North and south: regional footprints on the transition pathway towards a low carbon, global economy. *Appl Energy.* 2010;87(9):2945–2951.

17. Weeks JR. *Population: An Introduction to Concepts and Issues.* Belmont, CA: Wadsworth; 2012.

18. Rees WE. Revisiting carrying capacity. *Pop Environ.* 1996;17(3):195–215. http://link.springer.com/article/10.1007/BF02208489#page-1.

19. Shiklomanov IA. World fresh water resources. In: Gleick PH, Pacific Institute for Studies in Development Environment and Security, Stockholm Environment Institute, eds. *Water in Crisis: A Guide to the World's Fresh Water Resources.* New York: Oxford University Press; 1993:xxiv.

20. United Nations Environment Programme (UNEP). *Vital Water Graphics.* New York: UNEP; 2008. http://www.unep.org/dewa/vitalwater/article192.html.

21. United Nations. *Millennium Development Goals Report 2013.* New York: United Nations; 2013. http://www.un.org/millenniumgoals/pdf/report-2013/mdg-report-2013-english.pdf. Accessed October 6, 2013.

22. United Nations. Goal 7: ensure environmental sustainability. In: *We Can End Poverty: Millennium Development Goals and Beyond 2015.* New York: United Nations. http://www.un.org/millenniumgoals/environ.shtml.

23. World Health Organization. Fast facts: 2013 JMP report: progress on sanitation and drinking-water. http://www.who.int/water_sanitation_health/publications/2013/jmp_fast_facts/en/index.html. Accessed October 7, 2013.

24. Engelman R, Population Action International. *People in the Balance: Population and Natural Resources at the Turn of the Millennium.* Washington, DC: Population Action International; 2000.

25. World Health Organization, UNICEF. *Water for Life: Making It Happen.* Geneva, Switzerland: WHO and UNICEF; 2005. http://www.who.int/water_sanitation_health/monitoring/jmp2005/en/.

26. United Nations Water. Water statistics: graphs & maps. 2013. http://www.unwater.org/statistics_sec.html. Accessed May 3, 2013.

27. United Nations Water. Factsheet on water and food. 2013. http://www.unwater.org/downloads/water_for_food.pdf. Accessed May 17, 2013.

28. United Nations Water, Food and Agriculture Organization (FAO). *Coping with Water Scarcity: Challenge of the Twenty-First Century.* 2007.

29. Hunter LM, Labor and Population Program, Population Matters (Project). *The Environmental Implications of Population Dynamics.* Santa Monica, CA: Rand; 2000.

30. World Wildlife Fund. Sustainable agriculture. http://worldwildlife.org/industries/sustainable-agriculture. Accessed October 12, 2013.

31. Meyerson FAB, Population Growth and Deforestation: A Critical and Complex Relationship. Washington DC: Population Reference Bureau; 2004. http://www.prb.org/Publications/Articles/2004/PopulationGrowthandDeforestationACriticalandComplexRelationship.aspx. Accessed March 22, 2014.

32. United Nations, Food and Agriculture Organization (FAO). *Global Forest Resources Assessment 2010: Main Report.* FAO Forestry Paper 163. Rome, Italy: FAO. http://www.fao.org/docrep/013/i1757e/i1757e.pdf. Accessed October 6, 2013.

33. United Nations Environment Programme (UNEP). *Annual Report 2011.* New York: UNEP. http://www.un.org/ru/publications/pdfs/unep_annual_report_2011.pdf. Accessed October 6, 2013.

34. United Nations Environmental Programme (UNEP). *Forests: Costa Rica.* New York: UNEP. http://www.unep.org/forests/Portals/142/docs/Costa_Rica.pdf.

35. United Nations, Food and Agriculture Organization (FAO). *FAO Yearbook: Fishery and Aquaculture Statistics 2010.* Rome: FAO; 2011.

36. World Resources Institute. Decline in fish stocks. 1998–1999. http://www.wri.org/publication/content/8385. Accessed May 17, 2013.

37. United Nations, Secretary General's Advisory Group on Energy and Climate Change (AGECC). *Energy for a Sustainable Future: Summary Report and Recommendations.* New York: United Nations; April 2010.

38. Intergovernmental Panel for Climate Change (IPCC). *Climate Change 2007: Impacts, Adaptation and Vulnerability. Contribution of Working Group II to the Fourth Assessment Report of the Intergovernmental Panel on Climate Change.* Cambridge, UK: Cambridge University Press; 2007.

39. Houghton, JT, Intergovernmental Panel on Climate Change. Working Group I, *Climate Change 2001: The Scientific Basis: Contribution of Working Group I to the Third Assessment Report of the Intergovernmental Panel on Climate Change;* 2001.

40. Vitousek PM, Mooney HA, Lubchenco J, Melillo JM. Human domination of Earth's ecosystems. *Science.* 1997;277(5325):494–499.
41. World Wildlife Fund. Mountain gorilla: overview. http://worldwildlife.org/species/mountain-gorilla. Accessed October 10, 2013.
42. National Research Council. Committee on Incorporating Sustainability in the US Environmental Protection Agency. *Sustainability and the US EPA.* Washington, DC: National Academies Press; 2011.
43. World Commission on Environment and Development (WCED). *Our Common Future.* New York: WCED; 1987.
44. United Nations. *Report of the United Nations Conference on Environment and Development (Rio de Janeiro, 3–14 June 1992). Annex I, Rio Declaration on Environment and Development.* Geneva, Switzerland: United Nations; 1992.
45. United Nations. *Programme for the Further Implementation of Agenda 21: Adopted by the Special Session of the United Nations General Assembly, 23–27 June 1997, New York.* New York: United Nations Department of Public Information; 1997.
46. United Nations Environment Programme (UNEP). *Towards a Green Economy: Pathways to Sustainable Development and Poverty Eradication.* New York: UNEP; 2011.
47. United Nations Conference on Trade and Development (UNCTAD). *The Least Developed Countries Report 2010: Towards a New International Development Architecture for LDCs.* New York/Geneva, Switzerland: United Nations; 2010.
48. Mjøs OD, Chairman of the Norwegian Nobel Committee. The Nobel Peace Prize 2004: award ceremony speech. Oslo, Norway: December 10, 2004. http://www.nobelprize.org/nobel_prizes/peace/laureates/2004/presentation-speech.html. Accessed January 13, 2013.
49. Nobel Prize. Nobel Peace Prize for 2004. Oslo, Norway: October 8, 2004. http://www.nobelprize.org/nobel_prizes/peace/laureates/2004/press.html. Accessed October 12, 2013.

Chapter 11

Climate Change, Population, and Reproductive Health

Karen Hardee

INTRODUCTION

Climate change and its effects on natural and human systems are becoming an increasingly critical global issue. The potential effects of climate change stem from an accumulation of greenhouse gas emissions that trap heat in the atmosphere, and the associated global warming. Over the past century, between 1906 and 2005, "global average surface temperature rose 0.6 to 0.9 degrees Celsius (1.1 to 1.6°F)."[1] Furthermore, the rate of the increase in temperature has nearly doubled in the last 50 years (Figure 11–1). The global temperature in 2012 was above average for the 36th consecutive year.[2]

Rising temperatures are leading to sea level rise and greater variability in weather. No one weather-related event can be attributed to climate change. However, the increasing number and frequency of such events show a trend that extreme climate-related events (e.g., droughts in the U.S. Midwest, Europe, Asia, and parts of Australia; flooding in Bangladesh, Pakistan, and parts of Africa [Figure 11–2]; tornadoes; hurricanes, including Hurricane Sandy in the United States in 2012 [Figure 11–3]; and ice melts in the Arctic) are widespread and are increasing in frequency.[2,3]

The impacts of climate change are already being felt and are likely to get worse in the future.[3,4] No country is exempt from the effects of steadily rising temperatures. Yet, while industrialized countries have contributed most to

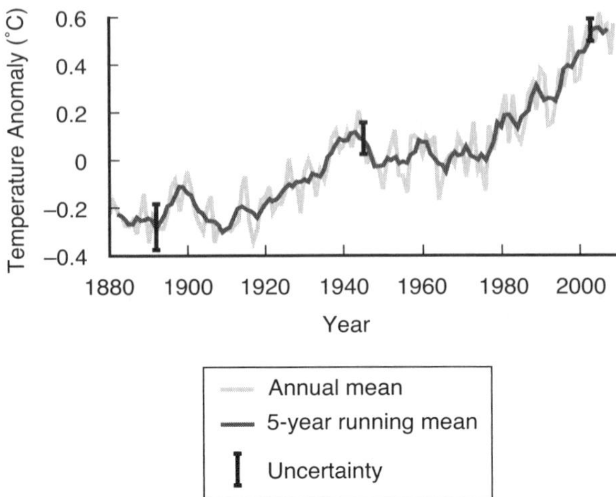

Figure 11–1 Rise in Global Temperatures

Reproduced from Earth Observatory, National Air and Space Administration (NASA). http://earthobservatory.nasa.gov/Features/GlobalWarming/page2.php. Accessed October 27, 2013.

Figure 11–2 Floodwaters Surrounding Houses in Dhaka, Bangladesh

© Stockbyte/Thinkstock

the buildup of greenhouse gases in the atmosphere, it is developing nations, where most of the world's people live, that will suffer the worst effects, even though these countries have contributed least to climate change. The flooding that left approximately one-fifth of Pakistan under water in 2010, for example, affected an estimated 20 million people.[5]

Figure 11–3 Damage from Hurricane Sandy

© craftandbrand/Thinkstock

Box 11–1 Human Activity and the Natural Environment

Rapid and widespread changes in the world's human population, coupled with unprecedented levels of consumption present profound challenges to human health and wellbeing, and the natural environment.

Reproduced from Royal Society. *People and the Planet.* London: Royal Society; 2012. http://royalsociety.org/uploadedFiles/Royal_Society_Content/policy/projects/people-planet/2012-04-25-PeoplePlanet.pdf. Accessed February 27, 2013.

The evidence that human activity is contributing to the increase in greenhouse gas emissions that is leading to climate change is overwhelming (Box 11–1).[6] In 2010, the U.S. National Research Council reiterated findings from the Intergovernmental Panel on Climate Change[*] (IPCC), saying that "a strong, credible body of scientific evidence shows that climate change is occurring, is caused largely by human activities, and poses significant risks for a broad range of human and natural systems."[7] If human activity is implicated in climate change, how do human population dynamics contribute? If, as Professor Joel Cohen[8] suggests, people are part of the problem and part of the solution, where do population, population dynamics, and

---

[*] "The Intergovernmental Panel on Climate Change (IPCC) is the leading international body for the assessment of climate change. It was established by the United Nations Environmental Programme (UNEP) and the World Meteorological Organization (WMO) in 1988 to provide the world with a clear scientific view on the current state of knowledge in climate change and its potential environmental and socio-economic impacts" (www.ipcc.ch).

reproductive health fit into climate science and into programming? Might changes in the size and composition of global population contribute to both "mitigation" of the emissions of greenhouse gases that contribute to climate change and to "adaptation" to the effects of climate change?

This chapter explores each of those questions. The first section presents definitions of climate change mitigation and adaptation. The second section examines human population dynamics as one of the drivers of climate change and reviews how population has been included in studies of climate change. The third section explores the relationships among climate change adaptation, population, and reproductive health. Finally, the chapter discusses challenges to incorporating population and reproductive health in climate change policies and programs.

DEFINITIONS OF CLIMATE CHANGE MITIGATION AND ADAPTATION

Mitigation

The IPCC defines mitigation as "a human intervention to reduce the sources or enhance the sinks of greenhouse gases."[9] Mitigation includes the range of policies, programs, and interventions to reduce emissions of greenhouse gases (GHG) into the atmosphere and to increase the trapping of GHG through sinks that hold carbon, for example, in forests and oceans (Box 11–2).

Box 11–2   Climate Change Mitigation

Climate change mitigation refers to efforts to reduce or prevent emission of greenhouse gases. It can refer to the use of new technologies or renewable energies, the adjustment of older equipment to be more energy-efficient, or the alteration of management practices or consumer behavior. Mitigation can be complex, as a plan for a new city, or simple, such as improving a cook stove design. Throughout the world, efforts range from developing high-tech subway systems to constructing bicycling paths and walkways. Other elements of mitigation include the protection of natural carbon sinks like forests and oceans and the creation of "new sinks through silviculture or green agriculture."

Reproduced from United Nations Environmental Program. Climate change mitigation. http://www.unep.org/climatechange/mitigation/. Accessed February 2, 2013.

Adaptation

Adaptation includes initiatives and measures to reduce the vulnerability of natural and human systems to actual or expected climate change effects. The IPCC defines a human system as follows:

> Any system in which human organisations play a major role. Often, but not always, the term is synonymous with "society" or "social system," e.g., agricultural system, political system, technological system, economic system; all are human systems.[9]

Adaptation is linked with vulnerability and resilience. Vulnerability is characterized as how a system is exposed to climate change and climate variation and its sensitivity to the effects of the climate change. Resilience refers to the ability of a system to absorb shocks and still maintain its basic structure, along with its capacity to adapt to external stressors and to change.[9] Studies on natural systems rather than on social systems continue to dominate research on impacts, vulnerability, and adaptation.[10,11]

POPULATION AS A DRIVER OF CLIMATE CHANGE

Attention to addressing climate change is based on climate models projecting greenhouse gas emissions into the future. The IPCC has produced periodic reports on climate science, starting with the First Assessment Report in 1990. The Fifth Assessment Report (AR5) expected in 2014. Population projections are central to emissions scenarios. The IPCC's 2007 Special Report on Emission Scenarios (SRES), builds on the IPAT identity, in which

$$\text{Impact} = \text{Population} \times \text{Affluence} \times \text{Technology}$$

This implies that the level of emissions results from the size of population, multiplied by affluence and the level of technology.[12] I = PAT was conceived in the 1970s by Ehrlich and Holdren to explain the factors explaining environmental impact.[13] When examining climate change, a variation known as the Kaya identity[14] is often used:

$$CO_2 \text{ emissions} = \text{Population} \times (\text{GDP/Population}) \times (\text{Energy/GDP}) \times (CO_2/\text{Energy})$$

In the Kaya identity equation, annual CO_2 emissions are the product of population, per capita income (or per capita economic production), the amount of energy used per unit of economic production (energy intensity), and the amount of CO_2 emitted per unit of energy produced (carbon intensity). Changes in one variable affect the others, and the effects of the variables can differ by region of the world. For example, population increase in a country with high per capita GDP would have a greater impact than population increase in a country with low GDP. Studies using the underlying $I = PAT$ equation have become more sophisticated in recent years—for example, through use of STIRPAT (STochastic Impacts by Regression on Population, Affluence, and Technology), which does "not assume that the effects of population growth on CO_2 have been proportional," among other models to project future effects of the drivers of climate change.[15(p2)]

How Population Has Been Included in Studies of Climate Change

O'Neill et al. review historical data on the link between population trends and CO_2 emissions from energy use (Box 11-3).[15] The first study of population and climate change that used the STRIPAT equation found roughly that a 1% increase in population size was associated with a 1% increase in CO_2 emissions.[16] This study included only population size as a variable and relied on cross-sectional data from 111 countries at one point in time (1989). Subsequent studies included more demographic variables, including urbanization and age structure, as well as time series data from multiple countries and multiple years; a comparison of the studies has also been undertaken.

The IPCC's 2007 SRES identified population growth, economic growth, technological change, and changes in patterns of energy and land use as the major driving forces of the growth in greenhouse gas emissions.[17] The SRES did not suggest a linear relationship between population and emissions, but instead emphasized that the relationships among the drivers of GHG emissions are complex.[15,18]

The 2007 SRES paid significant attention to population size, the only demographic factor currently included in emissions scenario modeling related to climate change. Yet the authors of the SRES acknowledged—and studies over more than two decades have shown—that other demographic factors, including urban–rural residence, age

---

Box 11-3 Population Size and Emissions

Population size is widely recognized as an important driving force of future emissions. Essentially, all quantitative emissions scenarios include changes in population size as one driver.

Reproduced from O'Neill BD, Liddle B, Jiang L, et al. Demographic change and carbon dioxide emissions. *Lancet.* Epub July 10, 2012, with permission from Elsevier.

structure, and household size, are associated with different patterns of energy consumption.[19-21] Projections of future population also suggest that total population size, aging, urbanization, and declining average household size will be important demographic trends in the coming decades.[22]

Population and Mitigation

Given that human population is among the drivers of climate change, a group of researchers asked what difference it would make to carbon emissions if the world's population in 2050 reached the United Nations' low, medium, or high levels in that year. The 2004 UN population projections, which were used in the analysis, range between roughly 8 billion and 11 billion people. The model that O'Neill and colleagues[15] developed used climate modeling techniques[†] and included urbanization, age structure, and household size in addition to population size and growth.[23(p17521)] The model showed that "if the world population were to follow a low rather than a medium growth path, worldwide emissions would be reduced by 1.4 gigatonnes of carbon (CtC) per year in 2050 and 1.5 GtC per year in 2100, or by about 15% and 40% respectively."[15(p5)] Figure 11–4 illustrates the result of their analysis for the world and for India and the United States. Their results showed that *aging* is the dominant demographic feature in the industrialized world and that *urbanization* is the predominant demographic feature related to climate change in developing countries.

In complementary research, Wheeler and Hammer analyzed the contribution that family planning, or access to contraception so that women are able to have the number of children they want to have, could make in addressing mitigation.[24] These researchers compared access to contraception with other interventions to reduce GHG emissions, such as wind and nuclear power, second-generation biofuels, and carbon capture and storage. Their analysis showed that access to contraception through family planning, in addition to girls' education, is "highly cost-competitive with almost all the existing options for carbon emissions abatement via low-carbon energy and forestry/agriculture." [24(p4)] In their analysis, a family planning cost of $17 was matched with each ton of carbon emissions abated, far less than other options.

---

† Their analysis uses the Population–Environment–Technology (PET) model, "a nine-region dynamic computable general equilibrium model of the global economy with a basic economic structure that is representative of the state of the art in emissions scenario modeling."

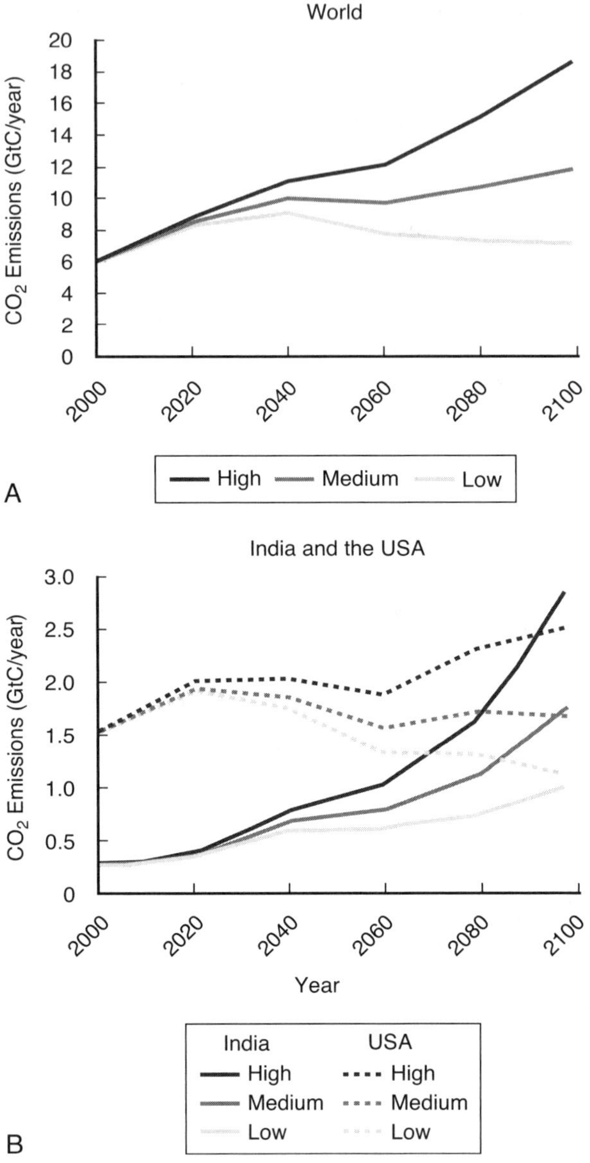

Figure 11–4 Carbon Emissions from Fossil Fuel Use, by Rate of Population Growth, for the World and for India and the United States

Reproduced from O'Neill BD, Liddle B, Jiang L, et al. Demographic change and carbon dioxide emissions. *Lancet.* Epub July 10, 2012. Reprinted with permission from Elsevier.

CLIMATE CHANGE ADAPTATION, POPULATION, AND REPRODUCTIVE HEALTH

While most global attention continues to focus on climate change mitigation, it has become increasingly clear over the past decade that adaptation to the ongoing effects of rising global temperatures is also critical (Box 11–4).

Demographic dynamics, including size, growth, urbanization, migration, and fertility, are linked with climate change adaptation. At a symposium on population footprints in 2011, participants agreed that, "population dynamics are relevant and heighten the challenges of adaptation."[25(p9)] Yet, as Schensul and Dodman explain in the book *The Demography of Adaption to Climate Change*, "current approaches to understanding climate change vulnerability . . . have masked important gaps in the understanding and practice of adaptation . . . includ[ing] superficial assessments of who is vulnerable and why, a focus on the physical environment over the social one as well as a lack of connection between the two and a static perception of vulnerability over time."[26(p19)]

Schensul and Doman noted that population dynamics in adaptation programs are important for three reasons. First, population projections provide information about the size and composition of the future populations, which is important for policy development and program planning. Second, the link between population dynamics and socioeconomic development is important for adaptation. Finally, there are links between adaptation and other demographic dynamics such as age structure, urbanization, migration, and fertility.

Awareness of the importance of demographic factors for climate change adaptation is beginning to emerge. Writing about the impacts of climate change for the United States, Karl et al. noted that demographic factors as well as aging infrastructure, buildings, and air pollution will exacerbate climate change impacts.[27] Studying adaptation in small communities in eastern Ontario, Canada, McLeman found that "adaptive capacity at the community level can change over time, and can be very sensitive to changes in the demographic composition of the community."[28(p312)] This author offers a typology of the interactions between population change and vulnerability to climate change that addresses conditions of population increase, decrease, and stability, and high volatility (for example rapid changes in population triggered by climatic or nonclimatic events).

Box 11–4 Population Dynamics Influence Climate Change Adaptation

Population dynamics, including urbanisation, migration, age structures, density and growth are significant because they influence resource use and consumption rates, are linked to health, the environment and the economy, and can affect a country's capacity to adapt to climate change.

Reproduced from University College London (UCL). *The UCL-Leverhulme Trust Population Footprints Symposium: Report.* London: UCL; May 25–26, 2011.

The Global Demographic Divide and Climate Change

Programs that address adaptation to climate change will need to be tailored to global demographic dynamics. Industrialized countries, which have demonstrated high past and present consumption rates, have certainly made the greatest contributions to climate change to date. Yet, ongoing rapid population growth in the developing world exacerbates scarcity of food and water, vulnerability to natural disasters and infectious diseases, and population displacement, which are all linked to climate change.[29-31] Fertility rates have fallen in much of the world; however, the global population continues to grow and the UN projects that more than half (27) of the world's 49 least-developed countries (LDCs) will at least double their current population by 2050. At the same time, aging of the population in other parts of the world has created a demographic divergence.[32] This divergence is illustrated by the Population Reference Bureau with data from Tanzania and Spain, two countries that had similar populations in 2012—48 million people in Tanzania and 46 million in Spain (Table 11-1). While Spain's population is projected to grow to 48 million in 2050, an increase of 2 million people, Tanzania could grow to 138 million people in that same year, nearly tripling the country's 2012 population.

Table 11–1 Key Population Indicators for Tanzania and Spain, 2012 and 2050

	Tanzania	Spain
Population (2012)	48 million	46 million
Projected population (2050)	138 million	48 million
Lifetime births per woman	5.4	1.4
Annual births	1.9 million	438,000
Percentage of the population younger than age 15	45%	15%
Percentage of the population age 65 or older	3%	17%
Percentage of the population age 65 or older in 2050	4%	33%
Life expectancy at birth	57 years	82 years
Infant mortality rate (per 1000 live births)	51	3.2
Annual number of infant deaths	98,000	1600
Number of adults ages 15–49 with HIV/AIDS	5.6%	0.4%

Reproduced from Population Reference Bureau. 2012 world population data sheet. 2012. http://www.prb.org/Publications/Datasheets/2012/world-population-data-sheet/data-sheet.aspx. Accessed February 16, 2013.

Urbanization, Migration, and Climate Change

The world is becoming increasingly urbanized, a trend that is consistent in all regions of the globe (Box 11–5 and Box 11–6). In its 2011 update on urbanization, the UN projected that all worldwide population growth between 2011 and 2050 will occur in urban areas. "Between 2011 and 2050, the world population is expected to increase by 2.3 billion, passing from 7.0 billion to 9.3 billion. . . . At the same time, the population living in urban areas is projected to gain 2.6 billion, passing from 3.6 billion in 2011 to 6.3 billion 2050."[33(p1)] These projections suggest that not only will all population growth over the next four decades occur in urban areas, but that urban areas will also continue to draw migrants from rural areas.

Discussions of urbanization, migration, and climate change often revolve around crisis narratives of millions of climate refugees migrating from the global South to the global North in search of relief and suggest that rural migration to cities will exacerbate urban poverty. While these narratives may be overblown, it is important to understand the dynamics of urbanization and migration in relation to climate change. Takoli offers a typology of migration that includes seasonal mobility generally related to agriculture, temporary migration related to income diversification, and permanent migration.[34]

Each of these types of migration can be linked with climate change. McGranahan et al. contend that policy discussions should focus on the fact that "mobility is and will be an increasingly fundamental feature of

---

Box 11–5 Urbanization

The level of urbanization is the share of the population living in urban rather than in rural settlements, and the rate of urbanization is the annual rate at which this urban share increases.

Reproduced from Poston D, Bouvier LF. *Population and Society: An Introduction to Demography.* Cambridge, UK: Cambridge University Press; 2010. Reprinted with the permission of Cambridge University Press.

Box 11–6 Urbanization in Developing Countries

Developing countries will be building the equivalent of a city of a million people every five days from now to 2050.

Reproduced from Royal Society. *People and the Planet.* London: Royal Society; 2012:7. http://royalsociety.org/uploadedFiles/Royal_Society_Content/policy/ projects/people-planet/2012-04-25-PeoplePlanet.pdf.

livelihoods, economies and settlement."[35(p27)] While the location and size of cities may be set, policies need to take into account unfolding climate conditions. In 2012, Hurricane Sandy, which hit the U.S. East Coast, highlighted the link between urban infrastructure, settlement patterns and people's housing and livelihoods. Urban planners need to think about how "to steer new urban development away from locations likely to be prone to future climate hazards."[35(p28)]

Bangladesh offers a stark illustration of the need to factor climate change into planning.[36] Figure 11–5 shows the possible effects on the country with sea level rise. Bangladesh is the world's seventh largest country, with an estimated population of 153 million people[32] and a very high population density of more than 1000 people per square kilometer. Bangladesh could lose almost one-fourth of its land by the end of the century, affecting millions of people.[37] Sea level rise is already putting added stress on land, water and food, all of which affect people.[38] Bangladesh, well aware of its predicament, has been a leader in addressing climate change. Saleem Huq, an expert on climate change adaptation with the International Centre for Climate Change and Development, calls climate change in Bangladesh the "ultimate storm": "We face [climate change] challenges in every corner of the country. The convergence of climate-related factors has created the ultimate storm here."[36(p1)]

Fertility, Reproductive Health, and Adaptation

Research linking population, fertility, and reproductive health with adaptation to climate change is now emerging.[18,29,31,39,40] Poor reproductive health contributes to one-third of the burden of disease among women[41] and can be exacerbated by climate change. Providing women and couples with the ability to have their desired number and spacing of children can contribute to promoting adaptive capacity within families and communities.

A recent study conducted in several regions of Ethiopia examined the relationships between fertility, reproductive health, and adaptation to climate change. First, researchers queried local inhabitants about whether they thought that population and fertility were related to climate change. Second, researchers explored whether respondents perceived family planning and reproductive health as potentially useful for increasing resilience to climate change impacts.[42]

Responses from different regions indicated an understanding of these linkages. Both women and men described the increasing challenges they face in adapting to climate change. Specifically, they recounted how rising temperatures, more frequent draughts, increased flooding, receding agricultural grazing land, and diminishing forests were making it more

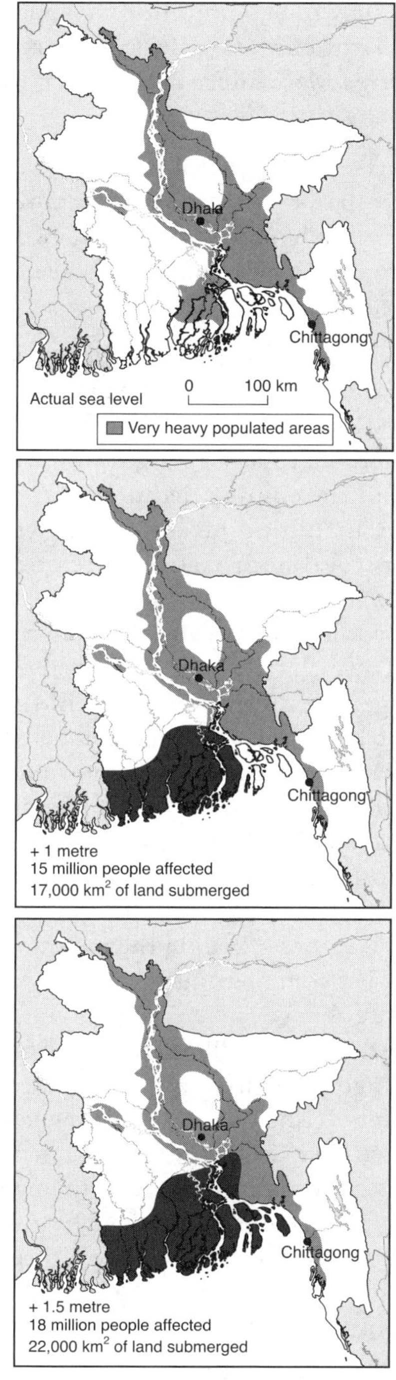

Figure 11–5 Potential Effects for Bangladesh of Sea Level Rise

Reproduced from United Nations Environment Program (UNEP). Impact of sea level rise in Bangladesh. http://www.unep.org/dewa/vitalwater/article146.html. Accessed January 8, 2014.

difficult for their families and communities to cope. Linking population pressure to the effects of climate change, they reported that families should consider having fewer children. As one young rural woman from Ethiopia explained:

> . . . if a family has limited children, he will have enough land for his kids and hence we can protect the forests. . . . In earlier years we had a lot of fallow lands, but now as a result of population growth we don't have adequate fallow land. Therefore, limiting number of children will help us to cope with the change in climate.[42]

An assessment of vulnerability to glacier ice melts in Asia also noted that "existing vulnerabilities in human health status, population pressure, degraded ecosystems and—especially—water stress make societies and ecosystems vulnerable to any changes in water availability as glacier melt accelerates in the coming decades."[43(P5)] The need to address population pressure, including meeting family planning needs, was one of nine cross-sectoral approaches recommended to addressing glacier melt.

Recent analysis of vulnerability using the Vulnerability and Resilience Index Measure (VRIM), which focuses on the adaptive capacity of countries, and the Global Change Assessment Model (GCAM), which is designed to provide scenarios related to climate change, compared outcomes using the United Nations mid-range population projection and a population projection assuming that all unmet need for family planning is satisfied (called the Universal Access to Family Planning scenario [UAFP]).[44] The analysis, which included data from Bangladesh, Nepal, Ethiopia, Kenya, Malawi, and Uganda, showed that resilience to climate change was between 2% and 10% higher under the UAFP scenario than in the mid-range population projection, suggesting a link between meeting women's needs for family planning and resilience to climate change.

When resilience scores using the VRIM are plotted on world maps that also show countries with high population growth, changes in agricultural production, and water stress, "[t]he 26 countries designated as population and climate change hotspots have an average population growth of 2.2 percent—a rate that, if unchanged, would result in a doubling of the population in 31 years."[45] In many of these countries, women have a high unmet need for family planning—that is, they say they want to either stop having children or want to wait at least 2 years before having another child, and yet are not using a method of contraception. Worldwide in 2012, 222 million women had an unmet need for family planning.[46]

POPULATION DYNAMICS AND CLIMATE CHANGE PROGRAMS

Given their importance, population dynamics are not being adequately factored into programs to address climate change. Advances in data availability are occurring, yet many challenges remain in generating data at local levels that can be used for adaptation planning.[47] While demographic data are generally available at national levels, generating and using data from subnational levels is also important. As Balk et al. note, "to facilitate a more in-depth analysis of climate change vulnerability, census data must be processed for very small areas in such a way that they effectively match the geographic distribution of hazards. . . . By the end of January 2012, 77 percent of all countries of the world had conducted their 2010 round of census, meaning that 87 percent of the global population had been enumerated."[48(p75)]

Data from census, household surveys, special studies and administrative statistics can be combined with geographic information systems (GIS) and remote sensing techniques to provide information useful for adaptation at local levels. Recurring surveys, such as the Demographic and Health Surveys (DHS), can be adapted to include questions on vulnerability to climate change and adaptive capacity. Jankowska et al. used DHS to model water availability, malnutrition, and livelihoods in Mali.[49] Moreland and Smith combined demographic projections with an economic model that incorporates the effects of climate change on agriculture and with a model of food requirements.[50] The resulting model is useful for policy advocacy to introduce population into discussions about food security. The model, tested in Ethiopia (Figure 11–6), reinforced the following:

- Climate change is expected to reduce agricultural yields in Ethiopia.
- Already, 60% of Ethiopians have insufficient food consumption.
- Almost 30% of children younger than 5 years of age in Ethiopia are underweight.
- Continued rapid population growth will add to the challenge of food insecurity.[51(p1)]

National Adaptation Programs of Action

Under the auspices of the United Nations Framework Convention on Climate Change (UNFCCC), 49 least-developed countries and small island states have been eligible to develop national adaptation programs of action (NAPAs). These plans are intended to link with the countries' national development plans. Analysis of NAPAs and national development

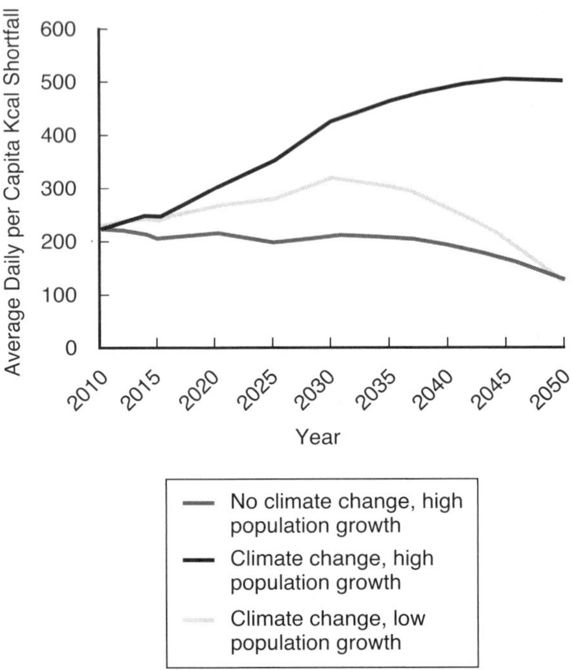

Figure 11-6 Food Gap in Ethiopia, 2010–2050

Reproduced from Moreland S, Smith E. *Modeling Climate Change, Food Security, and Population.* Chapel Hill, NC: MEASURE Evaluation; 2012. http://www.cpc.unc.edu/measure/publications/sr-12-69/.

plans, using population as an example, shows that critical issues such as population pressure and dynamics are falling through the cracks in both climate change adaptation programs and development assistance.[52] Analysis of NAPAs found that among 41 NAPAs completed by May 2009, fully 37 mentioned rapid population growth as a factor exacerbating the effects of climate change.[52,53] Yet only 6 NAPAs suggested that addressing population through family planning should be part of the country's adaptation strategy, and no projects to do so were funded.

While NAPAs are not the only climate adaptation programming, they give an instructive representation of such programming, which focuses on addressing climate change adaptation through technological solutions to climate change systems (e.g., building sea walls). Among the 448 projects proposed in 41 NAPAs, half address three issues: food security (21%), water resources (16%), and terrestrial ecosystems (15%). The social sectors are hardly addressed. The health sector accounts for about 7% of the total projects, and the education sector is not represented at all.

Possible Synergies for Climate Change Programs

While climate-related policies tend to neglect population, development policies and population policies are increasingly acknowledging the important role of climate change, as shown in case studies in Kenya and Malawi.[54] Yet, lack of coordination among policies and their implementation remains an issue. Community-based adaptation (CBA), a newly emerging type of climate adaptation programming, offers a potential for integrated programming that focuses on both natural and human systems.[55] CBA fosters community engagement, and projects are developed based on climate science and local knowledge about weather trends and ways to adapt to the trends.[56]

CBA is intended to meet community needs (Box 11–7). For example, if a community identifies lack of access to reproductive health services or to education as an issue related to its ability to adapt to climate change, a CBA project should theoretically be able to accommodate this issue. Both international organizations, including the United Nations Development Program (UNDP) and nongovernmental organizations (NGOs) such as CARE, have embraced CBA. Analysis of ongoing projects, however, suggests that the bias against social-sector programming continues. For example, although UNDP's CBA project activities are designed to increase the resilience of land and biodiversity resources to the impacts of climate change,[57] none of the projects include components related to reproductive health, despite high levels of unmet need for family planning services in a number of countries where UNDP is supporting CBA projects (Table 11–2).

Population, health, and environment (PHE) programs, designed to address the complex connections between people, their health, and their environment, offers a promising approach to climate change adaptation. PHE programs in the Philippines,[58] Ethiopia,[59,60] Cambodia,[61] Tanzania,[62]

Box 11–7   People-Centered Adaptation

The path to climate change adaptation must be people centered. A critical component of success is the well-being and rights of vulnerable people and communities. "Incorporating population dynamics into adaptation can help in understanding who is most vulnerable, why and how to target policies to decrease that vulnerability."

Reproduced from Schensul D, Dodman D. Populating adaptation: incorporating demographic dynamics in climate change adaptation policy and practice. In: Martine G, Schensul D, eds. *The Demography of Adaptation to Climate Change.* New York/London/Mexico City: UNFPA, IIED, and El Colegio de Mexico; 2013:1–23.

Table 11–2 Unmet Need for Contraception in 10 Countries with CBA Projects

Country	Unmet Need Among Married Women (%)	Year
Bangladesh	17.1	2007
Bolivia	20.2	2008
Guatemala	27.6	2002
Jamaica	11.7	2002–2003
Kazakhstan	8.7	1999
Morocco	10	2003–2004
Namibia	6.7	2006–2007
Niger	15.8	2006
Samoa	45.6	2009
Vietnam	4.8	2002

*Unmet need is the percentage of married women who are fecund and who desire to postpone or stop childbearing, but who are not currently using a contraceptive method.

With the exception of Jamaica and Samoa, unmet need figures are from Measure DHS StatCompiler (http://www.statcompiler.com/, accessed July 19, 2012); unmet need data for Samoa are from Samoa's 2009 DHS; and unmet need data for Jamaica are from the United Nations, 2012. TFR data are 2005–2010 figures from United Nations, 2011. From Bremner et al., forthcoming.

Nepal,[63] Madagascar,[64] and Rwanda,[65] among other countries, illustrate the effect of addressing natural resources, livelihoods, population, and health simultaneously.

In the Philippines, for example, programs that integrated coastal resources management (CRM) and reproductive health services in one municipality of Palawan Province, when compared to separate service programs (reproductive health in one municipality and CRM in another municipality), "generated higher impacts on human and ecosystem health outcomes compared to the independent coastal resources management and reproductive health interventions." Project beneficiaries also perceived the integrated programming approach as more realistically addressing their needs. Youth, for example, were encouraged to become stewards of the environment and of their sexuality.[58]

In Nepal, a PHE project in the Terai region tested whether forest user groups could include family planning and education with programming addressing conservation, health, and sustainable livelihood activity.[63] Over a two-year period, the project taught women and girls about reproductive health, family planning, and environmental issues. The project provided education and services to promote safer sex and condom use, conducted media campaigns, and promoted energy efficiency in homes. Contraceptive use in the project area rose from 43% to nearly 73%. The project helped shift attitudes toward the benefits of family planning from simply improved health to also contributing to resource management.

Similarly, in the coastal reserve Velondriake on the southwest coast of Madagascar, integrated programs sought to protect coral reefs, mangroves, seagrass beds, baobab forests, and other threatened habitats, as well as to provide family planning at the request of women in the 25 villages in the reserve. These programs both improved natural resource management and increased the use of family planning.[64]

Such an integrated approach is being tried in other locations. For example, Tanzania is experiencing climate change, including altered precipitation and increased sea surface temperatures that are affecting coastal communities. In preparing community-based climate change vulnerability and adaptation action plans for a project in the Bagamoyo district, attention was given to environmental impacts and to population, health, equity, and food security. Each of these issues arose during preparatory discussions with communities in that area.[62] In Rwanda, a project with coffee cooperatives that combined a health component with activities to improve livelihoods and business practices was well received by cooperative staff.[65] The health component focused on HIV prevention, reproductive health and family planning, maternal health, nutrition, water, and sanitation. Community activities addressed issues that linked coffee farming and family health.

In short, the PHE approach shows considerable promise. Nevertheless, PHE programs are currently ineligible for international climate change adaptation funding, including support for CBA programs.[55] Indeed, population itself is virtually absent in most adaptation assessment tools. Furthermore, the United Nations Development Program's guidelines for implementing gender-sensitive adaptation programming, issued in 2010, do not even mention reproductive health.[66]

Need for More Evidence on Reproductive Health and Climate Change

The evidence base linking reproductive health and climate change adaptation is growing but remains insufficient. More analyses and especially country studies are needed to show the effects of reproductive health programming on building resilience and helping people and communities adapt to climate change.

It has also been observed that "family planning policies would have a substantial environmental co-benefit,"[23(p17525)] and that if the unmet need for family planning were satisfied in all countries, world population growth would fall between the UN's low and medium projections.[67] Cohen (2010) calls for attention to universal secondary education, voluntary contraception, maternal health services, and smarter urban design and construction, noting that these efforts represent "effective responses to the twin challenges of reducing poverty and reducing greenhouse gas emissions."[8(p179)] Wheeler and Hammer found a strong synergy between the potential contributions of family planning and female education and emissions reductions.[24] Further, they concluded that given the benefits of both interventions to sustainable development and climate change, along with their low cost, "the evidence is strongly consistent with an allocation of some carbon mitigation resources to the population policy options" (p. 16).

CONCLUSION

Emerging evidence highlights both the link between macro population trends and climate change mitigation as well as the benefits of incorporating demographic dynamics into climate change adaptation programming. The number of people in the world and in each country, the areas where they live, their age structure and household composition, their fertility, their education and health status, their patterns of movement, their use of energy, and their consumption levels are all important and need to be understood to address climate change. Reproductive health and family programs can contribute to building resilience in women, thereby enabling them to thrive in the face of climate change. Increasing calls are being made to address population factors, often with a focus on reproductive health, as part of climate change programming, and with climate funding (Box 11–8).

Despite this mounting evidence, evidence-based policy making on this topic is fraught with controversy. Population has been inadequately included in climate studies and has been absent from climate negotiations at the global level, in part due to "a long history—from Malthus to Erlich to the recent resurgence of concerns linked to climate change and sustainable

---

Box 11–8   Political Leadership and Financial Commitment Needed

Reproductive health and voluntary family planning programs urgently require political leadership and financial commitment, both nationally and internationally. This is needed to continue the downward trajectory of fertility rates, especially in countries where the unmet need for contraception is high.

Reproduced from Royal Society. *People and the Planet.* London: Royal Society; 2012. http://royalsociety.org/uploadedFiles/Royal_Society_Content/policy/projects/people-planet/2012-04-25-PeoplePlanet.pdf. Accessed February 27, 2013.

development—of blaming population growth for the world's problems."[26(p2)] There is a need to reframe the discourse on population to delink it from its past associations.

Country studies can contribute because, if "the affected countries themselves identify [population] as a local priority [it] avoids the conflict that comes from framing population regulation as a way of reducing global greenhouse emissions."[68(p807)] For example, many countries themselves identified population as an exacerbating factor for climate change adaptation in their NAPAs. Furthermore, many countries are increasing the attention paid to population stabilization. At a conference on population footprints in 2011, Dr. Eliya Zulu, president of the African Institute for Development Policy, noted that developing countries are taking on the topic themselves. He argued that the primary concern for slowing population growth in Africa relates to development, and he noted that there is increasing support for family planning among African leaders because women themselves are saying they want fewer children. "We must therefore challenge the notion that family planning programmes are 'top down' and imposed on developing countries by the global North."[25(p21)] It is both possible and essential to discuss and address population and climate change topics and interactions in a just and ethical manner.[69]

Population and reproductive health issues have all too often been left out of climate change policies. It is important to fully understand the global architecture being built to respond to climate change, including grasping financing issues and the potential for using climate funding or a mix of climate and development funding for programs that address population factors and reproductive health. Given the complexity of climate change impacts on both people and the natural environment, programs that integrate social and natural systems or people and their environment are likely to show the most promise in meeting needs to increase resilience and promote adaptation. Population, health, and environment approaches offer lessons for community-based adaptation programming. Documenting the effects on climate change outcomes of programming that addresses population issues and reproductive health in addition to natural systems, the environment and livelihoods, will be critical, along with advocacy for inclusion of integrated programming in climate change policies and programs.

DISCUSSION QUESTIONS

1. Discuss the three principal ways that human population is related to climate change.
2. What is the difference between climate change mitigation and climate change adaptation?

3. Why have population dynamics been neglected in climate change programming? What would you recommend for more integrated programs at the international and country levels?
4. How is reproductive health linked to climate change mitigation and adaptation?
5. Identify the geographic areas most likely to be affected by climate change. What should be done to address current and future impacts?

REFERENCES

1. National Aeronautics and Space Administration (NASA). Earth Observatory resources page. Global warming. http://earthobservatory.nasa.gov/Features/GlobalWarming/page2.php. Accessed February 2, 2013.
2. National Oceanic and Atmospheric Administration (NOAA), National Climatic Data Center. *State of the Climate Global Analysis for Annual 2012*. Washington, DC: NOAA, National Climatic Data Center; December 2012. http://www.ncdc.noaa.gov/sotc/global/2012/13. Accessed January 31, 2013.
3. Intergovernmental Panel for Climate Change (IPCC). Special report on managing the risks of extreme events and disasters to advance climate change adaptation (SREX). 2012. http://www.ipcc-wg2.gov/SREX/. Accessed February 23, 2013.
4. Intergovernmental Panel for Climate Change (IPCC). *Climate Change 2001: Impacts, Adaptation, and Vulnerability, Contribution of Working Group II to the Third Assessment Report of the IPCC.* Cambridge, UK: Cambridge University Press; 2001.
5. Witte G. Pakistan floods affecting 20 million; cholera outbreak feared. *Washington Post.* August 15, 2010. http://www.washingtonpost.com/wpdyn/content/article/2010/08/14/AR2010081400427.html. Accessed February 1, 2013.
6. Intergovernmental Panel for Climate Change (IPCC). *Climate Change 2007: Impacts, Adaptation and Vulnerability. Contribution of Working Group II to the Fourth Assessment Report of the Intergovernmental Panel on Climate Change.* Cambridge, UK: Cambridge University Press; 2007.
7. National Research Council. *Advancing the Science of Climate Change.* Washington, DC: National Academies Press; 2010.
8. Cohen JE. Population and climate change. *Proc Am Philosoph Soc.* 2010;154(2):158–182.
9. Intergovernmental Panel for Climate Change (IPCC). Glossary of terms used in the IPCC Fourth Assessment Report. 2007. http://www.ipcc.ch/publications_and_data/publications_and_data_glossary.shtml. Accessed February 2, 2013.
10. Cannon T, Muller-Mahn D. Vulnerability, resilience and development discourses in context of climate change. *Natural Hazards.* 2010;55(3):621–635.
11. Malone EL. *Vulnerability and Resilience in the Face of Climate Change: Current Research and Needs for Population Information.* PNWD-4087. Pacific Northwest Division, Batelle; 2009.

12. Nakicenovic N, Alcamo J, Davis G, et al. *Special Report on Emissions Scenarios: A Special Report of Working Group III of the Intergovernmental Panel on Climate Change.* Cambridge, UK: Cambridge University Press; 2000.

13. Ehrlich PR, Holdren JP. Impact of population growth. *Science.* 1971;171: 1212–1217.

14. Kaya Y. *Impact of Carbon Dioxide Emission Control on GNP Growth: Interpretation of Proposed Scenarios.* Paper presented to the IPCC Energy and Industry Subgroup, Response Strategies Working Group, Paris; 1990.

15. O'Neill BD, Liddle B, Jiang L, et al. Demographic change and carbon dioxide emissions. *Lancet.* July 10, 2012. doi: 10.1016/S140-6736912):60958-1.

16. Dietz T, Rosa EA. Effects of population and affluence on CO_2 emissions. *Proc Natl Acad Sci USA.* 1997;94:175–179.

17. Fisher B, Nakicenovic N, Alfsen K, et al. Issues related to mitigation in the long-term context. In: Metz B, Davidson OR, Bosch PR, et al., eds. *Climate Change 2007: Mitigation of Climate Change. Contribution of Working Group III to the Fourth Assessment Report of the Intergovernmental Panel on Climate Change.* Cambridge, UK: Cambridge University Press; 2007:169–250.

18. Hoepf Young M, Mogelgaard K, Hardee K. *Projecting Population, Projecting Climate Change: Population in IPCC Scenarios. PAI Working Paper.* 2009;09-02.

19. Jones DW. Urbanization and energy use in economic development. *Energy J.* 1989;10:29–44.

20. Pachauri S, Jiang L. The household energy transition in India and China. *Energy Policy.* 2008;36:4022–4035P.

21. Yamasaki E, Tominaga N. Evolution of an aging society and effects on residential energy demand. *Energy Policy.* 1997;25:903–912.

22. Lutz W, Samir KC. Dimensions of global population projections: what do we know about the future population trends and structures? *Philosoph Trans Roy Soc.* 2010;365:2779–2791.

23. O'Neill BC, Dalton M, Fuchs R, et al. Global demographic trends and future carbon emissions. *Proc Natl Acad Sci.* 2010. doi: /10.1073/pnas.1004581107.

24. Wheeler D, Hammer D. *The Economics of Population Policy for Carbon Emissions Reduction in Developing Countries.* CGD Working Paper, 229. Washington, DC: Center for Global Development; 2010.

25. University College London (UCL). *The UCL-Leverhulme Trust Population Footprints Symposium: Report.* London: UCL; May 25–26, 2011.

26. Schensul D, Dodman D. Populating adaptation: incorporating demographic dynamics in climate change adaptation policy and practice. In: Martine G, Schensul D, eds. *The Demography of Adaptation to Climate Change.* New York/London/Mexico City: UNFPA, IIED, and El Colegio de Mexico; 2013:1–23.

27. Karl TR, Melillo JM, Peterson TC, eds. *Global Climate Change Impacts in the United States.* Cambridge, UK: Cambridge University Press; 2009.

28. McLeman R. Impacts of population change on vulnerability and the capacity to adapt to climate change and variability: a typology based on lessons from a hard country. *Pop Environ.* 2010;31:286–316.

29. Jiang L, Hardee K. How do recent population trends matter to climate change? *Pop Res Policy Rev.* August 23, 2010;1–26. doi: 10.1007/s11113-010-9189-7.

30. Royal Society. *People and the Planet.* London: Royal Society; 2012. http://royalsociety.org/uploadedFiles/Royal_Society_Content/policy/projects/people-planet/2012-04-25-PeoplePlanet.pdf. Accessed February 27, 2013.

31. UNFPA. *Facing a Changing World: Women, Population and Climate. State of the World Population Report 2009.* New York: UNFPA; 2009.

32. Population Reference Bureau. World population data sheet 2012. http://www.prb.org/Publications/Datasheets/2012/world-population-data-sheet/data-sheet.aspx. Accessed February 2, 2013.

33. United Nations. *World Urbanization Prospects: The 2011 Revision.* New York: UN Economic and Social Affairs Division; 2011.

34. Takoli C. Migration as a response to local and global transformation: a typology of mobility in the context of climate change. In: Martine G, Schensul D, eds. *The Demography of Adaptation to Climate Change.* New York/London/Mexico City: UNFPA, IIED, and El Colegio de Mexico; 2013:1–23.

35. McGranahan G, Balk D, Martine G, Tacoli C. Fair and effective responses to urbanization and climate change: tapping synergies and avoiding exclusionary policies. In: Martine G, Schensul D, eds. *The Demography of Adaptation to Climate Change.* New York/London/Mexico City: UNFPA, IIED, and El Colegio de Mexico; 2013:24–39.

36. Aulakh R. How bad can climate change get? Bangladesh already knows. *Toronto Star.com.* February 9, 2013. http://www.thestar.com/news/world/2013/02/09/bangladesh_faces_mass_migration_loss_of_land_from_climate_change.html. Accessed February 16, 2013.

37. Ericson JP, Vorosmarty CJ, Dingman SL, et al. Effective sea-level rise and deltas: causes of change and human dimension implications. *Glob Planet Change.* 2005;50:63–82.

38. Union of Concerned Scientists. Climate hotspot map: Ganges–Brahmaputra Delta, Bangladesh. http://www.climatehotmap.org/global-warming-locations/ganges-brahmaputra-delta-bangladesh.html. Accessed February 16, 2013.

39. Mogelgaard K. Reproductive health and solutions to climate change: connecting the dots. *Global Health Magazine.* Summer 2010, 18–20.

40. Stephenson J, Newman K, Mayhew S. Population dynamics and climate change: what are the links? *J Public Health.* 2010;32(2):150–156.

41. World Health Organization (WHO). *The world health report 2005: Make every mother and child count.* Geneva, Switzerland: WHO; 2005.

42. Rovin K, Hardee K, Kidanu A. Linking population, fertility and family planning with adaptation to climate change: views from Ethiopia. *Afr J Reprod Health.* 2013;17(3);15–29.

43. Malone EL. Changing glaciers and hydrology in Asia. In: *Addressing Vulnerability to Glacier Ice Melt Impacts.* Washington, DC: USAID; 2010:5.

44. Malone EL, Brenkert AL, Delgado A. *Climate Change Resilience and Universal Access to Family Planning. PAI Working Paper 2011.* Washington, DC: Population Action International; 2011.

45. Population Action International. Climate change. http://populationaction .org/topics/climate-change/?post_type=cpt_data. Accessed February 16, 2013.

46. Singh S, Darroch J. *Adding It Up: Cost and Benefits of Family Planning Services, Estimates for 2012.* New York: Guttmacher Institute; 2012.

47. Guzman HM, Schensul D, Zhung S. Understanding vulnerability and adaptation using census data. In: Martine G, Schensul D, eds. *The Demography of Adaptation to Climate Change.* New York/London/Mexico City: UNFPA, IIED, and El Colegio de Mexico; 2013:24–39.

48. Balk D, Guzman HM, Schensul D. Harnessing census data for environmental and climate change analysis. In: Martine G, Schensul D, eds. *The Demography of Adaptation to Climate Change.* New York/London/Mexico City: UNFPA, IIED, and El Colegio de Mexico; 2013:24–39.

49. Jankowska MM, Lopez-Carr D, Funk C, et al. Climate change and human health: spatial modeling of water availability, malnutrition, and livelihoods in Mali, Africa. *Appl Geography.* 2012;33:4–15. doi:10.1016/j.apgeog.2011.08.009.

50. Moreland S, Smith E. *Modeling Climate Change, Food Security, and Population.* Chapel Hill, NC: MEASURE Evaluation; 2012. http://www.cpc.unc.edu/measure/ publications/sr-12-69/. Accessed February 16, 2013.

51. Moreland S. Study summary: modeling food security, population growth and climate change. In: *Improving Access to Family Planning Can Promote Food Security in the Face of Ethiopia's Changing Climate.* Washington, DC: Futures Group; 2012. http://futuresgroup.com/files/publications/Ethiopia_4_page_domestic_ Feb_27_2012_pm.pdf. Accessed February 16, 2013.

52. Hardee K, Mutunga C. Strengthening the link between climate change adaptation and national development plans: lessons from the case of population in National Adaptation Programmes of Action (NAPAs). *Mitigation and Adaptation Strategies for Global Change.* 2009. doi: 10.1007/s11027-009-9208-3.

53. Bryant L, Carver L, Butler CD, Anage A. Climate change and family planning: least developed countries define the agenda. *Bull WHO.* 2009;87:852–857.

54. Mutunga C, Zulu E, De Souza RM. *Population Dynamics and Climate Compatible Development in Africa.* Washington, DC: Population Action International; 2012.

55. Bremner J, Hardee K, Mogelgaard K, D'Agnes H. Is there a link between population, health and environment (PHE) and climate change adaptation? Presented at the *Population Dynamics and Climate Change II: Building for Adaptation Workshop,* Mexico City, Mexico, October, 2010; forthcoming from UNFPA.

56. Reid H, Alam M, Berger R, et al. Community-based adaptation to climate change: an overview. In: IIED. *Participatory Learning and Action 60.* London: IIED; 2009:11–38.

57. United Nations Development Program (UNDP). Community-based adaptation. http://www.undp-alm.org/projects/spa-community-based-adaptation -project/about. Accessed October 26, 2012.

58. D'Agnes L, D'Agnes H, Schwartz JB, et al. Integrated management of coastal resources and human health yields added value: a comparative study in Palawan (Philippines). *Environ Conserv.* 2010. doi: 10.1017/S0376892910000779.

59. Hardee K, Deribe S, Teklu N, Edmonds J. The contribution of a PHE approach to climate change adaption in Ethiopia. *BALANCED Newsletter.* 2010;1(2):19–21.

60. Worku M. The missing links: poverty, population and the environment in Ethiopia. *Focus.* 2007;14.

61. Conservation International. *Health and Conservation in the Cardamoms in Cambodia.* Phnom Phen: Conservation International and CARE; 2008.

62. Torell E, Tobey J, Mahenge J, et al. Climate change along the north coast of Tanzania: community-based adaptation planning with a population, health, and environment (PHE) Lens. *BALANCED Newsletter.* 2010;1(2):12–15.

63. D'Agnes L, Oglethrope J, Thapa S, et al. Forests for the future: family planning in Nepal's Terai region. *Focus.* 2009;18.

64. Mohan V. *Providing Sexual and Reproductive Health Services for Communities in Velondriake, Southwest Madagascar: Project Development Plan 2009–2011.* 2009. https://blueventures.org.

65. Kitzantides I. Coffee and community: combining agribusiness and health in Rwanda. *Focus.* 2011;22.

66. United Nations Development Program (UNDP). *Gender, Climate Change and Community-Based Adaptation: A Guidebook for Designing and Implementing Gender-Sensitive Community-Based Adaptation Programmes and Projects.* New York: UNDP; 2010.

67. Moreland S, Smith E, Sharma S. *World Population Prospects and Unmet Need for Family Planning.* Washington, DC: Futures Group; 2010.

68. Campbell-Lendrum D, Lusti-Narasimhan M. Taking the heat out of the population and climate debate. *Bull WHO.* 2009;87:807. doi: 10.2471/BLT.09.072652.

69. Petroni S. Policy review: thoughts on addressing population and climate change in a just and ethical manner. *Pop Environ.* 2010. doi: 10.1007/s11111-009-0085-1.

Food Security, Population, and Reproductive Health

Richard E. White

INTRODUCTION

The continued growth of human population, combined with reduced growth in global food production, has heightened attention to the issue of food security.[1] Nearly 1 billion of today's more than 7 billion people have inadequate nutrition, yet population is projected to grow by another 2 billion by mid-century, posing a critical challenge to human welfare.

Food security entails multiple dimensions, each of which poses challenges to be met in the coming decades. Food security has a reciprocal relationship with population, with a critical divide existing between a demographic transition to a well-nourished society with a stable population and a demographic trap of a society impoverished by continued population growth. Another reciprocal relationship exists between food security and women's overall and reproductive health, due to the critical roles that women play in agriculture across much of the developing world and universally as mothers. Long-term food security requires stable populations, which is contingent upon the reproductive behavior of women as well as the development of a sustainable food system.

FOOD SECURITY

Food security exists when all people at all times have access to sufficient, safe, nutritious food to maintain a healthy and active life.[2] Food security is one of the central concerns of the Millennium Development Goals (MDGs),[3] and food security and sustainable agriculture also was one of seven critical issues identified at the 2012 United Nations (UN) Conference on Sustainable Development (Rio+20).[4]

Food insecurity wears many faces (Figure 12-1). The Integrated Food Security Phase Classification[5] identifies five phases, with extremes of chronic insecurity and famine (Box 12-1).

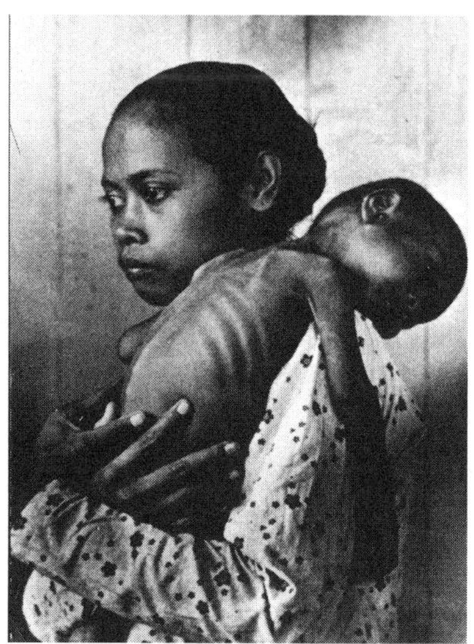

Figure 12–1 Malnutrition Is a Tragedy for Millions of Children and Their Families

Courtesy of World Health Organization and National Library of Medicine

---

Box 12–1   The Integrated Food Security Phase Classification (IPC)[5]

The IPC addresses the multidimensional nature of food security issues. It emphasizes household food security and draws on four commonly accepted conceptual frameworks for food security, nutrition, and livelihoods analysis: risk, sustainable livelihoods, the UN children's nutritional framework, and the four elements of food security discussed in the text. It further distinguishes between acute and chronic food insecurity and assesses acute situations at two points in time: the present and a foreseeable future time to facilitate possible intervention.

The framework identifies five phases of acute food insecurity: (1) none/minimal; (2) stressed; (3) crisis; (4) emergency; (5) humanitarian catastrophe/famine.

Data from the IPC: Integrated Food Security Phase Classification. What is the added-value of IPC. http://www.ipcinfo.org/ipcinfo-about/what-is-the-added-value-of-ipc/en/. Accessed March 27, 2013.

Chronic undernutrition is pervasive in some developing countries (Figure 12–2).[1,6] Of the 26 nations listed as the "world's hungriest countries," only Haiti (number 6) does not lie in Africa or Asia. Six of these countries are in Asia: Timor Leste (7), Yemen (10), Bangladesh (12), India (15), Pakistan (23), and Laos (25). The other 19 are in Africa, including the hungriest 5 countries: Democratic Republic of Congo, Burundi, Eritrea, Chad, and Ethiopia.[6]

Major indicators of food insecurity include (1) share of undernourished people, (2) share of underweight children younger than 5 years, and (3) mortality among children younger than 5 years. The hungriest three countries in the world have more than 60% malnutrition; Haiti, at 57%, is the only other country above the 50% malnutrition mark. In most of these countries, the percentage of underweight children (ranging from 15% to 45%) is less than the malnutrition rate in the whole population; in a few cases, it is half or less—for example, in Haiti it is one-third. In contrast, there are five countries—Timor Leste, Yemen, Bangladesh, India, and Madagascar—where the rate of underweight children is substantially higher than undernourishment in the general population; in India, the child rate is more than double the general undernourishment rate. For these countries, the data strongly suggest that the distribution of food is biased against children, and very likely their mothers. Surprisingly, however, the rate of under 5 mortality in these countries is less than 10%; Bangladesh has the lowest rate at 5%, which may relate to public health systems that minimize mortality from infectious disease. In other countries, child mortality rates range as high as 21%.

At the opposite end of the food security spectrum is famine. The most notorious famine in recent history occurred in China between 1958 and 1962. While its exact toll is unknown, millions starved to death. Assuming a

Figure 12–2 Life Holds No Joy for the Malnourished
Courtesy of World Health Organization and National Library of Medicine

2.25% annual growth rate in China from 1957 to 1962 yields a population about 78 million larger than actually recorded. Roughly half of this difference probably resulted from a high death rate and half from low birth rates.[7] Figure 12–3 shows the resulting deep reduction in world population growth.[8]

A key element of any famine is the economic inability of families to secure food. Appropriate government policies, for emergency assistance or job creation, can mitigate this condition.[9] In the Chinese case cited earlier, the government policy known as the Great Leap Forward caused a drop in food production, with the resulting deprivation then compounded by the Chinese government's failure to intervene—for example, by redistributing available food.

The UN Food and Agriculture Organization (FAO) uses the percentage of undernourishment (PoU), as its primary indicator of global food insecurity.[10] Derived from estimates of the average nutritional energy supply and the average daily energy requirement, this statistic has the virtue of availability across countries and over time. However, it underestimates the actual level of malnutrition in a population for several reasons: (1) it does not consider aspects of nutrition other than gross caloric availability; (2) it uses minimum requirements for daily activities, not including intense manual labor that is the lot of many poor people; and (3) it is a three-year average that cannot account for the impact of short-term spikes in food prices. Recognizing these issues, FAO has also developed a broader set of food security indicators.[11]

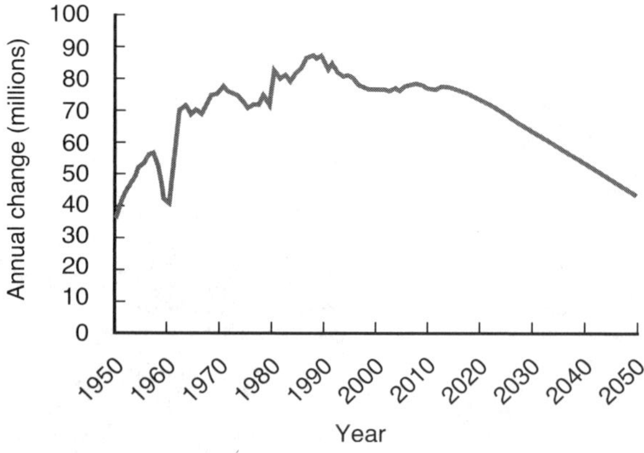

Figure 12–3 Annual World Population Change, 1950–2050

Reproduced from the United States Census Bureau, International Programs. World population, 2013. http://www.census.gov/population/international/data/worldpop/graph_annualchange.php. Accessed January 8, 2014.

Factors Impacting Food Security

The FAO[12] identifies four dimensions of food security: food availability, affordable access, household and physiological utilization, and temporal stability. To this list, others have added food sovereignty.[13]

Availability

At the largest scale, the issue of food security relates to the global availability of food and how it compares with human nutritional needs. For the 2010–2012 period, the global "average dietary food supply adequacy," expressed as a percentage, was 121% (up from 114% in 1990–1992).[11(TableV01)] Individual country values ranged from 73% for Burundi to 153% for Israel; the average for developing countries was 117%, while that for developed countries was 134%. The value of the index implies that the gross global food supply could feed roughly 8.5 billion people. The persistence of hunger in the world at the present time, therefore, results primarily from the distribution of nutritional resources, not the absolute quantity of food available.

Statistics on gross food production, particularly of grains, fail to capture an important dimension of nutrition: micronutrients. A result is the phenomenon of "hidden hunger," which occurs in an estimated 2 billion people worldwide, owing to a lack of critical vitamins and minerals, even in a diet that provides adequate caloric intake. Iron-deficiency anemia degrades maternal and infant health, slows physical and cognitive development in children, and diminishes adult work capacity. Iodine deficiency impacts maternal health and fetal development, particularly reducing mental capacity. Vitamin A deficiency impairs the immune system and leads to impaired vision and blindness. Zinc and folate deficiencies also negatively affect maternal health and fetal development.[14]

Access

MDG #1 calls for the eradication of extreme poverty and hunger, with specific targets of halving the proportion of people living on less than $1 per day and the proportion of people suffering from hunger between 1990 and 2015. Figure 12–4 shows progress made globally toward meeting these goals, in addition to the under-5 mortality rate, also relevant to this chapter.[10(p11, Figure5)] Barring retrogression by 2015, the poverty-reduction target has been achieved already. Until the "Great Recession" of 2008–2009, the proportion of undernourished people also had been on track to meet MDG #1. The disparity between poverty and undernourishment trends, however, suggests that while increasing income may be necessary to enable people to improve their diets, it is not sufficient.

Figure 12–4 also shows that improvement in under-5 mortality has lagged behind improvement in undernourishment, suggesting that children, and perhaps their mothers, have made less progress than men. Further, while the *proportion* of undernourished people has fallen, the actual number of the undernourished, because of population growth, decreased only 13%, from 980 million in 1990–1992 to 852 million in 2007–2009. Estimates for 2010–2012 indicate that the Great Recession prevented further progress in this area.

The global data illustrated in Figure 12–4 mask large regional differences, as shown in Figure 12–5.[10(p9, Figure2)] Asia and the Pacific (dominated by China and India) and Latin America and the Caribbean continue to make progress toward meeting the hunger MDG. Again, because of population growth, progress has been slower toward the more ambitious goal articulated at the 1996 World Food Summit (WFS) of halving the *number* of hungry people.[15] Nevertheless, the roughly 25% reduction in this region represents more than 200 million people. In contrast, in sub-Saharan Africa, the number of hungry people has grown by roughly 64 million, even though their proportion in the overall population has fallen. In the Near East and

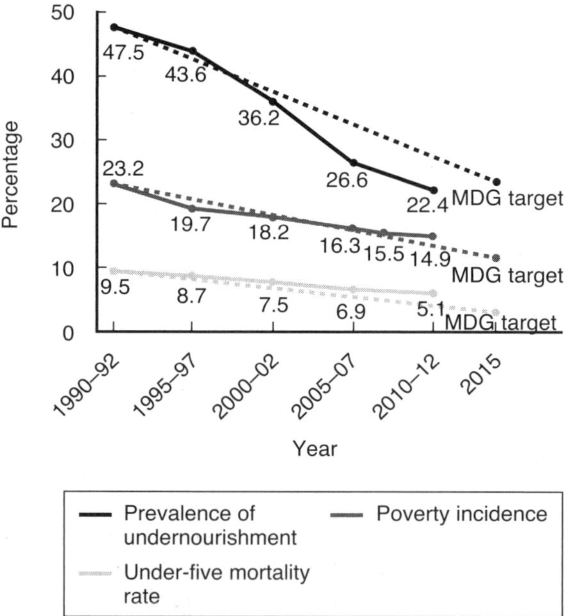

Figure 12–4 Poverty, Undernourishment, and Child Mortality in the Developing World

Reproduced from *The State of Food Insecurity in the World 2012: Economic Growth Is Necessary but Not Sufficient to Accelerate Reduction of Hunger and Malnutrition.* Rome: UN Food and Agriculture Organization; 2012. http://www.fao.org/docrep/016/i3027e/i3027e.pdf. Accessed March 27, 2013.

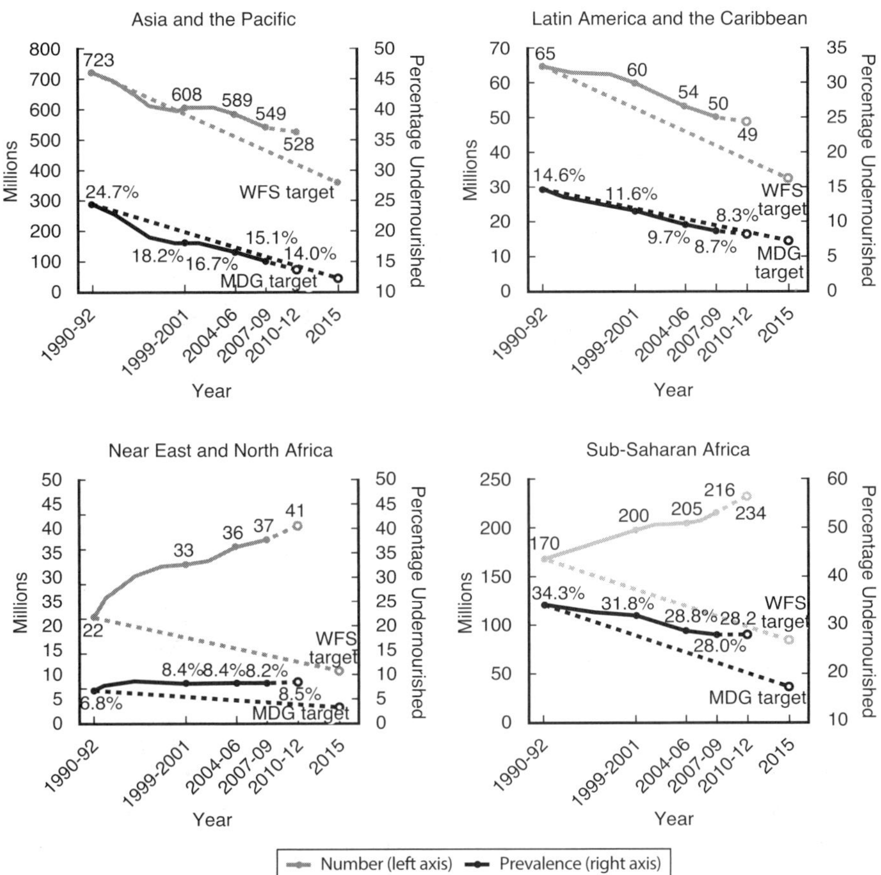

Figure 12–5 Hunger Trends in the Developing Regions

Reproduced from *The State of Food Insecurity in the World 2012: Economic Growth Is Necessary But Not Sufficient to Accelerate Reduction of Hunger and Malnutrition.* Rome: UN Food and Agriculture Organization; 2012. http://www.fao.org/docrep/016/i3027e/i3027e.pdf. Accessed March 27, 2013.

North Africa, the proportion of hungry people grew in the 1990s and has merely stabilized since, while population growth has nearly doubled the number to more than 40 million. These data illustrate how population growth is exacerbating the challenge of food security.

Economic growth in developing countries has positive impacts on the diversity of nutrition, enabling more people to gain greater access to sources of animal protein, as well as to fruits and vegetables; both trends reduce micronutrient deficiencies. Figure 12-6 illustrates the improved access to fruits and vegetables, but also reveals that in Africa availability is still only about three-fourths of the recommended 400 grams per person

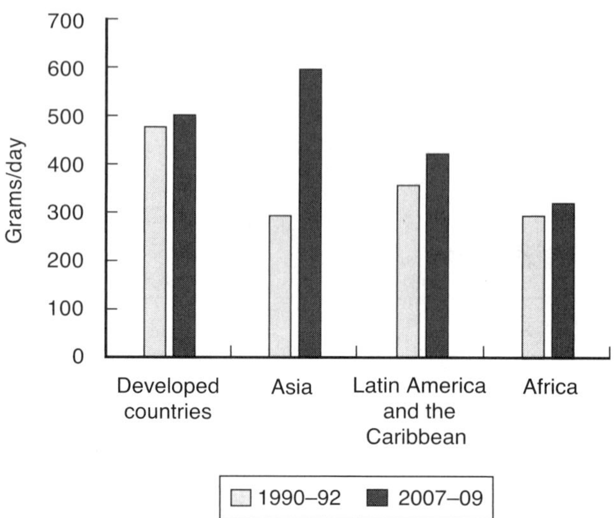

Figure 12–6 Consumption of Fruits and Vegetables Is Increasing but Remains Insufficient in Some Regions

Reproduced from *The State of Food Insecurity in the World 2012: Economic Growth Is Necessary but Not Sufficient to Accelerate Reduction of Hunger and Malnutrition.* Rome: UN Food and Agriculture Organization; 2012. http://www.fao.org/docrep/016/i3027e/i3027e.pdf. Accessed March 27, 2013.

per day.[10(p20, Figure13)] The FAO[10] stresses that not all economic growth reaches the poorest, for whom better nutrition is most critical. Agricultural growth provides rural incomes that more directly improve nutrition, and "nutrition-sensitive" policies are important for ensuring that growing national incomes yield improved nutrition for the neediest.

Although food access is a particularly acute issue in many parts of the developing world, developed countries are not exempt from this concern, because the poor everywhere are challenged to secure adequate nutrition and must devote a large fraction of their income to securing food. For example, some urban areas of the United States are "food deserts," where residents have limited local access to fresh food and frequently buy meals from fast-food establishments (Figure 12-7)—circumstances that contribute to obesity.[16] Obesity, an alternative form of malnutrition that derives at least in part from easy access to an excess of food in an unbalanced diet, affects more than 1.4 billion adults around the world, many in places also afflicted by undernutrition.[17] For example, such disparities exist between urban and rural populations in African countries.[18]

A critical element of food access that gross food availability fails to capture is the ability of poor people to acquire food. This may result from inability of subsistence farmers to grow sufficient food, at least under some conditions.

Figure 12–7 Midsection of an Overweight Man Sitting on a Park Bench with Takeout Food

© Digital Vision/Thinkstock

More often, it results from inadequate financial resources for consumers to compete in the marketplace. Economic access to food is a critical element in famine,[9] but clearly applies to chronic undernourishment as well. Angus and Butler bluntly state: "Food follows the money."[19(p74)]

Utilization

Even when adequate nutrition is available and accessed, malnutrition can result from inadequate use of the available food and the inability of individuals to absorb nutrients. Non-utilization can include the following components: loss of nutrients through waste or food preparation; inadequate knowledge or application of nutrition, childcare, or sanitation techniques; cultural practices that may limit groups or individuals within a family from consuming an adequate diet; and the state of health of individuals within the family.[20,21] Notably, cultural practices in some countries can lead to relative deprivation of women and children.

Stability

A key element in the definition of food security is the phrase "at all times." Reduction in malnutrition globally has occurred since 1990, especially because of rapid economic growth in China and other parts of Asia

(Figures 12–4 and 12–5).[10] The Great Recession, however, slowed progress and hints at challenges for the stability of food supply and the requisite increase of about 60% by midcentury. Challenges to stable food security come from price spikes, the unsustainability of industrial agriculture, resource constraints, and climate change.

Price Spikes

The FAO PoU statistic[10] fails to capture the full effects of short-term price spikes, which have impacts beyond mere nutrition. A model for the dependence of social unrest on food prices, including the impacts of market speculation,[22] indicates that a surge of social unrest, like that occurring in the Middle East in 2009 and again in 2011 (the so-called Arab Spring), directly reflects the impacts of surging food prices on vulnerable populations. Ominously, the study projects that the general upward trend of world food prices implies further unrest in the future.[23]

Underlying recent price spikes were other factors that threatened the temporal stability of food supplies. The 2007–2008 food price spike followed surging oil prices driven by disparity between demand and supply. If renewed economic growth once again causes oil demand to exceed immediate supply, food prices could spike again. The 2010–2011 spike followed a drought in Russia in 2010. Similar results may follow if more frequent or persistent droughts expected from climate change occur in major food-producing regions, such as Great Plains of the United States, which suffered severe drought in 2012.

Industrial Agriculture

Another threat to long-term food security is the unsustainability of industrial agriculture.[24,25] Industrialization of agriculture, particularly in developed countries, has led to enormous increases in production through the substitution of fossil energy for human and animal labor, plus augmentation of natural fertilizers with synthetic nitrogen and mined phosphorus, in addition to synthetic pesticides and herbicides and, most recently, genetically engineered crops. These innovations, however, have imposed diverse negative effects on the environment and on social systems: reduction of agricultural and wild biodiversity; pollution of surface and groundwater with pesticide runoff, excessive nutrients, and animal wastes; loss of soil and of organic matter in remaining soil; increased incidence of crop and animal diseases; disruption of agricultural communities; negative health consequences; and economic costs of agricultural subsidies.[26(p291)] "Sustainable intensification" of global agriculture as a response to these impacts is a critical strategy for achieving long-term food security.[27]

Resource Constraints

Long-term stability of agricultural production depends on the stability of critical inputs: land, water, soil, and energy.

Much of the growth in agricultural productivity in the 20th century came from increases in land under cultivation, as well as large increases in crop yields due to irrigation. Figure 12-8 illustrates the constancy of land for rainfed agriculture and the growth of irrigated agriculture since 1961.[29(p14, Figure1)] Superposed on that is the trend of agricultural land *per capita*, which has fallen by roughly half, although *irrigated* land *per capita* has fallen by less than 10%.[28,29] Further increases in cultivated land area would exacerbate climate change through deforestation, even as agricultural land is lost to other uses and degradation, as well as biofuel production.[30] Whether increasing the amount of land under cultivation can contribute significantly to meeting the food security challenge is unclear.

Maintaining even present levels of irrigation, unfortunately, is improbable owing to unsustainable withdrawals of groundwater. Countries that are over-pumping aquifers account for more than half the world's human population, including several of the world's most productive regions, such as the Great Plains of the United States, the North China Plain, and multiple states in India.[31] Conversely, great potential exists for improving

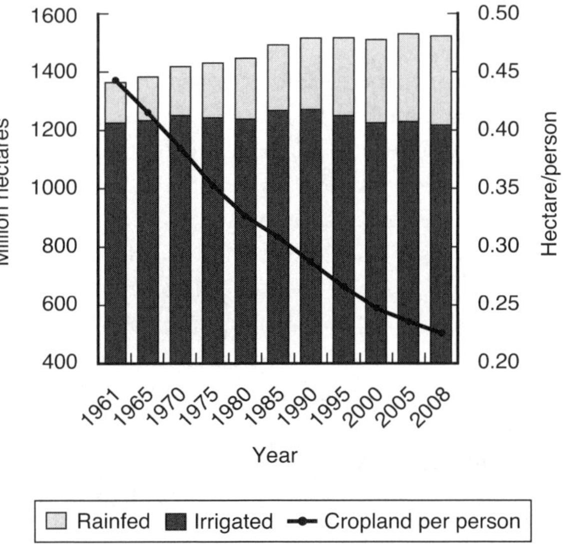

Figure 12–8 Evolution of Land Under Irrigated and Rainfed Cropping (1961–2008), with Variation of Cropland Area per Person

Reproduced from *The State of the World's Land and Water Resources for Food and Agriculture: Managing Systems at Risk. Summary Report.* Rome: UN Food and Agriculture Organization; 2011:14.

water productivity[29(p29)] and for significant improvements in agricultural productivity through expanded irrigation in other regions, such as sub-Saharan Africa.[29(p39)]

Another agricultural resource under stress is soil. Application of industrial methods in developed countries has led to depletion of the nutrients in soil faster than they are replaced by natural processes.[32] In developing countries, growth in population and nearly proportional growth in the number of grazing animals in semi-arid landscapes is leading to desertification and persistent Dust Bowl conditions, particularly in the African Sahel and northwest China and western Mongolia.[1(p48)]

A final resource constraint that could affect the stability of agricultural production is the finite supply of fossil fuel, which is the primary energy source for industrial agriculture.[4,26] Increasing application of new drilling technology in the United States has led to a sharp increase in natural gas supplies and reversal of the long-term decline in oil production, leading the International Energy Agency to project increasing supplies of this fossil fuel to 2035.[33] These developments may delay energy bottlenecks that could greatly increase food prices, but the energy-intensive technologies behind the increased gas and oil production also require higher prices to ensure profitability.[34] Continued reliance on fossil fuels, of course, will exacerbate climate change.

Climate Change

Fossil fuel emissions continue to track toward the high end of projections made by the Intergovernmental Panel on Climate Change (IPCC),[35] even as extreme weather events impose increasing costs around the world. The impacts of climate change include increasing the likelihood of droughts (Figure 12–9 and Figure 12–10), such as those that have afflicted major crop-producing regions such as Russia (2010) and the United States (2011, 2012), and the extreme events of the 2012–2013 "Angry Summer" in Australia. As discussed earlier, commodity shortages on the world market have led to food price spikes that contributed to social unrest, in addition to direct impacts on nutrition among the poor. Further, studies have documented decreased agricultural yields already attributed to the warming climate.[4,36] Warming and associated trends are certain to continue for several decades as the result of heat-trapping gases already in the atmosphere. They will become more severe in the absence of concerted political action to curb emissions that would mitigate further climate change in the latter part of this century. To quote the 2014 IPCC assessment: "All aspects of food security are potentially affected by climate change, including food access, utilization, and price stability."[37]

Figure 12–9 Dry Riverbed, Uganda, Africa

© Arjen de Ruiter/Shutterstock.com

Figure 12–10 Dead Cow in African Drought

© 1a_photography/iStock/Thinkstock

Direct methane and nitrous oxide emissions from agriculture contribute roughly 10% to 12% of global greenhouse gas emissions; deforestation contributes a comparable but more uncertain amount.[38] These figures, however, do not include contributions from fossil fuel energy consumed in the food system. Thus global agriculture perversely threatens long-term food security through climate change, even as it strives to increase short-term productivity.

The Right to Food and Food Sovereignty

The 1948 United Nations Universal Declaration of Human Rights includes the right to food. Only the 1996 World Food Conference, however, began an effort to bring this right to bear. One outcome was appointment of the Special Rapporteur on the Right to Food, whose website[39] articulates a definition similar to that of the FAO for food security, but stresses both physical and economic access. The Rapporteur's definition goes even further by incorporating cultural appropriateness and stressing each state's obligations for enabling food production or food purchase by its citizens.

A related dimension of food security has to do with control. In the United States and globally, the agricultural system is dominated by a small number of corporations.[40] On-farm income represents just a small fraction of the agricultural economy, with large profits going to corporate investors. These same corporations exert disproportionate political influence (e.g., on the U.S. farm bill) and likewise profit from food aid policies. The governing paradigm is to generate the greatest return per dollar invested, which depends critically on industrial scale and in general does not correspond to the peak productivity of land, maximum energy efficiency, or availability of food for the largest number of people.[41] These circumstances have generated movements for *food sovereignty*, "the rights of nations and peoples to control their own food systems, including their own markets, production modes, food cultures and environments,"[13(p2)] and for *slow money*,[42] which promotes investment in small agricultural enterprises with the intention of promoting greater production of nutritious food for local populations, rather than maximum financial return.

Woody Tasch, the author of the book *Slow Money*, identifies the corporate imperative to maximize profits as one of the root causes of food insecurity: "The problems we face with respect to soil fertility, biodiversity, food quality, and local economies are not primarily problems of technology. They are problems of finance. In a financial system organized to optimize the efficient use of capital, we should not be surprised to end up with cheapened food, millions of acres of GMO corn, billions of food miles, dying Main Streets, kids who think food comes from supermarkets, and obesity epidemics side-by-side with persistent hunger."[42(pxvi)] Angus and Butler[19] echo these comments, while proponents of food sovereignty identify the concept and the associated social movement as "a critical alternative to the dominant neoliberal model for agriculture and trade."[13(p2)]

Projections of Food Security

The Green Revolution is rightfully credited with significant improvements in the human condition during the second half of the 20th century.

Innovations in the genetic strains of staple crops, combined with application of fossil-fuel–intensive farming techniques, including mechanized application of artificial fertilizer and pumped irrigation, led to dramatic increases in yields and in overall food production.[43] However, as discussed later in this chapter, the increased availability of food contributed to the acceleration of population growth. As a consequence, *per capita* food production has stagnated since the 1980s, as shown in Figure 12–11.[44] Moreover, the distribution of available agricultural bounty is uneven.

Brown describes the multiple obstacles to increasing food supply to meet the mid-range projected global population of 9 billion by midcentury: the progressive reduction in arable land per person, the increase in demand for meat products and corresponding diversion of grain supplies to animals; the diversion of agricultural land to biofuel production; topsoil loss; water scarcity; plateauing yields of staple crops in the most productive countries; climate change; and increasing purchases by food-challenged countries of land in other countries (primarily in developing ones) to the detriment of the local populations. He does not cite potential limitations in fossil fuel availability.[1]

These arguments engender skepticism that global food production can grow by the 60% or so required to meet the projected needs of 9 billion people in 2050. However, given the present surplus of food, the prospect for greater agricultural yields in some of the most undernourished portions of the world, and the prospect for significant reduction of food waste, greater attention to economic access and food sovereignty have the potential to increase food security, even if gross per capita food production falls.[19]

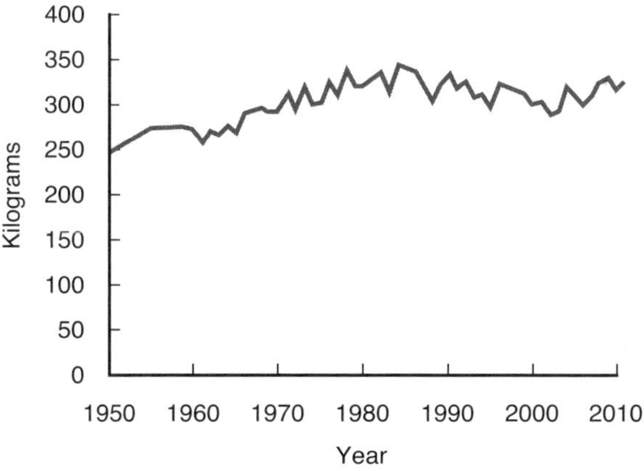

Figure 12–11 Grain Production per Person, 1950–2010

Reproduced from Earth Policy Institute. World grain production per person, 1950–2011. http://www.earth-policy.org/datacenter/xls/book_fpep_ch1_3.xlsx. Accessed March 27, 2013.

A guide to the associated challenges comes from the Agrimonde Project.[45] This effort in foresight modeled alternative scenarios for meeting global nutritional needs, which might be dubbed "business as usual," with a primary focus on economic growth, and "sustainable development," seeking to include ecological balance along with economic growth. Both scenarios are able to achieve adequate agricultural production to feed 9 billion people in 2050. The "business as usual" scenario (which tacitly assumes abundant fossil fuel energy) relies on expanded cultivated area and biotechnology for significant crop yield increases, and it sacrifices ecological impacts for economic growth. The "sustainable development" scenario relies on significant technological and cultural shifts. It posits a transition to locally based agriculture, expanding even more the land area under cultivation (primarily at the expense of rangelands), minimizing fossil fuel inputs and food wastage, shifting away from using grain as animal feed, and increasing international trade without undermining markets in developing countries. The two contrasting scenarios provide provocative insight into the long-term challenges of food security—in particular, the need to ensure food access by poor populations.

Sub-Saharan Africa is "ground zero" for food security. As noted earlier, it is the locus of the majority of the most food-insecure countries. This region was largely bypassed by the Green Revolution, and the agricultural sector there is relatively undeveloped, leading to low yields, even compared to developing countries in other regions. Moreover, it is the region with the highest human fertility levels and population growth rates. Torrey[18] has reviewed present food production and demographic trends. In terms of gross food production, sub-Saharan Africa today produces roughly enough food to meet its residents' caloric needs, but wastage and poor distribution systems lead to widespread undernourishment. Torrey has estimated the caloric needs by midcentury, allowing for reduced human fertility, changing demographic profiles, and urbanization. Meeting the needs of a population projected to increase by two-thirds even in low-fertility scenarios will require a proportionately larger increase in food supply, owing to the caloric needs of a population made up of a larger fraction of adults. Increases in agricultural productivity will be essential to meet this demand, but could have synergistic effects by improving the larger economy. The critical role of women in this effort has been stressed by Bremner.[46]

Strategies

Food and nutrition represent a key theme as the UN looks beyond the 2015 target date for the MDGs.[47] One articulation of extended sustainable development goals includes "sustainable food security.[47](pp305–307) End hunger and achieve long-term food security—including better nutrition—through sustainable systems of production, distribution and consumption."[48](pp305–307)

Recognizing that "business as usual" is not sustainable, the FAO has provided a policy maker's guide[27] for the needed transition toward "sustainable intensification" of global agriculture, "producing more from the same area of land while reducing negative environmental impacts and increasing contributions to natural capital and the flow of environmental services."[27](Chapter1) Approaches to achieve this goal include[26] (1) agroecology, "the application of ecological science to the study, design, and management of sustainable agriculture,"[49] particularly among the smallholder farmers in developing countries,[50] who feed 80% of the world's people; (2) organic agriculture, as practiced in developed countries; and (3) improved crop cultivars, achieved by conventional plant breeding guided by modern genetic mapping, and perhaps by genetically modified organisms (GMOs), although the sustainability of GMOs remains controversial.[51,52]

Technological strategies identified for achieving sustainable agricultural output include[53,54] stopping the expansion of global agricultural area; closing the gaps between potential and actual yields for existing crops; increasing agricultural resource efficiency; increasing production with improved cultivars, especially in developing countries; increasing food delivery by shifting diets and reducing waste; and expanding aquaculture. Notably, these strategies do not address food access and utilization—aspects of food security that relate less to technology than to financial and social systems, not least the persistence of deep poverty.

FOOD SECURITY AND POPULATION GROWTH

The classic discussion of population and food security appears in the work of Thomas Malthus (1766–1834), who famously argued that food production could expand roughly arithmetically, whereas population could grow geometrically, ensuring that food insecurity for the poor is inevitable.[55](p75) As discussed in the previous section, global food production has more than kept pace with population, in spite of enormous population growth.

There is a reciprocal relationship between nutrition and population. Growing food security feeds population growth, particularly during the demographic transition, when improved nutrition and health care lead to an imbalance between births and deaths. However, excessive population growth during the demographic transition may trap countries in poverty, thereby exacerbating food insecurity. The critical issue is whether birth rates decline quickly enough for a country to complete the demographic transition rather than to become caught in the demographic trap.

Food Security as a Cause of Population Growth

Hopfenberg[56] has argued that for humans, as for other animals, population growth is driven by food supply, without significant impact from other variables. Hopfenberg and Pimentel[57] go even further, arguing that the paradigm of increasing food supply to feed a growing population is misguided: population growth will be limited only by limiting food supply. This mechanistic approach leads to the moral conundrum of sacrificing the world's poor to achieve population stability and associated limitation of environmental degradation.

The growth in population following the Green Revolution broadly supports Hopfenberg's contention. The work of Fogel[58] provides causal texture to his thesis. Data from developed countries since the Industrial Revolution show that greater food security and the corresponding improvements in nutrition reduce childhood stunting (low height for age) and wasting (low weight for age), with lifelong consequences in reduced morbidity and mortality. Moreover, per capita food consumption grew approximately 0.6% per year during the last several decades of the 20th century,[58(p54)] including improvements in consumption of high-quality protein (primarily from animal products).[58(p137, fn3.21)] Consequently, it is reasonable to infer that improvements in nutrition did, indeed, contribute significantly to population growth in the late 20th century.

However, factors other than nutrition also contribute to population growth. For example, although some two-thirds of the health improvements, and consequent longevity, in American men through the 20th century resulted from improved nutrition, other factors such as public health initiatives and improved health care must account for the remaining one-third of health improvements.[59] Similarly, as discussed further later in this chapter, extension of family planning services can lead to lower birth rates.

Population Growth as a Threat to Food Security

The world presently produces more than enough food to feed the entire human population, but many of the world's poor lack economic access to adequate nutrition. The "demographic trap" describes a situation in which rapid population growth outruns economic development, leaving countries trapped in poverty and food insecurity.[55(p441)]

Specific data illustrate the connection between demographic pressure and food insecurity in some countries. Seven of the 26 hungriest nations[6] also appear in the list of the top 20 failing states.[60] Each of the 7 has a total fertility rate (TFR) of 4.4 or higher; 5 of them have population growth rates of 2% or more. High rankings in terms of both demographic pressure and hunger contribute significantly to their high failing-state index values, supporting the contention that population pressure is trapping these countries in malnutrition and civil unrest.

Detailed models have projected food demand and food supply under three population/economic growth scenarios and five subregions of Africa.[61] Most of the conclusions, however, address the extreme cases of (1) low population growth with high income and (2) high population growth with limited income. Interestingly, the former case yields the highest demand for food, presumably owing to improved economic access, but also increases food production faster than population growth. Conversely, even though the opposite case yields the lowest food demand, population growth outstrips food production—a scenario with grim human implications. These data reinforce the potential threat of the demographic trap and the FAO call for nutrition-sensitive economic growth.[10]

The tragic case of Rwanda illustrates how these trends can play out in a specific country.[62] Although multiple trends contributed to the genocide of 1994, one factor was population growth that overwhelmed the national capacity to produce food. Fourfold population growth between 1948 and 1992 resulted in increasing subdivision of available land, which in turn reduced both fallow time and manure fertilization, and drove peasants to cultivation of increasingly marginal (and eroding) landscapes. During the period of increasing civil unrest in 1991–1992, two-thirds of the population failed to obtain the minimum food energy requirement of 2100 calories per person per day, and the communities (communes) that experienced violence were correlated with nutritional deficits. No violence occurred where the average food energy was more than 1500 calories per person per day; relatively little violence occurred in the most deprived communes (6 of the 21 communes with less than 1000 calories per person per day), but violence occurred in 23 of the 29 communes with intermediate nutrition. A brief cease-fire ended catastrophically with the assassination of the president, triggering the genocide that killed hundreds of thousands and led 2 million people (approximately 25% of the population) to flee the country, at least temporarily. More stable government and economic growth have improved conditions in Rwanda, but great food security challenges persist in the country.[63]

Demographic Transition or Demographic Trap?

Successful navigation of the demographic transition and avoidance of the demographic trap depend on timely reduction in fertility. Three factors play key roles here: awareness of women that fertility reduction is a choice; objective advantages to smaller family size; and availability of acceptable means of fertility reduction.[64] The first factor depends largely on women's educational status; the second on economic variables, including empowerment of women; and the third on access to reproductive health services.

A powerful comparison of two progressive states in India (Kerala and Tamil Nadu) shows that education and empowerment of women there have led to even larger fertility reductions than the coercive measures adopted in China.[9(Chapter9)] Furthermore, comparisons between these two states and other Indian states where birth rates remain high reveal that education of women and their participation in the work force are the *only* two elements that have significantly impacted birth rates. Multivariate analysis of a global data set confirms that literacy rate for women aged 15 and older has the strongest influence on reductions in total fertility rates.[64] Others have also documented the importance of women's education in decreasing birth rates.[65(p376),66(p241)]

Improved access to contraception has led to avoidance of an estimated 230 million births per year, and an additional avoidance of some 270,000 annual maternal deaths.[67] These results suggest that extending modern reproductive health services—in particular, meeting the unfulfilled need for contraception of an estimated 222 million women worldwide—is likely to accelerate reductions in both fertility and maternal mortality. Yet extending contraception services to these women would still leave millions more women without access to family planning.[67] Even so, gradual fulfillment of the unmet need for family planning would reduce total fertility rates below the replacement level after 2030 and approach the levels of the UN's low population estimate for 2050.[68]

Even where food security prevails, particularly in industrialized countries, ongoing population growth imposes growing environmental impacts, such as climate change and biodiversity loss. Population stabilization, therefore, is a necessary goal for achievement of global sustainability. Although poverty reduction and improved food access could greatly reduce food insecurity in the world today, long-term stability of food security requires cessation of population growth. Efforts in this direction require engagement of the world's women.

WOMEN, FOOD SECURITY, AND REPRODUCTIVE HEALTH

Women and Food Security

Women as food producers have an important role to play in improving food security. Conversely, malnutrition among women exacerbates food insecurity both for them and for their families. The case studies that follow illustrate the positive outcomes realized from empowering women.

Women as Food Producers and Caregivers

In many parts of the world, particularly in Africa, women play a critical role in smallholder farming. Regionally, women constitute 20% to 50% of the agricultural labor force; in some individual African and Middle Eastern countries, women's share of this work force exceeds 60%. Given equal resources, women farmers are as productive as their male counterparts, but traditional gender roles greatly limit their access to land, technology, education, and capital, so that women farmers typically produce 10% to 20% lower yields. Folding this estimate together with country data on percentages of women farmers leads to the estimate that providing equal access for women to agricultural resources could increase food production in developing countries by 2.5% to 4%. Furthermore, such action would reduce the number of hungry people by 12% to 17%—roughly 100 to 150 million—many of whom are women and their children.[69]

Because women are typically the primary caregivers for their families, improving their status would further directly improve the lot of their children, leading to better nourishment, health, education, and lifetime productivity, thereby paying societal dividends for generations. Improved education and status for women also leads to their choosing to have fewer children.

Nutritional Shortfall and Agricultural Productivity

The "hunger trap"[70] describes a situation in which hungry people lack sufficient nutritional energy to participate successfully in the economy. The consequences of this situation include the perpetuation of hunger, often preferentially afflicting women and girls, leading to greater numbers of underweight and malnourished children, who are even less able to earn sufficient income to escape poverty. Indeed, the statistics on women in agriculture rest on the numbers of women who actually participate in the agricultural economy.[69] To the extent that hungry women might want to work in gardens or fields, but are nutritionally unable to do so, these figures are underestimates.

The impact of undernutrition on general labor productivity is estimated at 6% to 10% of per capita GDP, with losses in heavy labor from iron deficiency alone exceeding those from general protein-energy malnutrition.[71] Such productivity losses from hunger surely affect the agricultural productivity of women and increase the food insecurity of their families. In addition, these losses likely contribute to the differences in agricultural productivity between women and men.

Case Studies

Winifred Omoding provides an encouraging example of farming success from Uganda.[72] Forced into farming simply to feed her family through nearly two decades of political violence, she has become an enthusiastic and successful farmer. Blessed with a three-hectare farm inherited by her husband, she received assistance though a microloan cooperative that she and other women founded with guidance from an aid organization. Further, Omoding was able to buy seeds (with a microloan) and receive advice from a new agricultural research institute. The result was restoration of the land's failing productivity and bounty, which enabled her to feed her large family (nine children), with surplus to sell through a marketing network created by the research institute. Continued prosperity, however, is vulnerable to climate change, to ongoing dependence on donor groups for some of her sales, and to the need for the national government to make infrastructure investments. Omoding's large family is symptomatic of the challenge faced by Uganda and other developing countries with high fertility rates, although her children will be well educated and very likely to have much smaller families.

An encouraging institutional example is the Self-Employed Women's Association (SEWA) in India.[73] This organization serves more than 1 million poor women, 94% of whom are in the unorganized labor sector, more than half of them poor farmers. With goals of full employment and self-reliance for its members, SEWA provides a spectrum of services otherwise scarcely available to members: banking; insurance and legal services; health and child care; housing and infrastructure; and capacity building. The women farmers of SEWA earn revenue far beyond earlier imagination, achieving food security for themselves and their families, as well as producing sustainably grown, affordable, and healthy food that reduces food insecurity in their local communities.

Both of these cases include examples of the kinds of institutional supports that custom frequently denies to women farmers. More stories of individuals and institutions addressing food security issues can be found through the Worldwatch Institute's Nourishing the Planet Program[74] and through Food Tank.[75]

Food Security and Reproductive Health

A key element linking food security, population, and reproductive health is maternal nutrition. Put simply, nutrition influences fertility. For instance, women who are very slender stop having menstrual periods. This happens whether the woman has a low body mass index (BMI: a measure of the ratio of a person's weight to height) due to exercise, anorexia nervosa (a psychiatric condition where people eat very little because they perceive that they are

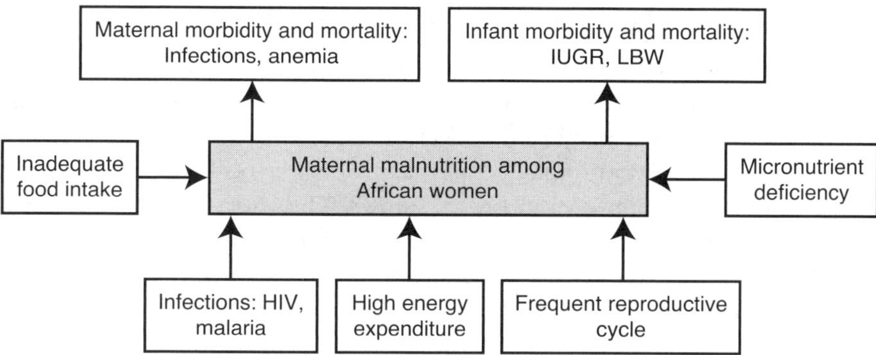

Figure 12–12 Effect of Maternal Malnutrition on Maternal and Infant Health

IUGR, intrauterine growth restriction; LBW, low birth weight

Reproduced from Lartey A. Maternal and child nutrition in sub-Saharan Africa: challenges and interventions. *Proc Nutr Soc.* 2008;67:105–108.

obese even when they are very thin), or lack of nutrition. Thus, if a woman is too thin, she does not ovulate. Whatever the mechanism, it protects women from conceiving when they would be unable to carry a fetus to viability.[76]

Malnourished mothers who carry their babies to term give birth to stunted children, and malnourished children are more likely to succumb to death by age 5. Figure 12–12[77] illustrates for Africa the complex web of causes and effects related to maternal and child morbidity and mortality; the relative impacts of the different factors may differ in other parts of the world, but the consequences are the same. The consequences of undernutrition in mothers and children include 3.5 million deaths annually, 35% of the disease burden for children between birth and age 5, and 11% of the global total of "disability adjusted lost years" (a standardized measure of the relative impact of disease and premature death). Short stature in mothers caused by malnutrition is a risk factor for caesarean delivery, and anemia creates risk of hemorrhage. Together, these two factors account for at least 20% of maternal deaths in childbirth.[78]

Such nutritional effects may lower birth rates, so widespread malnutrition could lower projections of population growth, albeit in a most inhumane manner. At the same time, they may create urgency for families to have more children to replace those who perish—a factor in the demographic trap.

CONCLUSION: BEYOND REPRODUCTIVE HEALTH

The achievement of food security represents a daunting challenge for humankind. The preceding discussion points to three sets of investments that will contribute to reaching this goal. Ultimately, however, food security represents just one aspect of a sustainable world.

Investments in Reproductive Health and Family Planning

Many discussions of food security tacitly or explicitly assume the accuracy of the mid-range UN demographic projection, which now forecasts a global population of 9.6 billion by 2050.[79] A critical question concerns the fertility projections. Like virtually all other global problems, food security becomes less challenging if population growth slows. The shortfall in availability of contraceptive services across the world— particularly in the developing world—is well documented, making it clear that there is ample room for improving women's access to voluntary reproductive health and family planning services. Past episodes of human rights violations in the name of "population control," however, created a legacy of distrust that hampers humane expansion of such services.[65] Such services must be viewed in the context of a larger program to improve the global status of women.[9,65,66]

Investments in Smallholder Programs and Education for Women

One component of such a program should be education of girls. Another component needs to be greater access of women smallholders to land, capital, and improved agricultural knowledge and techniques. Because of their currently restricted access to such resources, women farmers are less productive than men. Increasing their productivity would mostly serve local needs, not least those of their own families. Improvement in women's productivity would directly address the issue of food security and indirectly reduce fertility levels. The outcome of such a concerted program would be a virtuous cycle of improved agricultural productivity, improved maternal and child nutrition, reduced child mortality, and lower birth rates, which would further improve nutritional levels in succeeding generations.

One caveat may be relevant here. Globally, agricultural investments are largely driven by the motivation of yielding corporate profits, which then become distributed to shareholders, almost always in developed countries or among the elites in developing countries. This system has failed to distribute equitably the existing bounty of global agriculture among the world's people. It is unlikely that success in meeting the MDG of halving world hunger by 2015—and continuing on to abolish hunger in the future—will be possible as long as decisions are driven almost exclusively by the goal of maximizing economic growth per se. National and international decision makers must weigh actual human welfare, most significantly food security, not merely gross economic productivity. To meet the challenge of ensuring food security, global agricultural productivity must rise in terms of yields per acre and per unit of energy input, not necessarily in terms of yield per dollar invested. Broadening the grounds for economic decision making—most importantly in terms of calories and micronutrients delivered to people, especially to women and children—would go far toward addressing the right to food and food sovereignty.

Effective Investments in Nutrition Programs

The existing international nutrition system is "fragmented and dysfunctional."[80] The scale of basic nutritional assistance (as opposed to emergency assistance) pales in comparison to the need: even if perfectly targeted at very young children in the 20 most food-insecure nations, the current level of expenditures would provide little more than $2/child/year, much less than the $5–$10/child/year required for an effective large-scale program. In comparison, the $2.2 billion per year devoted to HIV/AIDS programs is roughly seven times greater than the funding devoted to basic nutritional assistance. Further, the negative impact of food aid and agricultural protectionism on rural agricultural productivity costs low- and middle-income countries more than $8.6 billion in lost GDP. Frequently placing food on sale in the target countries, food aid programs also limit their impact on food insecurity by commonly leaving the nutritional resources beyond the resources of the most economically vulnerable populations.

Nutritional interventions exist that could reduce stunting, micronutrient deficiencies, and child deaths in the short term, but few actions that clearly would improve maternal health have had large-scale assessments. Investments in women's education, financial status, and empowerment are needed to eliminate stunting in populations in the long term.[81]

Food Security and Sustainability

Educating and empowering women, particularly with general health and reproductive health services, holds promise for stabilizing population growth by mid-century. Such an achievement would moderate the food security challenge as well as the environmental impacts of population growth. In the case of nutrition, as is true for climate change and other issues, population stabilization is necessary for sustainability. By itself, stabilizing population will not solve the issue of providing universal food access, which requires structural change in the food system. The movements for the right to food, food sovereignty, and food justice point to the larger issue of distribution and the imperative to empower all with the positive freedom[9(Chapter9)] to access adequate nutrition, not merely with the negative freedom from hunger.

A sustainable economy requires transitioning to a dynamic equilibrium in which humankind neither takes from nor deposits into the environment more than natural processes can accommodate.[82] Ongoing technological innovation could then improve quality of life, but without a quantitative increase in mineral extraction or waste disposal beyond the capacity of the environment to regenerate itself. Such a situation could include progressive improvements in nutrition, both by enabling farmers everywhere to achieve high yields and by minimizing food waste, while preserving the essential nutrient flows through the biosphere. Stability of such a steady-state economy requires equitable distribution of resources, particularly those essential to meeting human needs—food and water. As stressed by Tasch[42] as well as by Dietz and O'Neill,[82] this will require restructuring of the financial system as well.

Such a program for radical reform will not be implemented quickly. Sustainable development, however, requires that it happen. Without it, the challenges of reproductive health, population, and food security are unlikely to be met.

DISCUSSION QUESTIONS

1. How is food security related to population growth?
2. How is food security related to reproductive health?
3. Discuss the five dimensions of food security and explain how they can be addressed.
4. Compare the idea of food sovereignty with the neoliberal model of agriculture.
5. Discuss the ways that food security can be regarded as a women's issue.

REFERENCES

1. Brown L. *Full Planet, Empty Plates: The New Geopolitics of Food Scarcity.* New York: W. W. Norton; 2012:122.
2. *An Introduction to the Basic Concepts of Food Security.* Rome: UN Food and Agriculture Organization; 2008. http://www.fao.org/docrep/013/al936e/al936e00.pdf.
3. We can end poverty: 2015 Millennium Development Goals. http://www.un.org/millenniumgoals/. Accessed March 27, 2013.
4. *Food Security and Sustainable Agriculture.* Rio 2012 Issues Briefs, No. 9. December 2011. http://www.uncsd2012.org/content/documents/227Issues%20Brief%209%20-%20FS%20and%20Sustainable%20Agriculture%20CLCnew.pdf. Accessed March 27, 2013.
5. What is the added-value of IPC. IPC: Integrated Food Security Phase Classification. http://www.ipcinfo.org/ipcinfo-about/what-is-the-added-value-of-ipc/en/. Accessed March 27, 2013.

6. Food insecurity indicators in world's hungriest countries. Earth Policy Institute. http://www.earth-policy.org/datacenter/xls/book_fpep_ch1_17.xlsx. Accessed March 27, 2013.

7. Great Chinese Famine. *Wikipedia.* February 26, 2013. http://en.wikipedia.org/wiki/Great_Chinese_Famine. Accessed March 27, 2013.

8. World population. United States Census Bureau, International Programs. http://www.census.gov/population/international/data/worldpop/graph_annualchange.php. Accessed March 27, 2013.

9. Sen A. *Development as Freedom.* New York: Anchor; 1999:163–188.

10. *The State of Food Insecurity in the World 2012: Economic Growth Is Necessary but Not Sufficient to Accelerate Reduction of Hunger and Malnutrition.* Rome: UN Food and Agriculture Organization; 2012. http://www.fao.org/docrep/016/i3027e/i3027e.pdf. Accessed March 27, 2013.

11. Food security indicators. In: *The State of Food Insecurity in the World 2012.* Rome: UN Food and Agriculture Organization; 2012. http://www.fao.org/publications/sofi/food-security-indicators/en/. Accessed March 27, 2013.

12. *An Introduction to the Basic Concepts of Food Security.* Rome: UN Food and Agriculture Organization; 2008. http://www.fao.org/docrep/013/al936e/al936e00.pdf. Accessed March 27, 2013.

13. Wittman H, Desmarais A, Wiebe N. The origin and potential of food sovereignty. In: Wittman H, Desmarais A, Wiebe N, eds. *Food Sovereignty.* Oakland, CA: Food First Books; 2010:1–14.

14. Kennedy G, Nantel G, Shanty P. The scourge of "hidden hunger": global dimensions of micronutrient deficiencies. *Food Nutr Agricult.* 2003;32:8. Rome: UN Food and Agriculture Organization. ftp://ftp.fao.org/docrep/fao/005/y8346m/y8346m01.pdf. Accessed April 14, 2013.

15. *Rome Declaration on World Food Security and World Food Summit Plan of Action.* Rome: UN Food and Agriculture Organization; November 13, 1996. http://www.fao.org/docrep/003/w3613e/w3613e00.htm. Accessed: March 27, 2013.

16. Beaulac J, Kristjansson E, Cummins. A systematic review of food deserts, 1966–2007. *Prev Chronic Dis.* 2009;6(3):A105. http://www.ncbi.nlm.nih.gov/pubmed/19527577. Accessed March 27, 2013.

17. Obesity and overweight. World Health Organization, Media Center. http://www.who.int/mediacentre/factsheets/fs311/en/index.html. Accessed March 27, 2013.

18. Torrey B. Population dynamics and future food requirements in sub-Saharan Africa. In: Pinstrup-Andersen P, ed. *United Nations University: African Food System and Its Interactions with Human Health and Nutrition.* Ithaca, NY: Cornell University Press; 2010:182–198.

19. Angus I, Butler, S. *Too Many People? Population, Immigration, and the Environmental Crisis.* Chicago, IL: Haymarket; 2011.

20. Definition of food security. USAID Policy Determination. http://transition.usaid.gov/policy/ads/200/pd19.pdf. Accessed March 27, 2013.

21. Food utilization. World Food Programme. http://www.foodsecurityatlas.org/idn/country/fsva-2009/chapter-4-food-utilization. Accessed March 27, 2013.

22. Lagi M, Bertrand KZ, Bar-Yam Y. The food crises and political instability in North Africa and the Middle East. New England Complex Systems Institute. 2011. http://necsi.edu/research/social/food_crises.pdf. Accessed March 28, 2013.

23. Lagi M, Bar-Yam Y, Bar-Yam Y. Update July 2012: the food crises: the US drought. New England Complex Systems Institute. July 23, 2012. http://necsi.edu/research/social/foodprices/updatejuly2012/. Accessed March 28, 2013.

24. Tilman D. The greening of the Green Revolution. *Nature.* 1998;396:211–212.

25. Kimbrell A. Seven deadly myths of industrial agriculture. In: Kimbrell A, ed. *Fatal Harvest: The Tragedy of Industrial Agriculture.* Washington, DC: Island Press; 2002:49–63.

26. White RE. Fossil fuel and food security. In: Kahn S, ed. *Fossil Fuel and the Environment.* Rijeka, Croatia: InTech; 2012:279–304. http://www.intechopen.com/books/fossil-fuel-and-the-environment/fossil-fuel-and-food-security.

27. *Save and Grow: A Policymaker's Guide to the Sustainable Intensification of Smallholder Crop Production.* Rome: UN Food and Agriculture Organization; 2012. http://www.fao.org/ag/save-and-grow/index_en.html. Accessed March 27, 2013.

28. World irrigated area and irrigated area per thousand people, 1961–2009. Earth Policy Institute. http://www.earth-policy.org/datacenter/xls/book_fpep_ch6_4.xlsx. Accessed March 27, 2013.

29. *The State of the World's Land and Water Resources for Food and Agriculture: Managing Systems at Risk, Summary Report.* Rome: UN Food and Agriculture Organization; 2011:14, Figure 1. http://www.fao.org/nr/water/docs/SOLAW_EX_SUMM_WEB_EN.pdf. Accessed March 27, 2013.

30. Smedshaug CA. *Feeding the World in the 21st Century: A Historical Analysis of Agriculture and Society.* London/New York: Anthem Press; 2010.

31. Countries overpumping aquifers in 2012. Earth Policy Institute. http://www.earth-policy.org/datacenter/xls/book_fpep_ch6_6.xlsx. Accessed March 27, 2013.

32. Pimentel D, Pimentel MH, eds. *Food, Energy, and Society.* 3rd ed. Boca Raton, FL: CRC Press; 2008: Chapter 15.

33. World energy outlook 2012: executive summary. International Energy Agency. 2012. http://www.iea.org/publications/freepublications/publication/English.pdf. Accessed March 28, 2013.

34. Guilford MC, Hall CA, O'Connor P, Cleveland CJ. A new long term assessment of energy return on investment (EROI) for U.S. oil and gas discovery and production. *Sustainability.* 2011;3(10):1866–1887. http://www.mdpi.com/2071-1050/3/10/1866. Accessed April 30, 2013.

35. Peters GP, Andrew RM, Boden T, et al. The challenge to keep global warming below 2 °C. *Nature Climate Change.* 2013;3:4–6. http://www.nature.com/nclimate/journal/v3/n1/full/nclimate1783.html. Accessed March 27, 2013.

36. Lobell DB, Schlenker W, Costa-Roberts J. Climate trends and global crop production since 1980. *Science.* 2011;233:616–620.

37. Field CB, Barros VR, Mastandrea MD, et al. *Working Group II: climate change 2014: impacts, adaptation, and vulnerability: summary for policymakers.* Intergovernmental

panel on Climate Change; 2014. http://ipcc-wg2.gov/AR5/images/uploads/ IPCC_WG2AR5_SPM_Approved.pdf. Accessed March 31, 2014.

38. Smith P, Martino D, Cai Z, et al. Agriculture. In Metz B, Davidson OR, Bosch PR, et al., eds. *Climate Change 2007: Mitigation. Contribution of Working Group III to the Fourth Assessment Report of the Intergovernmental Panel on Climate Change.* Cambridge, UK/New York: Cambridge University Press; 2007. http://www.ipcc-wg3.de/ assessment-reports/fourth-assessment-report/.files-ar4/Chapter08.pdf. Accessed April 19, 2013.

39. Right to food. Olivier De Schutter: United Nations Special Rapporteur on the Right to Food. http://www.srfood.org/index.php/en/right-to-food. Accessed March 28, 2013.

40. The global food crisis: ABCD of food: how the multinationals dominate trade. *The Guardian; Poverty Matters Blog.* http://www.guardian.co.uk/global -development/poverty-matters/2011/jun/02/abcd-food-giants-dominate -trade. Accessed April 30, 2013.

41. McKibbin B. *Deep Economy: The Wealth of Communities and the Durable Future.* New York: Times Books; 2007: Chapter 2.

42. Tasch W. *Inquiries into the Nature of Slow Money: Investing as if Food, Farms, and Fertility Mattered.* White River Junction, VT: Chelsea Green; 2008.

43. Evenson RE, Gollin D. Assessing the impact of the Green Revolution, 1960–2000. *Science.* 2003;300:758–762.

44. World grain production per person, 1950–2011. Earth Policy Institute. http://www.earth-policy.org/datacenter/xls/book_fpep_ch1_3.xlsx. Accessed March 27, 2013.

45. Paillard S, Dorin B, Le Cotty T, et al. Food security by 2050: insights from the Agrimonde Project. *European Foresight Platform.* EFP Brief No. 196. September 2011. http://www.foresight-platform.eu/wp-content/uploads/2011/10/ EFP-Brief-No.-196_Agrimonde.pdf. Accessed March 28, 2013.

46. Bremner J. Population and food security: Africa's challenge. *Issue Brief.* Population Reference Bureau; February, 2012. http://www.prb.org/pdf12/ population-food-security-africa.pdf.

47. Food security and nutrition. http://www.worldwewant2015.org/food2015. Accessed July 4, 2013.

48. Griggs D, Stafford-Smith M, Gaffney O, et al. Sustainable development goals for people and planet. *Nature.* 2013;495:305–307.

49. De Schutter O, Vanloqueren G. The new Green Revolution: how twenty-first century science can feed the world. *Solutions.* 2011;2(4):33–44.

50. McIntyre BD, Herren HR, Wakhungu J, Watson RT, eds. *Agriculture at the Crossroads: International Assessment of Agricultural Knowledge, Science and Technology for Development. Synthesis Report.* Washington, DC: Island Press; 2009.

51. Fedoroff NV, Battisti DS, Beachy RN, et al. Radically rethinking agriculture for the 21st century. *Science.* 2010;327:833–834.

52. Benbrook C. Innovations in evaluating agricultural development projects. In: *State of the World, 2011: Innovations That Nourish the Planet.* New York: W. W. Norton; 2011:169–172.

53. Godfray HCJ, Beddington JR, Crute IR, et al. Food security: the challenge of feeding 9 billion people. *Science.* 2010;327:812–818.

54. Foley JA, Ramankutty N, Brauman KA, et al. Solutions for a cultivated planet. *Nature.* 2011:478;337–342.

55. Weeks JR. *Population: An Introduction to Concepts and Issues.* 10th ed. Belmont, CA: Thomson Wadsworth; 2008.

56. Hopfenberg R. Human carrying capacity is determined by food availability. *Pop Environ.* 2003;25:109–117. http://www.panearth.org/WVPI/Papers/CarryingCapacity.pdf.

57. Hopfenberg R, Pimentel, D. Human population numbers as a function of food supply. *Environ Develop Sustain.* 2001;3:1–15. http://www.panearth.org/WVPI/Papers/HumanPopulationNumbers.pdf.

58. Fogel RW. *The Escape from Hunger and Premature Death, 1700–2100: Europe, America, and the Third World.* New York: Cambridge University Press; 2004.

59. Kim JM. *The Economics of Nutrition, Body Build, and Health: Waaler Surfaces and Human Physical Capital.* PhD dissertation, Department of Economics, University of Chicago; 1996. http://www.300luanvan.neu.edu.vn/Upload/9711196.pdf.

60. Top 20 failing states, 2012. Earth Policy Institute. 2012. http://www.earth-policy.org/datacenter/xls/book_fpep_ch2_7.xlsx. Accessed March 28, 2012.

61. Thomas KJA, Zuberi T. *Demographic Change, the IMPACT Model, and Food Security in Sub-Saharan Africa.* United Nations Development Programme, Working Paper WP 2012-003. February, 2012. http://web.undp.org/africa/knowledge/WP-2012-003-thomas-zuberi-impact.pdf.

62. Gasana J. Remember Rwanda? *World Watch.* 2002;15(5):24–33.

63. Rwanda: comprehensive food security and vulnerability analysis and nutrition survey, December 2012. World Food Program. http://www.wfp.org/content/rwanda-comprehensive-food-security-vulnerability-analysis-nutrition-survey-dec-2012. Accessed June 17, 2013.

64. Lutz W, Qiang R. Determinants of human population growth. *Philosoph Trans R Soc Lond B Biol Sci.* 2002;357:1197–1210. http://www.ncbi.nlm.nih.gov/pmc/articles/PMC1693032/pdf/12396512.pdf. Accessed June 16, 2013.

65. Connelly M. *Fatal Misconception: The Struggle to Control World Population.* Cambridge, MA: Belknap; 2008.

66. Engelman R. *More: Population, Nature, and What Women Want.* Washington, DC: Island; 2008.

67. Ahmed S, Li Q, Liu L, Tsui A O. Maternal deaths averted by contraceptive use: an analysis of 172 countries. *Lancet.* 2012;380:111–125. http://www.thelancet.com/journals/lancet/article/PIIS0140-6736(12)60478-4/fulltext.

68. Moreland S, Smith E, Sharma S. *World Population Prospects and Unmet Need for Family Planning.* Washington, DC: Futures Group; 2010. http://futuresgroup.com/files/publications/World_Population_Prospects.pdf.

69. *The State of Food and Agriculture 2010–11: Women in Agriculture, Closing the Gender Gap for Development.* Rome: UN Food and Agriculture Organization; 2011. http://www.fao.org/docrep/013/i2050e/i2050e.pdf.

70. The hunger trap. United Nations World Food Programme. http://one.wfp .org/policies/introduction/other/documents/pdf/hungertrap.pdf. Accessed April 1, 2014.

71. Sanchez P, Swaminathan MS, Dobie P, Yuskel N. *Halving Hunger: It Can Be Done.* UN Millennium Project, Task Force on Hunger. London: Earthscan; 2005:30. http://www.unmillenniumproject.org/reports/tf_hunger.htm. Accessed March 28, 2013.

72. Vince G. From one farmer, hope—and reason for worry. *Science.* 2010;327: 798–799.

73. Nierenberg D. Agriculture: Growing Food—and Solutions. In: *State of the World 2013: Is Sustainability Still Possible.* Washington, DC: Worldwatch Institute; 2013:190–200.

74. Nourishing the planet. Worldwatch Institute: vision for a sustainable world. March 29, 2013. http://www.worldwatch.org/nourishingtheplanet. Accessed March 29, 2013.

75. Food Tank: the food think tank. http://www.foodtank.org. Accessed March 28, 2013.

76. Baird DT, Cnattingius S, Collins J, et al. Nutrition and reproduction in women. *Hum Reprod Update.* 2006;12:193–207. http://humupd.oxfordjournals.org/ content/12/3/193. Accessed March 28, 2013.

77. Lartey A. Maternal and child nutrition in sub-Saharan Africa: challenges and interventions. *Proc Nutr Soc.* 2008;67:105–108. http://journals.cambridge.org/ action/displayAbstract?fromPage=online&aid=1681164. Accessed March 28, 2013.

78. Black RE, Allen LH, Bhutta ZA, et al. Maternal and child undernutrition: global and regional exposures and health consequences. *Lancet.* 2008;371:243–260. http://www.thelancet.com/journals/lancet/article/PIIS0140-6736(07)61690-0/ fulltext. Accessed March 28, 2013.

79. World population prospects: the 2012 revision. United Nations, Department of Economic and Social Affairs. http://esa.un.org/wpp/Documentation/ publications.htm. Accessed July 5, 2013.

80. Morris SS, Cogill B, Uauy R, for the Maternal and Child Undernutrition Study Group. Effective international action against undernutrition: why has it proven so difficult and what can be done to accelerate progress? *Lancet.* 2008;371:608–621. http://www.thelancet.com/journals/lancet/ article/PIIS0140-6736(07)61695-X/fulltext#article_upsell. Accessed March 28, 2013.

81. Bhutta ZA, Ahmed T, Black RE, et al., for the Maternal and Child Undernutrition Study Group. What works? Interventions for maternal and child undernu-trition and survival. *Lancet.* 2008;371:414–440. http://www.thelancet.com/ journals/lancet/article/PIIS0140-6736(07)61693-6/fulltext#article_upsell. Accessed March 27, 2013.

82. Dietz R, O'Neill, D. *Enough Is Enough: Building a Sustainable Economy in a World of Finite Resources.* San Francisco: Berrett-Koehler; 2013.

POLICIES

Chapter 13

Global Population and Reproductive Health Policies

John F. May

INTRODUCTION

World population reached a new threshold of 7 billion persons in 2011 and is expected to exceed 9.3 billion persons by the year 2050.[1] Global population has many subsets: people live in about 240 different countries and geopolitical entities. Among and within the various continents and countries, population diversity stems from different tempos of the demographic and epidemiological transitions. Moreover, disparate policies and politics address population issues. Recent decades have shown that population and reproductive health issues can polarize political camps, as they have in the United States.[2]

This chapter first presents key population issues and challenges. Next, it describes the first policy interventions in population and family planning. It then examines how population became a global issue. The next section explores how explicit population policies developed within both countries and international organizations. The chapter then explains how and why family planning evolved into reproductive health. The final section explores future prospects for population policies and reproductive health assistance.

CURRENT POPULATION ISSUES AND CHALLENGES

The unprecedented size and growth of the human population is a major global issue. The fact that nearly all of this growth is taking place in the world's poorer countries is one aspect of this serious topic. Another is the multitude of people who continue to lack access to contraceptive supplies and services.

Global Fertility

Concerning fertility or live births, the countries of the world can be categorized into three major groups. Approximately 16% of the world population lives in countries where women have at least 4 children on average; these are mainly the 49 least-developed countries (LDCs). A second batch of countries, home to 38% of the world population, has total fertility rates (TFR) between replacement level (TFR of 2.1) and 4 children per woman; this diverse group of nations is in the process of completing the demographic transition. Finally, a third category of countries, encompassing 46% of the world population, has reached replacement-level or even lower fertility levels; these countries include the United States, Canada, most of Europe, Iran, and several Asian nations. Obviously, the demographic prospects of these three groups of countries will differ dramatically.[3(PP26-28)] In addition to major differences in their demographic components, population issues vary greatly by country, depending on geography, resource bases, and specific politics.

Demographic Diversity

Current demographic issues include the completion of the demographic, epidemiologic, and migratory transitions; rapidly growing or contracting populations; major changes in age structures; population aging and the youth bulge; rapid urbanization, including the formation of slums, and rural depopulation; intensifying migratory flows that affect both sender and recipient countries; and, last but not least, issues of climate change, food supply, national and international security, inequity, gender imbalance, and poverty. Remarkably, all of these diverse, and sometimes opposing, trends are taking place *at the same time* in the different parts of the world. Moreover, the situation of the populations in the poorest countries—the "bottom billion"—is particularly challenging and requires specific policies.[4(PP3-13)]

INITIAL POPULATION INTERVENTIONS

The first organized efforts in population policy were geared toward reducing high levels of mortality. These interventions were quite successful, particularly in developing countries (many of which were still under colonial rule after World War II). Mortality reduction was achieved through wide-scale exogenous medical techniques (e.g., vaccinations, oral rehydration, nutrition supplements), as well as through improved access to clean water and sanitation. In addition, governments and international donors, such as the World Health Organization (WHO), designed programs designed to control and, when possible, eradicate specific diseases (e.g., smallpox). These programs are considered to be the first successes at coordinating global public policy (Figure 13–1).

These interventions contributed to improved survival, particularly child survival, in large swaths of the developing world. This decline in mortality accelerated the rate of natural population growth. By the early 1950s, United Nations (UN) demographers realized that populations were growing at unprecedented rates, particularly in Asia.[3(pvi)] Fertility levels were not

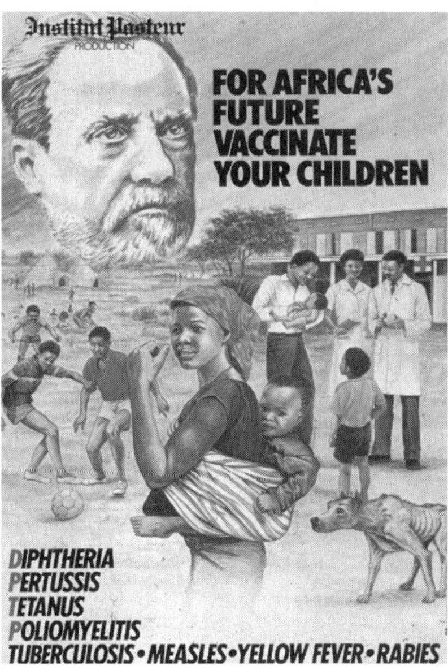

Figure 13–1 Through Widespread Immunization, the First Population Programs Focused on High Mortality

Courtesy of National Library of Medicine.

falling rapidly in response to the sharp declines in mortality; the lag between decreasing mortality levels and the decline of fertility dragged on in many developing countries. Moreover, population momentum further accelerated the population growth.

Concern spread about an impending *population explosion*.[5] Echoing Malthus's earlier alarms,[6] international leaders called for action to stem rapid population growth. Several vanguard countries, such as Japan in the late 1940s, had already enacted fertility reduction measures, but eugenic considerations dominated this policy. Subsequently, other countries, such as India, launched family planning programs, although the Indian program struggled for many years. Egypt paid attention to its population challenge in the early 1950s and started a family planning program in the 1960s, but did not act decisively until the 1980s.

The year 1952 began the global effort to curb rapid population growth. In that year, India enacted the world's first national population policy, with the goal of mainstreaming a national family planning program. That year, two important international population institutions were born: the International Planned Parenthood Federation (IPPF) and the Population Council.[3(pvii)]

The 1960s were important for global family planning efforts. In 1960–1962, Pakistan, the Republic of Korea (South Korea), Egypt, Sri Lanka, and Fiji initiated national family planning programs. Indonesia, Singapore, Taiwan, and Thailand followed suit shortly thereafter. Other developing countries launched family planning programs and adopted population policies in the late 1960s and early 1970s.[7(p185)] Private foundations in the United States (e.g., Ford, Mellon, Rockefeller, and the Population Council) supported these efforts by funding the dissemination of contraceptive technology and other technical assistance.

By the 1970s and 1980s, population policies and programs were becoming more comprehensive and elaborate (see Box 13–1 for definitions). In some countries, they had expanded beyond the delivery of family planning services and supplies. These early experiences shaped the international deliberations on population and family planning (in later years, reproductive health) over the next several decades. Family planning gradually reached most parts of the developing world,[8] with the exception of some least-developed countries, particularly in sub-Saharan Africa.[9]

GLOBALIZATION OF POPULATION ISSUES

This section explains how population issues became international concerns. Such global recognition was achieved by organizing major

---

Box 13-1   What Are Population Policies?

Population policies are defined as actions taken explicitly or implicitly by public authorities to prevent, delay, or address imbalances between demographic changes, on the one hand, and social, economic, environmental, and political goals, on the other hand.[3(pp1-2)] Population policy interventions include four instruments or mechanisms—namely, the availability of information, laws and regulations, taxation and subsidy mechanisms, and direct investments, including the offer of services.[10] These instruments, which help foster couples' choices, can be implemented through both the public (e.g., governmental) and private sectors.

A crucial issue is the choice of the proper *policy levers*—that is, the mechanisms for policy implementation. In an attempt to influence fertility, governments can enact normative laws and regulations[11] and/or adopt institutional reforms.[12] Incentives may come in the form of payments for children (e.g., in Western Europe), tax exemptions for childcare costs (e.g., in the United States), or other financial incentives (e.g., money given to individuals to accept sterilization as has been promoted by some countries in Asia).

---

international conferences, creating new institutions, strengthening international assistance, collecting demographic data, and monitoring other trends. Despite this international commitment, the Millennium Development Goals—the global development framework adopted in 2000—did not take population issues into account. However, this oversight was corrected in 2005.

International Population Conferences

The United Nations was created in 1945. A quarter of a century later, international conferences had become important in the formulation of global social and economic policies. For example, an international conference devoted to human environmental issues was held in Stockholm in 1972; its success paved the way for subsequent international conferences.[13,14(p168)]

The United Nations Population Commission sponsored the first two modern scientific conferences dedicated to population. These meetings took place in Rome and Belgrade in 1954 and 1965, respectively. They were technical rather than political gatherings, and ideologies played a minimal role in the deliberations.[15(p247)]

Thereafter, between 1974 and 1994, the United Nations organized three decennial world conferences to discuss population issues, held in Bucharest, Mexico City, and Cairo, respectively.[16(p1),17(pp725-748),18] These conferences became political events because they involved government officials, not just population experts. National and international media provided extensive press coverage. The conferences were important vehicles for disseminating new ideas and approaches, sometimes promoting the convergence of opinions,[19] globalizing population discourse, and fostering international population and reproductive assistance. They facilitated the transmission of population and family planning experiences across national borders.[20]

Bucharest Conference (1974)

The Bucharest Conference took place in 1974, designated World Population Year by the United Nations (Figure 13–2), in the context of the Cold War and amid concerns over rapid demographic growth. Here Third World (an older name for less-developed countries) countries' desire for a new international economic order coalesced with diverse nationalistic and religious opposition to family planning. Pitted against each other were development proponents and family planning proponents. The former, led by socialist and non-aligned developing countries, were convinced that socioeconomic development was the only remedy to mitigate rapid demographic growth in developing countries. The latter, led by the United States, Canada, and several European countries, were joined by large developing countries such

Figure 13–2 East African Community, Circa 1974

A stamp printed in East African Community, dedicated to World Population Year, shows women receiving family planning advice.

as Mexico and Egypt. They viewed family planning programs as prerequisites to socioeconomic development.[20(p108)]

The main achievement of this conference was the World Population Plan of Action (WPPA).[21] Despite their opposing views, participants eventually agreed on this document, the first international declaration on population. The WPPA covered the inter-relations between population, development, and the international economy. It also stated the right of couples to choose their family size as well as the responsibility of governments to provide the means to exercise this right. Finally, the document insisted on accelerated economic and social development, including expanded international assistance. Noteworthy is the fact that the Bucharest conference was the first to be conducted at the ministerial level.

Mexico City Conference (1984)

The decennial follow-up to Bucharest was held in Mexico City in 1984. Convened to assess progress on the WPPA, this conference produced two documents: the Mexico Declaration on Population and Development and the Mexico Recommendations. For the most part, delegates agreed on the need to reduce fertility and design population policies. By this time, southern countries had abandoned their development-only position and were actively promoting family planning services (Figure 13–3).

Despite this unanimity, the reversal of longstanding U.S. population policy clouded the conference's atmosphere. The newly elected Republican administration challenged the necessity of public interventions on population variables.[22] Not only did the U.S. delegation question the effectiveness of family planning programs, but it also championed the private sector's role in fostering economic growth.[23] Thus began the wave of economic liberalism, which assumed that market rules and private-sector initiatives would solve social problems.

Cairo Conference (1994)

In 1994, the United Nations convened the International Conference on Population and Development, also known as the Cairo conference.[24] Continuing the WPPA discourse, the Cairo conference produced the 20-year ICPD Programme of Action. This document was innovative because it proposed the first quantitative goals for access to family planning services, reduction of maternal and infant and child mortality levels, and universal access to primary education (priority being given to female education). Despite some contentious issues, such as free access to induced abortion, adolescent access to contraception, and women's empowerment,

Figure 13–3 Latin American Family Planning Poster Asking How Many Children a Family Can Afford

Courtesy of United Nations Population Fund

nearly a global consensus emerged from Cairo—a consensus that holds today.[18,25,26]

The 1994 ICPD remains the most recent global population conference. ICPD recommendations were revisited by the United Nations General Assembly in 1999 (ICPD+5), 2004 (ICPD at 10), and 2009 (ICPD at 15). In 2014, the United Nations is expected to devote a special session of the General Assembly to population issues.

Other Pertinent UN Conferences

The United Nations has organized numerous other conferences on subjects related to those discussed at the conferences of Bucharest, Mexico City, and Cairo. One of the most important was the 1968 International Conference on Human Rights in Tehran.[3(p109)] This conference established that individuals and couples have the right to *freely* and *responsibly* decide the number and spacing of their children. In 1972, United Nations Conference on the Human Environment met in Stockholm. This successful meeting paved the way for subsequent international conferences addressing a broad range of social problems.[3(p105),14(p168)]

Twenty years later, the Earth Summit—formally the 1992 Conference on Environment and Development—took place in Rio de Janeiro. The Earth

Summit defined renewable or sustainable development, a major topic for subsequent conferences on long-term development. In 1995, the Conference on Women and Development took place in Beijing. Both the Beijing Conference and the 1995 World Summit for Social Development held in Copenhagen restated women's central role in development.[3(p108)]

Aftermath of the Population Conferences

The population conferences stressed the prominent role of the state in implementing their recommendations. They outlined actions necessary to achieve a balance between population growth and available resources. Because relationships between population and socioeconomic development are complex, these conferences had to address most development issues, including health, nutrition, education, status of women, labor force, gender, urbanization, environment, and political reform. The conferences also recommended that technical population expertise should be provided by multilateral assistance, bilateral contributions, and South–South cooperation.

Successive population conferences revealed the diminishing role of the United States. American leadership on population and development was slowly subsumed by the leadership of the United Nations, and later by European countries and Japan. The reversal of previous American policy at the Mexico City conference dealt a major blow to the U.S supremacy in this area.[17(pp734-735)] Domestic abortion politics, leading to periodic and highly partisan defunding of the United Nations Population Fund (UNFPA) and the International Planned Parenthood Federation (IPPF), have weakened the American role in global population and reproductive health.[1(pp197-198)]

New Population Institutions

The internationalization of population issues was fostered by the creation of new population institutions. Institutions addressing population issues can be domestic or international agencies, professional associations, or research institutions.[27-30] Some population institutions have specific mandates—for example, dissemination of data and public education (e.g., Population Reference Bureau), or advocacy functions (e.g., Population Action International, formerly the Population Crisis Committee). A population institution can fulfill several roles, such as undertaking research and implementing programs (e.g., the Population Council).

Early Country-Level Efforts

In the 1950s, and especially during the 1960s and the 1970s, many governments in developing countries created national institutions to tackle population and family planning issues. Unfortunately, these institutions were often weak, in minor ministries, and without effective national leadership. Governments had the choice of creating new population departments or tasking existing governmental ministries with additional population-related responsibilities. Many countries did establish stand-alone national population commissions or councils. Sometimes a dual approach prevailed, with the strategic functions of population programs being entrusted to the Ministry of Planning and the health aspects (e.g., the delivery of family planning services) being given to the Ministry of Health. In sum, institutional arrangements varied by the administrative culture of the country along with its historical, often colonial, legacy.

International Agencies

In addition to country-level efforts, a vast network of international agencies concerned with population and family planning issues emerged after World War II. The Economic and Social Council of the United Nations established the Population Commission in 1946, with a Population Division serving as its Secretariat. This intergovernmental body advises the Economic and Social Council on population-related issues, trends, and development strategies. The Population Commission meets every year and serves as a forum for countries to communicate about their policies and progress. Following the 1994 ICPD, the Population Commission was renamed the Commission on Population and Development, acquiring a key role in monitoring and assessing the implementation of the ICPD Programme of Action.

Initially, the World Health Organization (WHO) was reluctant to embark on large-scale family planning programs. WHO's Department of Reproductive Health and Research as well as its longstanding Human Reproduction Program now carry out contraceptive research as well as technical guidance for family planning and reproductive health programs worldwide.

To fill in the gap left by WHO on the programmatic front, a new specialized UN agency, the United Nations Fund for Population Activities (UNFPA), was created in 1967; it began operations in 1969. (Later, this agency was renamed the United Nations Population Fund.) UNFPA continues to support global population, reproductive health, and family planning programs. Its efforts are joined by other specialized UN agencies, such as the International Labour Organization (ILO), the World Bank, and several regional development banks (e.g., the African Development Bank).

Important bilateral funding and technical contributions complement these efforts. In particular, the U.S. Agency for International Development

(USAID) and several European countries have played key roles. Among the European bilateral donors, one should mention the Nordic countries, the Netherlands, the U.K. Department for International Development (DfID), and the German *Gesellschaft für Internationale Zusammenarbeit* (GIZ). Japan's efforts, both multilateral and bilateral, have also been important, even if Japan has often relied on third parties (e.g., multilateral institutions) for program implementation.

The activities of numerous nongovernmental organizations (NGOs), funded by a mix of public and private money, complement the work of the multilateral and bilateral international population organizations. Among these NGOs, the most important is London-based IPPF, founded by eight national family planning associations in 1952. IPPF currently has member-associations in more than 140 countries.

Numerous and very influential U.S. institutions deserve special mention. These encompass NGOs, scientific organizations, professional networks, women's coalitions, and foundations. For example, the Bill and Melinda Gates Foundation works in the areas of health, immunization (e.g., polio vaccine), HIV/AIDS, reproductive health, and urban poverty. Another example is the William and Flora Hewlett Foundation, which has an active population and development program. While these agencies are responsible to their boards of directors or donors, they nonetheless enjoy a semi-autonomous status. However, the proliferation of these organizations inevitably has contributed to fragmentation of efforts and programs in population and reproductive health.

While some agencies, such as USAID, have focused on family planning issues (when permitted by the particular presidential administration), other institutions or countries have neglected this area altogether. It is interesting to note that in 2008 eight European countries accounted for more than 75% of the regular contributions to UNFPA. Together with Japan, these countries covered more than 80% of the organization's regular budget (in 2008, the United States had not resumed its funding to UNFPA). By comparison, the predominantly Catholic European countries (Austria, Belgium, France, Ireland, Luxembourg, Spain, and Italy) financed only 10% of UNFPA's regular budget.[31]

Data Collection and Monitoring

The global drive to gather demographic data began after World War II, when most developing countries organized regular population and housing censuses. At that time, the United Nations coordinated international rounds of censuses to help governments initiate these major operations.

The U.S. Census Bureau and USAID often supported these efforts with both funding and technical assistance.

The results from these post–World War II censuses raised the alarm about rapid population growth. Consequently, world population conferences and new population institutions further boosted data collection endeavors. The new institutions also monitor governments' views and policies regarding population trends.

Census data cannot provide all the information that is required for policy design and development planning; they need to be complemented with other data sources. The World Fertility Survey (WFS), carried out between 1972 and 1984, was the first attempt to use a standard questionnaire on human fertility in both industrialized and developing countries.[32] Other rounds of surveys later conducted included the Contraceptive Prevalence Surveys (CPS) from 1977 to 1985, the World Health Surveys (WHS), and more recently the Demographic and Health Surveys (DHS).

The DHS program, which was launched in 1984, has been implemented at 4- to 6-year intervals in more than 60 developing countries. It has assembled an impressive body of demographic and health statistics data that are internationally comparable.[33] The core questionnaire covers fertility preferences, family planning (knowledge, past and current use, method mix, and means of supply of methods), excision (female genital mutilation [FGM]), prenatal care, assistance during labor, vaccination, prevalence and treatment of diarrhea, breastfeeding and complementary feeding, knowledge of and attitudes toward sexually transmitted infections (especially HIV/AIDS), and infant and child mortality. These surveys provide invaluable data for program design.

Civil registration is the third main source of demographic information. Vital statistics are derived from the system that records births, marriages, divorces, and deaths for administrative and legal purposes. In industrialized countries, the vital registration system functions rather well. In developing countries, however, vital statistics are often of poor quality and the coverage of events is incomplete, especially with respect to mortality. Efforts to improve vital statistics in developing countries are costly and not as well funded by external donors as are censuses and surveys.

As early as 1963, the United Nations also invited governments around the world to report about their population dynamics and policies (questionnaires were regularly sent to each of the United Nations member and nonmember states). These population policy inquiries (i.e., the series of *United Nations Inquiry Among Governments on Population and Development*) are conducted every five years. As of 2011, the UN had conducted 10 such enquiries. The biennial UN report, *World Population Policies*, publishes the results.[34] Ideally, the perceptions of governments about their demographic challenges will prompt them to seek international population and reproductive assistance.

Widespread population projections have also emerged, particularly given that they can now be produced easily with readily available software. These have become powerful tools to enhance the policy dialogue on population and development issues and are now a main driver of international population and reproductive health assistance.

Millennium Development Goals

International concern about the widening socioeconomic gap between rich and poor countries increased in the 1990s. Consequently, world leaders and the international development community resolved to promote human development more forcefully. Several global conferences helped to forge a wide consensus to set and achieve global development goals. In 2000, the United Nations Millennium Summit approved a set of 8 development goals (with 19 targets and 60 indicators) called the Millennium Development Goals (MDGs). UN member states and international development institutions committed to attain these benchmarks by 2015.[35]

The MDGs helped reframe the international development agenda, but reproductive rights were omitted because of their political sensitivity.[36,37] In fact, opposition to reproductive rights threatened the adoption of the entire MDG framework, so the only way to reach a consensus was to eliminate the reproductive health goal altogether. Nevertheless, reproductive health services are vital to the full achievement of MDG 5 (Improve maternal health), dubbed the "mother of all MDGs." This glaring omission was finally addressed in 2005 with the addition of indicators specifically related to reproductive health. Particularly important among these are the contraceptive prevalence rate and level of unmet need for family planning.[38]

Paradoxically, the MDGs do not specifically mention demographic variables, yet achievement of seven out of eight goals depends on demographic outcomes.[39] MDG 1 (to eradicate poverty) can be achieved only with smaller family sizes. Poor households, which have the most children, also have the greatest difficulties securing access to education, health, and food, as well as having less access to monetary employment. MDG 2 (to achieve primary education) cannot be reached in 2015 if school-age populations continue to double every 20 years or so, as they are doing in countries with very high fertility. Gender equality (MDG 3) is crucial in countries where *reproductive rights* remain nominal. MDGs 4 and 5 (child mortality and maternal survival, respectively) have a direct impact on demographic outcomes and cannot be reached when half of all pregnancies are at risk because they are too early, too late, too many, and too close. MDG 6 (combat HIV/AIDS, malaria, and other diseases) can be implemented only with adequate health facilities and personnel—difficult to muster when population is increasing rapidly. Finally, MDG 7 (environmental sustainability) implies a reduction in rates of population growth, a crucial step in reducing the demographic pressure on ecosystems.

MDG 8 (develop a global partnership for development) is particularly important for coordinating efforts in international population and reproductive health assistance. At the global level, such coordination is spearheaded in principle by the United Nations. At the national level, coordination has occurred in some countries. In Bangladesh, for instance, a Population and Health Consortium was established in 1987 as a semi-formal grouping of the World Bank, its co-financing partners, and several UN agencies. The Consortium helped to create harmony among donors and, more importantly, between the government and the donor community.[3(p118)]

Because they provided an overarching framework, the MDGs helped coordinate international efforts in population—and later, reproductive health—assistance. However, the real challenge is to mainstream population and reproductive health issues within the MDG framework. This task remains unfulfilled to this day, becoming more acute as the global community deliberates the post-2015 international development framework.

FORMALIZATION OF POPULATION POLICIES

Following their initial focus on mortality and fertility, the scope of population policies broadened. This process took place against the backdrop of three different policy paradigms.[40] The first period, the *population control approach* (1965–1974), corresponded to initial efforts aimed specifically at curtailing fertility. The second period, the *population planning approach* (1974–1981), focused on integrating population issues with overall development endeavors and contributed to more comprehensive national population policies. The third period, the *competitive pluralism in population policy approach* (1981–1994), saw the adoption of broader population policies in addition to other policy initiatives.

Although this framework was developed in 1990,[40] its third and last period lasted far beyond that time. The era of *competitive pluralism,* which involves both public and private sectors and multiple approaches and numerous stakeholders, persists.

The 2009 inauguration of U.S. President Barack Obama rekindled hope for better international cooperation in family planning efforts.[41,42] American involvement in the organization of the global 2012 Summit on Family Planning in London supported this optimism.

Complexity of Current Issues

The complexity of current issues has broadened the scope of population policies. The HIV/AIDS epidemic, for example, has mustered a great deal of energy and resources, unfortunately to the detriment of family planning programs.[43] Results of UN surveys on population policies also reveal other preoccupations of governments. In 2009, a majority of governments (87%) viewed the HIV/AIDS epidemic as their most significant demographic issue. Within developing countries, high levels of child mortality (before age 5) and high levels of maternal mortality constituted the second and third most important concerns. Developing countries also worried about their expanding working-age populations, for whom they needed to create jobs. Developed countries, in contrast, were concerned with population aging and HIV/AIDS, and, to a lesser extent, with low fertility levels.[34(p7)] In addition, even before climate change became a dominant preoccupation, population policies had started to address environmental concerns.

In developing countries, an overarching goal of population policies is to integrate demographic variables in different sector policies and to create changes in social norms (e.g., through awareness-raising campaigns). Population policies include measures to modify fertility, mortality, or migration trends, as well as efforts to influence the aspirations of individuals and families (Figure 13–4). At the operational level, policies have been implemented through programs and projects, along with data collection and monitoring efforts. In developed countries, population policies are often implemented through social policies, which can be explicit or implicit. In all situations, finding the right institutional implementation arrangements is the most critical challenge.

Population policies do not operate in a vacuum, but rather are implemented within the specific administrative settings of nations. A lack of strong institutions, as in sub-Saharan Africa, or the fragmentation of the institutional policy actors, as in some developed countries, may hinder successful policy implementation. Specific bottlenecks may also impede the implementation of population policies, such as insufficient infrastructure (e.g., lack of health facilities or schools). Moreover, policies are implemented against the backdrop of contextual variables, such as education and

Figure 13–4 African Family Planning Poster

Courtesy of the Cameroon National Planning Association for Family Welfare, International Planned Parenthood Federation

employment levels, health, urbanization, gender roles, cultural norms, and religious beliefs. Finally, funding shortfalls may hinder the overall implementation. Governments may not have enough resources, especially for interventions that require large human capital investments, such as education and health. Externally funded policies and programs often suffer from the volatility of international funding.

International Institutions

The international institutions active in the area of population and family planning (and later reproductive health) have not been immune from the dominant population paradigms described earlier. Moreover, disagreements and rivalries have arisen as some agencies have favored specific strategies. The most striking contrast has been between USAID, which funds the largest family planning program in the world, and UNFPA, the largest international organization specialized in the field of population and development. USAID privileged (and still does to a large extent) the supply of services and put less emphasis on demand creation. USAID concentrates its efforts on making family planning programs and information, education, and communication programs (IEC) available, sometimes complemented by actions to change behavior. In contrast, the United Nations, through UNFPA, embraces a broader approach to population questions, which are seen in the context of global development. UNFPA has argued for the inclusion of all sectors (e.g., education, gender, legal reform) in the implementation of population and reproductive health policies. The adoption of the Cairo agenda in 1994 further broadened the scope of the agency. However, in the early 2000s, UNFPA launched an initiative to "reposition" family planning within its reproductive health programs.

Development banks have also provided population and reproductive assistance. The first World Bank population loan was granted to Jamaica in 1970. To build its own capacity, the World Bank started its population programs in small countries before moving to larger countries, such as Bangladesh. The rapid expansion of the World Bank population portfolio was fostered under the leadership of Robert McNamara, its president from 1968 to 1981. Since then, the World Bank portfolio of loans and grants for health, nutrition, and population (HNP) has grown. This organization is currently the largest contributor to the HNP sector for countries with medium or low incomes.[44,45(p15)] The World Bank also finances many projects dedicated to education, water, sanitation, urban planning, social protection, and women's empowerment, among other causes, each of which has direct or indirect effects on demographic variables.[45(p16)] However, after the publication of its seminal *1984 World Development Report* (WDR) devoted to population issues,[46] the World Bank did not push the population agenda as forcefully. While admitting that demographic growth remains an important issue, this organization did not spend much political capital to promote family planning programs. Nonetheless, there is currently a renewed interest for population and reproductive health issues within the World Bank, especially for sub-Saharan Africa and more specifically for the Sahel. Population issues are coming back around for discussion about the possibility of African countries benefiting from a demographic dividend.

THE ROAD TO CAIRO: FROM FAMILY PLANNING TO REPRODUCTIVE HEALTH

In the 1950s and 1960s, only a few Asian and Caribbean countries acted to curb high levels of fertility. By the late 1960s, many of the larger Asian countries had adopted population policies that promoted family planning services. Programs in East and Southeast Asia were quite successful in lowering birth rates. However, earlier programs in India and Pakistan did not succeed in curtailing high levels of fertility.

The failure of some Asian programs gave rise to increasing pessimism about how family planning *by itself* could effectively reduce fertility. Apparently, just supplying family planning services was insufficient for triggering fertility declines. Clearly, individual couples had not rushed to the new networks of family planning clinics, built at great costs.

Demographers were convinced that more was needed to stimulate demand for smaller families. Calls began for measures "beyond family planning," actions to stimulate the demand for smaller families or to reward individuals or couples for limiting their family sizes. Policy levers ranged from indirect measures, such as improving girls' access to education or reducing under-5 mortality rates, to monetary incentives for using contraceptives or limiting births. Sometimes, authorities resorted to payment to people to undergo sterilization. They even threatened the financial security of couples that would not comply with the new fertility norms promoted by governments. In some countries, authorities went as far as to a ration access to housing and other benefits, based on family size. Policy makers in India became so frustrated by the failure of voluntary family planning efforts that they turned increasingly to such solutions, culminating in the coercive sterilization campaigns during the so-called Emergency Period (1975–1977). In 1979, Chinese authorities resorted to similarly coercive policies, with the *One-Child Policy*, although fertility had already dropped to fewer than three children per woman. At the time, Western population assistance officials, relieved that Asian programs had finally gotten some traction, paid scant attention to these abuses.

The shortcomings of the *all family planning approach*, combined with an increasing awareness of programs' abuses, raised concerns and led to an international drive to assert human rights, especially women's rights. From a narrow definition of family planning, the paradigm shifted to the broader concept of reproductive rights. The Cairo conference endorsed this change,[47,48(pp63–82)] despite heated debates around the issues of induced abortion and adolescent sexuality.

The prevailing assumption was that the optimal fertility of individuals and couples (hopefully, close to replacement level) could best be obtained if all persons could exercise their reproductive rights. Therefore, individual women and couples had to become the central focus of family planning programs. Many governments perceived reproductive rights as less intrusive and threatening than the occasionally heavy-handed family planning programs of the past. At the same time, it was fully recognized that fertility is dependent on many other health and development variables—namely, levels of infant, child, and maternal mortality; the physical integrity of women;

educational attainments of young girls; needs of adolescents; and levels of poverty.[49(pp3–34)]

The definitions of reproductive rights and health draw on the broader definition of health proposed by the WHO. The 1994 ICPD proclaimed, "Reproductive rights embrace certain human rights that are already recognized in national laws, international human rights documents and other relevant UN consensus documents. These rights rest on the recognition of the basic right of all couples and individuals to decide freely and responsibly the number, spacing and timing of their children and to have the information and means to do so . . ." With respect to reproductive health, the ICPD stated, "Reproductive health is a state of complete physical, mental and social well-being in all matters relating to the reproductive system and to its functions and processes. It implies that people have the capability to reproduce and the freedom to decide if, when and how often to do so."[24]

However, implementation has proved most challenging. First, to achieve reproductive rights, one needs to fulfill many prerequisite conditions (e.g., legal, social, economic). In this respect, experience has shown the importance of women's empowerment, as well as broader choices for women through expanded access to education, health services, skills development, and employment. Second, the breadth of the reproductive health concept implies that specific programs address many different dimensions.

The Cairo agenda includes three main areas. First, the components of the reproductive health (RH) service package cover family planning, maternal, infant, and women's health care, infertility prevention and treatment, abortion, treatment of reproductive tract and sexually transmitted infections, breast and cervical cancer, and harmful practices such as early marriage and female genital mutilation. The second group of additional RH components includes adolescent and men's RH, gender-based violence, and gender. The additional components include child health/child survival, elderly women/menopause, and nutrition.[50(p5)] Specific services regarding HIV/AIDS have been added to this already long list. Finally, reproductive health programs now include interventions to address the issue of vesicovaginal fistula. To address all these dimensions, reproductive health programs need to cover prenatal care, safe delivery, and postnatal care, as well as education and counseling on human sexuality and responsible parenthood.

The implementation of this tall order was less than optimal and funding did not match the rhetoric. First, many countries did not invest in reproductive health services in a systematic or comprehensive way, and very seldom have their efforts been sustained over time. Some countries, such as China and Vietnam, only nominally subscribed to the changes enacted in Cairo.

Second, population programs since 1994 have strived to satisfy the personal needs of people and couples. By doing so, however, they have lost sight of the big demographic picture as well as the huge logistical requirements needed to serve burgeoning populations. Third, the broad reproductive rights agenda with its many different components has been parceled out to a large number of different implementers (e.g., NGOs). These organizations picked up isolated aspects of the overall reproductive health agenda without adopting a holistic approach. As a result, the implementation of the reproductive rights agenda remained fragmented and unfocused. A major component of the population and reproductive assistance became fragmented as well.

Despite increases in contraceptive coverage in recent years,[51] an estimated 222 million women who want to avoid pregnancy do not use an effective method of contraception. Demand should rise dramatically in the next decades as record numbers of young people enter their prime reproductive ages. The unmet need for family planning is particularly acute in the LDCs, so it becomes urgent to set key priorities within the larger framework of reproductive health services. Among these priorities, it appears that five key elements should belong to any effective reproductive health program: family planning, maternal health, children's health, prevention of HIV/AIDS, and violence against women.

REPOSITIONING FAMILY PLANNING

About a decade after the Cairo conference, several donors, including UNFPA and USAID, repositioned family planning within their programs. Today, the unfinished agenda of family planning and reproductive health calls for renewed efforts. In 2012, the Bill and Melinda Gates Foundation and the U.K. Department for International Development (DfID) convened the London Summit of Family Planning. This helped to put family planning back at the heart of the global development agenda after 20 years of neglect because of competing agendas, the belief that the population problem is solved,[52] and a kind of donors' fatigue with respect to family planning. The London Summit's goal was to raise pledges of approximately $8 billion to provide family planning services to 120 million women through 2020.

In recent history, family planning programs have been woefully underfunded. Today, many women are still not able to attain the family planning method of their choice, and too often they must resort to abortion when services are not available. Commodities stock-outs remain far too frequent. Greater access to family planning is not only an issue of public health but also a matter of human rights and, in the long run, of sustainable economic growth.

The primary challenge of the aftermath of the London Summit will be in the implementation. Without going into the details of who or which agencies will implement the goals of the Summit, four crucial elements appear necessary for the renewed efforts on family planning to succeed.[53]

First, there must be a sense of urgency. The 16% or so of the world population that lives in countries where fertility remains more than 4 children per woman needs rapid access to contraception. Most of the 49 LDCs fall into this category; their aggregate population is expected to more than triple during the 21st century. Among these countries, Bangladesh managed to reduce its fertility levels through strong leadership and door-to-door family planning programs. Most other LDCs with very high fertility—in particular, those in the Sahel—need action immediately, as they face formidable developmental and environmental challenges.

Second, family planning must remain voluntary—everywhere and for everyone. Since the 1994 ICPD, family planning programs have adhered closely to this ideal. These programs have strived to provide high-quality services, with the proper counseling and methods of choice at affordable prices. Nonetheless, much room for improvement remains. Supply-side approaches must be combined with demand-creation efforts. Moreover, synergies between HIV/AIDS and family planning programs should be explored further.

Third, the renewed focus on family planning will be successful only if broader development goals are also considered. Gender-sensitive policies, including universal female education as well as women's labor force participation and legal autonomy, are necessary. Much more needs to be done to combine family planning programs with conditional or unconditional cash transfers and income-generating activities. Moreover, it is critically important to eliminate child marriage: doing so is a prerequisite for rapid fertility declines. New mechanisms for services provision must also be tried. The results-based financing system, in which providers are incentivized to provide good services, has proved very effective in Rwanda and other countries. Such new approaches should be scaled up.

Fourth, policy makers must keep aware of macro-demographic considerations, and the linkages between population growth and other development sectors. In particular, they need to realize that capturing the potential benefits of a demographic dividend will require a rapid acceleration of the fertility transition. Population policies and programs also need to better define their contribution to obtaining global public goods. To achieve this, more evidence-based policies are required. Recent decades have yielded a wealth of new information on population policies and programs. However, much of this research has yet to be translated into actionable policies.

Another neglected dimension is evaluating policy interventions aimed at modifying demographic components. Both the global decline in mortality and the contraceptive revolution were triggered by the supply of vaccines and contraceptives, respectively. More importantly, they occurred because profound socioeconomic changes took place, such as increased female literacy and labor force participation. The intellectual challenge is to disentangle the respective impact of these programmatic interventions from the broader improvements of socioeconomic conditions. This information would be invaluable for the design and funding of future policies and programs.

In recent years, more rigorous evaluation tools have become available, such as randomized controlled trials (RCTs).[54(p14)] Unfortunately, these have rarely been used to measure the effectiveness of family planning programs, with the exception of the experiment conducted in Matlab, a rural region East of Dhaka in Bangladesh as early as 1977 (this effort is still ongoing). More analyses of this nature are needed because funding is harder to obtain; governments and donors alike are becoming more eager to measure rigorously the outcomes of their contributions.

NEW PRIORITY AREAS

A host of new population and development issues have emerged around the world during the recent decades, and they present policy makers with major challenges. As mentioned earlier, these issues include, but are not limited to, the completion of the demographic, epidemiological, and migratory transitions; HIV/AIDS; rapidly growing or contracting populations; changing age structures;[55] the phenomena of population aging and youth bulge; rapid urbanization and rural depopulation as well as the formation of slums; intensifying migratory flows; and the concerns around climate change, food supply, security, inequity, gender imbalances, and poverty.

Several of these new issues relate to the priority groups that have been identified for policy interventions—namely, women, adolescents, elderly people, and migrants.[53(pp249-254)] For the most part, these groups have either been left out of mainstream development policies or have not benefited from socioeconomic progress. Because these disparate groups will be key for sustainable development in the long run, they call for specific interventions.

In addition to specific groups, some regions of the world require immediate and specific attention. Concerning high fertility and rapid population growth, high-priority geographical areas are Western and Middle Africa, particularly in the Sahel region. Other regions, affected by very low fertility levels and rapid population aging, need urgent attention, particularly Russia and

Eastern Europe as well as a number of countries in Asia (e.g., Japan, South Korea, Singapore, and parts of China).

In recent decades, the international policy climate has not been conducive for decisive action on population and reproductive health issues. The conservative backlash against Western feminist and liberal attitudes, along with the abortion debate (particularly pronounced in the United States), has left profound scars. Consequently, frank discussions on international population and reproductive health assistance have been muffled. Moreover, population is seldom considered when addressing climate change, other environmental issues, poverty, or inequity.

Nevertheless, there are signs that the population factor could return to the forefront of global policy concerns. First, pivotal goals, such as mitigating climate change, alleviating poverty, and reducing inequity, are undeniably linked to demographic factors. Second, linkages between age structure and economic growth have received renewed scrutiny, as illustrated by the ongoing interest and work around the demographic dividend within major development institutions (e.g., the World Bank and African Development Bank). Third, the consultations leading to the post-2015 development agenda replacing the MDGs (see www.worldwewant.org and www.ngosbeyond2014.org) as well as analyses conducted by the Population Council[56] and the Guttmacher Institute[57] elucidated population concerns. However, what will replace the MDGs remains unclear; the importance of population and reproductive health issues in a new framework remains unknown.

Apparently, there is an emerging consensus about the need to provide services to underserved populations—for example, the 120 million women targeted at the London Summit on Family Planning. Prominent global institutions have also endorsed the need to fully integrate population and reproductive health within broader international development. However, three crucial elements are missing. First, there is no sense of urgency, despite the fact that high-fertility situations beg for rapid interventions. Second, coordination between the myriad development actors and stakeholders is lacking. Third, country leadership is often weak.

CONCLUSION

The world is demographically more fragmented than ever before. Today's major population challenges are to address high-fertility situations, which foster positive population momentum, as well as situations of very low fertility (below replacement level), which compound rapid population aging.

International migratory flows are increasing as well, especially as the world economy recovers from the 2008 Great Recession. At the same time, the world population is becoming more urbanized (two-thirds of the world population will live in urban areas in 2050). Such rapid urbanization often leads to proliferation of slums, especially in low-income countries.

In this context, traditional population policies and interventions need to be reframed. In the past, such interventions were aimed at reducing high levels of mortality and especially fertility. However, as many new demographic issues have emerged (e.g., sub-replacement fertility, population aging, and increasing migratory flows), one size no longer fits all. Policies need to adapt to a variety of new and sometimes very different challenges, including climate change, food crises, water shortages, and the political instability brought about by distorted age structures (e.g., the youth bulge in some Arab countries).

While there is still a need to provide family planning and reproductive health services to 222 million underserved couples, many countries in Eastern Europe and Asia also want to increase their sagging fertility rates. Population aging is taking place at an unprecedented speed, careening countries and policy makers into uncharted demographic territory. Expanding migratory movements also will need new policies, informed by new data collection and analysis and emphasizing migrant rights.

Since the 1994 Cairo conference, the overall framework of population policies and population assistance has been more bottom-up than top-down; the major focus of population and reproductive health interventions has been fulfilling the rights and aspirations of individuals. This human rights approach needs to be replicated in other policy areas, to address priority groups such as women, adolescents, elderly people, and migrants. In addition, population assistance needs to address the issues of climate change, poverty, inequity, and security. For policy designers, the most difficult challenge of all will be to reconcile the rights of individuals with their responsibilities as global citizens.

DISCUSSION QUESTIONS

1. Should countries with sub-replacement level fertility implement policies to raise fertility rates? How effective are these policies likely to be?
2. More than two decades after the ICPD meeting in Cairo, debate continues about moving from family planning to reproductive health. Argue for and against the reproductive health paradigm.
3. How did the issue of population become globalized?

4. What needs to be done at an international and national levels to address the unmet need for family planning services?

5. Initially, the Millennium Development Goals did not address reproductive health. Do you think that their replacement will include reproductive health?

REFERENCES

1. United Nations. *World Population Prospects: The 2010 Revision*. New York: United Nations, Department of Economic and Social Affairs, Population Division; 2010.

2. Hoff DS. *The State and the Stork: The Population Debate and Policy Making in US History*. Chicago/London: University of Chicago Press; 2012.

3. May JF. *World Population Policies: Their Origin, Evolution, and Impact*. New York, NY: Springer; 2012.

4. Collier P. *The Bottom Billion: Why the Poorest Countries Are Failing and What Can Be Done About It*. Oxford, UK: Oxford University Press; 2007.

5. Ehrlich PR. *The Population Bomb*. New York: Ballantine Books; 1968.

6. Malthus TR. In: Flew A, ed. *An Essay on the Principle of Population, As It Affects the Future Improvement of Society with Remarks on the Speculations of Mr. Godwin, M. Condorcet, and Other Writers*. London, UK: Penguin Books; 1976. (Reprint of the 1798 edition).

7. Tsui AO. Population policies, family planning programs, and fertility: the record. In: Bulatao RA, Casterline JB, eds. *Global Fertility Transition*. New York: Population Council; 2001. *Pop Develop Rev.* 27(suppl):184–204.

8. Robinson WC, Ross JA. Family planning: the quiet revolution. In: Robinson WC, Ross JA, eds. *The Global Family Planning Revolution: Three Decades of Population Policies and Programs*. Washington, DC: World Bank; 2007:421–449.

9. Bongaarts J, Casterline J. Fertility transition: is sub-Saharan Africa different? In: McNicoll G, Bongaarts J, Churchill EP, eds. *Population and Public Policy: Essays in Honor of Paul Demeny*. New York: Population Council; 2012. *Pop Develop Rev.* 38(suppl):153–168.

10. Mosley WH, Jamison DT, Henderson DA. The health sector in developing countries: problems for the 1990s and beyond. *Ann Rev Public Health.* 1990;11:335–358.

11. Heckel NI. Population laws and policies in sub-Saharan Africa: 1975–1985. *Int Family Plan Perspect.* 1986;12(4):122–124.

12. McNicoll G. Institutional determinants of fertility change. *Pop Develop Rev.* 1980;6(3):441–462.

13. Finkle JL, McIntosh CA. United Nations Population conferences: shaping the policy agenda for the twenty-first century. *Stud Family Plan.* 2002; 33(1):11–23.

14. McIntosh CA, Finkle JL. International population conferences. In: Demeny P, McNicoll G, eds. *The Encyclopedia of Population*. Vol. 1. New York: Macmillan Reference USA; 2003:168–170.

15. Notestein FW. World Population Conference Rome, August 31–September 10. *Pop Index.* 1954;20(4):241–248.

16. Singh S, Darroch JE, Ashford LS, Vlassof M. *Adding It Up: The Costs and Benefits of Investing in Family Planning and Maternal and Newborn Health.* New York: Guttmacher Institute and United Nations Population Fund; 2009.

17. Chasteland JC. De 1950 à 2000: La communauté internationale face au problème de la croissance de la population mondiale. In: Chasteland JC, Chesnais JC, eds. *La population du monde: Géants démographiques et défis internationaux.* 2nd ed. Paris: Institut national d'études démographiques; 2002:717–753.

18. Chasteland JC. World population growth and the international community from 1950 to the present day. In: Caselli G, Vallin J, Wunsch G, eds. *Demography: Analysis and Synthesis. A Treatise in Population Studies.* Vol. IV. San Diego: Academic Press/Elsevier; 2006:457–486.

19. Kantner JF, Kantner A. *The Struggle for International Consensus on Population and Development.* New York, NY: Palgrave Macmillan; 2006.

20. Finkle JL, Crane BB. The politics of Bucharest: population, development and the new international economic order. *Pop Develop Rev.* 1975;1(1):87–114.

21. United Nations. *Report of the United Nations World Population Conference, Bucharest, 19–30 August 1974.* New York: United Nations, Department of Economic and Social Affairs; 1975.

22. Finkle JL, Crane BB. Ideology and politics at Mexico City: The United States at the 1984 International Conference on Population. *Pop Develop Rev.* 1985;11(1):1–28.

23. United States. Policy statement of the United States of America at the International Conference on Population. *Pop Develop Rev.* 1984;10(3):574–579.

24. United Nations. *Report on the International Conference on Population and Development, Cairo, September 5–13, 1994.* New York: United Nations, Department of Economic and Social Affairs, Population Division; 1995.

25. DeJong J. The role and limitations of the Cairo International Conference on Population and Development. *Soc Sci Med.* 2000;51(6):941–953.

26. McIntosh CA, Finkle JL. The Cairo Conference on Population and Development: a new paradigm? *Pop Develop Rev.* 1995;21(2):223–260.

27. Caldwell JC. Population organizations: professional associations. In: Demeny P, McNicoll G, eds. *The Encyclopedia of Population.* Vol. 2. New York: Macmillan Reference USA; 2003:742–744.

28. Chamie J. Population organizations: United Nations system. In: Demeny P, McNicoll G, eds. *The Encyclopedia of Population.* Vol. 2. New York: Macmillan Reference USA; 2003:749–752.

29. Haaga J. Population organizations: research institutions. In: Demeny P, McNicoll G, eds. *The Encyclopedia of Population.* Vol. 2. New York: Macmillan Reference USA; 2003:744–748.

30. MacDonald AL. Population organizations: national and international agencies. In: Demeny P, McNicoll G, eds. *The Encyclopedia of Population.* Vol. 2. New York: Macmillan Reference USA; 2003:739–742.

31. UNFPA. *Annual Report 2008*. New York: United Nations Population Fund; 2009. http://www.unfpa.org/webdav/site/global/shared/documents/publications/2009/annual_report_2008.pdf. Accessed April 3, 2013.

32. Cleland J, Verma V. The World Fertility Survey: an appraisal of methodology. *J Am Stat Assoc*. 1989;84(407):756–767.

33. Ayad M, Barrère B. Présentation des enquêtes démographiques et de santé. *Population*. 1991;46(4):964–975.

34. United Nations. *World Population Policies 2009*. New York: United Nations, Department of Economic and Social Affairs, Population Division; 2010.

35. UNDP. *Human Development Report 2003: Millennium Development Goals: A Compact Among Nations to End Human Poverty*. New York/Oxford, UK: Oxford University Press; 2003.

36. Crossette B. Reproductive health and the Millennium Development Goals: the missing link. *Stud Family Plan*. 2005;36(1):71–79.

37. Campbell White A, Merrick TW, Yazbeck AS. *Reproductive Health. The Missing Millennium Development Goal: Poverty, Health, and Development in a Changing World*. Washington, DC: World Bank; 2006.

38. United Nations. *Report of the Secretary-General on the Work of the Organization*. General Assembly, Official Records, Sixty-second Session, Suppl. 1. New York: United Nations; 2007.

39. Haslegrave M, Bernstein S. ICPD goals: essential to the Millennium Development Goals. *Reprod Health Matt*. 2005;13(25):106–108.

40. Finkle JL, Crane BB. The politics of international population policy. In: *Proceedings of the Expert Group Meeting on the International Transmission of Population Policy Experience, New York, June 27–30, 1988*. New York: United Nations, Department of International Economic and Social Affairs; 1990:167–182.

41. Gillespie D, Maguire ES, Neuse M, et al. International family planning budgets in the "new US" era. *Lancet*. 2009;373(9674):1505–1507.

42. Speidel JJ, Sinding S, Gillespie D, et al. *Making the Case for U.S. International Family Planning Assistance*. Washington, DC: United States Agency for International Development; 2009.

43. Shiffman J, Berlan D, Hafner T. Has aid for AIDS raised all health funding boats? *J Acq Immune Defic Syndr*. 2009;52(suppl 1):S45–S48.

44. Global Health Council. *Banking on Reproductive Health: The World Bank's Support for Population, the Cairo Agenda and the Millennium Development Goals*. Washington, DC: Global Health Council; 2004.

45. World Bank. *Population and the World Bank: Adapting to Change*. Rev. ed. Washington, DC: World Bank; 2000.

46. World Bank. *World Development Report 1984*. Washington, DC: World Bank; 1984.

47. Goldberg M. *The Means of Reproduction: Sex, Power, and the Future of the World*. New York: Penguin Press; 2009.

48. Halfon S. *The Cairo Consensus: Demographic Surveys, Women's Empowerment, and Regime Change in Population Policy*. Lanham, MD: Lexington Books; 2007.

49. Finkle JL, McIntosh CA, eds. The new politics of population: conflict and consensus in family planning. *Pop Develop Rev*. 1994;20(suppl).

50. UNFPA. *Implementing the Reproductive Health Vision: Progress and Future Challenges for UNFPA.* Evaluation Findings, issue 15. New York: United Nations Population Fund, Office of Oversight and Evaluation; 1999.

51. Singh JS. *Creating a New Consensus on Population: The Politics of Reproductive Health, Reproductive Rights and Women's Empowerment.* 2nd ed. London/Sterling: Earthscan; 2009.

52. Blanc AK, Tsui AO. The dilemma of past success: insiders' views on the future of the international family planning movement. *Stud Family Plan.* 2005;36(4):263–276.

53. May JF. Family planning is back. *Blog Post.* Washington, DC: Center for Global Development; 2012. http://www.cgdev.org/blog/family-planning-back-john-may. Accessed April 5, 2013.

54. Banerjee AV, Duflo E. *Poor Economics: A Radical Rethinking of the Way to Fight Global Poverty.* New York: Public Affairs; 2011.

55. Denton EH. How changes in age structure can impact policy making. *SAIS Rev Int Affairs.* 2011;31(2):15–34.

56. Bongaarts J, Cleland J, Townsend JW, et al. *Family Planning Programs for the 21st Century: Rationale and Design.* New York: Population Council; 2012.

57. Singh S, Darroch JE. *Adding It Up: The Costs and Benefits of Contraceptive Services.* New York: Guttmacher Institute & United Nations Population Fund; 2012.

Chapter 14

Toward the Future

Deborah R. McFarlane

INTRODUCTION

The world population is projected to reach 8.0 billion in 2025 and 9.0 billion in 2041.[1] Table 14–1 shows that the U.S. Census Bureau projects a global population of nearly 9.4 billion by the middle of the 21st century.

Table 14–1 World Population, 2015–2050

Year	World Population (Mid-year)	Annual Growth Rate	Annual Population Change
2015	7,253,260,112	1.060	76,859,527
2016	7,330,119,639	1.042	76,401,577
2017	7,406,521,216	1.022	75,717,812
2018	7,482,239,028	1.001	74,891,046
2019	7,557,130,074	0.978	73,941,616
2020	7,631,071,690	0.957	73,043,998
2021	7,704,115,688	0.937	72,177,655
2022	7,776,293,343	0.915	71,188,943
2023	7,847,482,286	0.894	70,117,888
2024	7,917,600,174	0.871	68,984,094

(*Continues*)

Table 14-1 World Population, 2015–2050 (*Continued*)

Year	World Population (Mid-year)	Annual Growth Rate	Annual Population Change
2025	7,986,584,268	0.851	67,928,189
2026	8,054,512,457	0.831	66,928,599
2027	8,121,441,056	0.811	65,867,468
2028	8,187,308,524	0.791	64,780,263
2029	8,252,088,787	0.772	63,669,522
2030	8,315,758,309	0.753	62,606,908
2031	8,378,365,217	0.735	61,617,218
2032	8,439,982,435	0.718	60,619,901
2033	8,500,602,336	0.701	59,605,959
2034	8,560,208,295	0.684	58,559,456
2035	8,618,767,751	0.668	57,560,668
2036	8,676,328,419	0.653	56,614,065
2037	8,732,942,484	0.637	55,644,617
2038	8,788,587,101	0.622	54,645,572
2039	8,843,232,673	0.606	53,611,906
2040	8,896,844,579	0.591	52,605,797
2041	8,949,450,376	0.577	51,634,225
2042	9,001,084,601	0.562	50,626,087
2043	9,051,710,688	0.548	49,575,885
2044	9,101,286,573	0.533	48,491,249
2045	9,149,777,822	0.518	47,436,461
2046	9,197,214,283	0.505	46,409,472
2047	9,243,623,755	0.491	45,353,264
2048	9,288,977,019	0.477	44,270,686
2049	9,333,247,705	0.463	43,169,270
2050	9,376,416,975		

Data from U.S. Bureau of the Census, International Data Base. World population 1950–2050. http://www.census.gov/population/international/data/worldpop/table_population.php.

Continued Population Growth

Although global growth rates are expected to decline, human population will continue to grow for the foreseeable future. Nearly all of that growth will occur in the poorer countries of the world, with scant resources to provide for increasing numbers of people (Box 14–1). At the same time, the populations of a few wealthy countries will decrease during the 21st century, accompanied by concerns about aging populations, immigration policies, and national identities. Both population growth and rising human consumption continue to drive environmental degradation, thus contributing to climate change.

Population, Consumption, and the Environment

The increasing human population and its growing resource demands are severely taxing the earth's carrying capacity. The absolute increase in numbers, along with the expanding world economy, have profoundly altered the connection between people and their environment. This skewed relationship persists, accompanied by alarming environmental repercussions. Both population growth and the consumption patterns of rich countries, in particular, are relentless in degrading ecosystems, including fresh water resources, oceans, wetlands, forests, fisheries, biodiversity, the atmosphere, and even the climate. Environmental depletion is as old as human civilization, but the current scale and velocity of natural degradation are unprecedented.[2,3]

Box 14–1   Africa Leads in Population Growth

With a projected growth of 1.3 billion from 2013 to 2050, Africa will add more people to the world population than any region. "Virtually all of that growth will be in the 51 countries of sub-Saharan Africa, the region's poorest. The 1.3 billion growth will exceed that of population-giant Asia. But even this projection assumes that birth rates will decline steadily in all countries of sub-Saharan Africa because of an increase in the use of family planning. If birth rates do not decline steadily, future projections of population growth will have to be increased."

Reproduced from Haub C, Kaneda T. *2013 World Population Data Sheet.* Washington, DC: Population Reference Bureau; 2013. http://www.prb.org/Publications/ Datasheets/2013/family-planning-worldwide-2013.aspx.

Demographic Divide

The demographic divide between the earth's poor and rich countries endures.[4] This fissure means large gaps in fertility and health. On one side of the divide are mostly poor countries with relatively high birth rates and low life expectancies. On the other side are mostly wealthy countries with low birth rates and high life expectancies—conditions that are creating rapidly aging populations and, in some cases, population declines.[5,6]

This demographic split also extends to individual countries, where there are vast differences in fertility and health. In Uganda, for example, women from the poorest fifth of families have twice the number of children as those from the wealthiest fifth.[4]

Policy Interventions

Coercive measures related to individual fertility are completely unacceptable. They violate human rights precepts, and they should never be considered as policy interventions. Nevertheless, action needs to be taken to manage population growth.

Decrying coercion, major policy interventions employed to address rapid population growth are promoting voluntary contraceptive use and raising the status of women. Modern contraceptive technology is highly effective for preventing pregnancy, yet some religious dominations—particularly the Catholic Church—still object to its practice. Raising the status of women, particularly educating girls, is directly associated with lower birth rates, but changing gender roles is also contentious in many settings.

REPRODUCTIVE HEALTH

Maternal Mortality

Women's reproductive health status differs greatly throughout the world. For example, maternal mortality ratios vary from 4 in Sweden to 1100 in Chad.[7] "Of the hundreds of thousands of women who die during pregnancy or childbirth each year,"[8] 99% live in the developing world[9] (Box 14-2). The majority of these deaths are from severe bleeding, infections, eclampsia, obstructed labor, and the sequelae of unsafe abortions—all causes for which there are highly effective interventions.[8]

Box 14–2 Geography of Maternal Mortality

Ninety percent of the world's maternal deaths occur in Africa and Asia.

Data from United Nations Population Fund (UNFPA). Safe motherhood: stepping up efforts to save mothers' lives. http://www.unfpa.org/public/home/mothers. Accessed December 20, 2013.

Sexually Transmitted Infections

Next to complications of pregnancy and childbirth, sexually transmitted infections (STIs) are the leading cause of health problems for women of reproductive age. Yet in many low- and middle-income countries, the diagnosis and treatment of STIs is sorely lacking. As a result, "the burden of STIs is greatest in low-income countries."[10(pp1–2)]

Access to Family Planning

Many women and their partners still lack access to family planning services. Globally, an estimated 222 million women do not have the means to delay pregnancy and childbearing. Certainly, modern contraceptive technology is up to the task. It is safe, effective, and reliable, but, unfortunately, out of reach for many of the world's poor citizens. If the unmet demand for modern contraception in the developing world were fully met, an estimated 54 million unintended pregnancies would be averted annually, including 22 million unplanned births, 25 million induced abortions, and 7 million miscarriages.[11]

Unsafe Abortion

Unsafe abortion is largely a consequence of unmet need for family planning services. Criminalizing abortion does not decrease its incidence, but it does affect the safety of this procedure. Where abortion is illegal, it is much more likely to be unsafe; therefore, risks to women's health can be significant with high rates of morbidity and mortality. Globally, the proportion of abortions that are unsafe is believed to have increased during the last two decades.[12–16]

Policy Interventions

Effective programs and policies demand clear goals, so delineating the parameters of reproductive health is important. The ICPD definition of this concept is the most widely accepted:

> Reproductive health is a state of complete physical, mental and social well-being and not merely the absence of disease or infirmity, in all matters relating to the reproductive system and to its functions and processes.[17]

While this definition is visionary, it lacks specificity in terms of actual services and measureable objectives.

An operational definition of reproductive health services must include family planning services and other basic health interventions (e.g., the diagnosis and treatment of STIs, postabortion care, and adolescent services) that enhance maternal and child health.[18,19] Certainly, other health services can be included, but the essential services need explication. If the goal of reproductive health is to improve the health of women and their families, then raising the status of women, particularly by educating girls, is also an important reproductive health strategy.

Many policy analysts and women's health advocates believe that reproductive and sexual health rights are central issues for women's empowerment as well as cornerstones of economic development. Controlling fertility provides women and their families with economic and health options that are not otherwise available to them. The United Nations Population Fund (UNPFA) has posited that to achieve sustainable and equitable economic development, people must be able to exercise control over their sexual and reproductive lives.[19,20]

SYNERGY BETWEEN POPULATION AND REPRODUCTIVE HEALTH

The most effective policy interventions for reducing population growth also improve reproductive health. Contraceptive access improves the lives of women and their families. Raising the economic and social status of women, particularly educating girls (Figure 14–1), is associated with lower fertility and better health outcomes. The concept of reproductive health,

Figure 14–1 Educating Girls Yields Many Benefits

© Nolte Lourens/Shutterstock.com

however, embraces more services than just family planning and is closely tied to human rights.

MOVING FORWARD

Moving forward in global population and reproductive health requires collective action—a daunting task given the current climate of global reproductive health politics and international economic insecurity.[21] Even so, there is too much at stake to shy away from the challenge. Global collaboration, policy action, resources and leadership, and research and evidence are required to move forward.

Collaboration Among Policy Communities

The linkages discussed in this text demand cooperation on many levels. First, the population and reproductive health communities must work together. More than two decades after the ICPD, the disputes between the population perspective and some women's health advocates are dated and unproductive. Each view brings valuable perspectives to policy development. From the population perspective, for example, demographic data are essential for large-scale planning and implementation of any reproductive health program. From the feminist reproductive health viewpoint, the insistence on high-quality care has strengthened family planning programs, while safeguarding human rights.

The environmental ramifications of population growth and unsustainable consumption, as discussed in the *Sustainability, Population, and Environmental Degradation* chapter, also demand collaboration between the population and reproductive health communities and the sustainability movement.[22] At the ground level, this is occurring; the *Climate Change, Population, and Reproductive Health* chapter discusses examples of population, health, and environment (PHE) programs[23] and the *Food Security, Population, and Reproductive Health* chapter discusses the interaction of reproductive health and food security.[24]

This collaboration needs to occur at other levels as well. At the global and national levels, separate policy communities (sometimes called policy silos) address environmental sustainability, population and reproductive health, food security, and climate change. In recent years, the United Nations has hosted separate forums for each of these policy arenas. In fact, all of these issues are linked. The design of effective and integrated policies demands discussion and collaboration between policy communities. The Millennium Development Goals (MDGs) have been a good start for such interdisciplinary policy discussions.

Policy Action

Global action is needed on many fronts. As the international community works to update and replace the MDGs, reproductive health and family planning goals should be boldly stated, not inferred and later updated as they were with the MDGs. Undoubtedly, more attention to these issues in 2000 would have promoted a faster decline in global maternal mortality—an MDG that was not met in 2015.[5]

Good policy design must incorporate valid causal mechanisms of how objectives will be attained, and it must employ metrics to evaluate progress toward policy objectives.[25] If there is to be realistic policy planning, reproductive health services must be delineated. Certainly, circumstances and needs vary among nations or regions. Nevertheless, essential services and their linkages to epidemiologic and demographic profiles can be spelled out. This method would allow priorities to be established, commensurate with available or likely resources.

Resources and Leadership

More and sustained financial support is necessary. At the London Family Planning Summit in 2012, both wealthy countries and poorer nations, along with private donors, pledged new support for family planning services. Unlike the country-specific fiscal commitments made at the 1994 ICPD, these pledges need to be honored, expanded, and updated.

As the world's largest economy and an original stronghold of international population assistance,[26,27] the United States should resume a leadership role, including, but not restricted to, greater fiscal contributions (Box 14–3). In recent decades, domestic politics, mainly surrounding abortion, have diminished the United States' stature in this field.[28] The history

---

Box 14–3　Reawakening the U.S. Interest in Global Population

Perhaps some combination of climate change, higher energy costs, crippling traffic in America's major cities, deteriorating national parks, and the religious right's continued effort to chip away at women's reproductive rights will awaken the population debate from its forty-year slumber. But the rightward tilt of the country suggests that neither a revival of Keynesian economics—nor a stable Population Keynesian challenge to the sacred cow that economic growth requires a steadily rising population—is likely.

Reproduced from Hoff DE. *The State and the Stork: The Population Debate and Policy Making in US History.* Chicago: University of Chicago Press; 2012.

of birth control in the United States is relevant here; it remains difficult for American politicians to address policy areas that are related to sexuality, including contraception.[29]

Although the United States remains the largest single contributor to population and reproductive health, the magnitude of its generosity is tempered by several factors. In terms of the country's population assistance contribution per gross national income, the American share is nowhere near first in the world; it was in ninth place in 2003 and much of that assistance was support for HIV/AIDS programs.[30,31]

Evidence and Research

To assess whether policies and programs have been successful, evidence is needed. Good policy evaluation, like policy design, needs metrics both to assess past efforts and to plan new directions.[32] During the past half-century, demographic techniques have advanced significantly, and con-traceptive technology has improved. Nevertheless, much remains to be accomplished.

One of these areas is the integration of health services. At an intuitive level, integrating reproductive health with other health services certainly makes sense. However, the cost-effectiveness and health impacts of inte-grated reproductive health care have not been fully explored, much less implemented. Unfortunately, much of this discussion remains at the ideo-logical and political levels. Scientific rigor must be employed here so that valid evidence is obtained, based on which effective policies and programs can be designed.[33]

Abortion is an area where the scientific evidence is clear, but in many settings, the politics surrounding this issue are difficult to surmount. To protect women's lives and their families' well-being, the role of abortion in both maternal mortality and fertility decline needs to be confronted in policy circles. The *Abortion and Reproductive Health* and *Benefits of Family Planning* chap-ters explain that as fertility aspirations decline, both contraceptive use and abortion rates rise. As access to contraception improves, abortion levels fall. Where abortion is illegal, it is more likely to be performed in unsafe condi-tions.[16,34] Currently, unsafe abortion accounts for at least 13% of maternal mortality worldwide.[35] Addressing this issue as a public health priority could save tens of thousands of lives each year.

Another research challenge for international family planning efforts is the development of new contraceptive methods. "Dual protection from preg-nancy and HIV/AIDS infection remains elusive, while the pandemic of HIV/AIDS continues to devastate sub-Saharan Africa and Southeast Asia."[36(pix)]

There is also "increasing recognition that a wider range of modalities is needed to address the changing contraceptive needs of the populations of the world across the reproductive life cycle."[36(pix)] Nevertheless, contraceptive development "has not been a major priority of the research community and pharmaceutical industry."[36(pix)]

Policy making, policy implementation, and policy evaluation take place in real time with ongoing needs and unexpected crises. At any time, there are competing demands for the attention of policy makers and the public. This situation strengthens the argument for good evidence and research. A continued focus on the synergy of population and reproductive health and how these issues are linked to other global issues is essential for moving forward.

PROSPECTS FOR THE FUTURE

Recent decades have seen population and reproductive health lose both momentum and funding. Relative to need, worldwide funding has not kept pace. Figure 14–2 shows that U.S. international assistance for population and reproductive health has varied greatly over the years. When adjusted for inflation (e.g., using 1974 constant dollars), American funding was actually

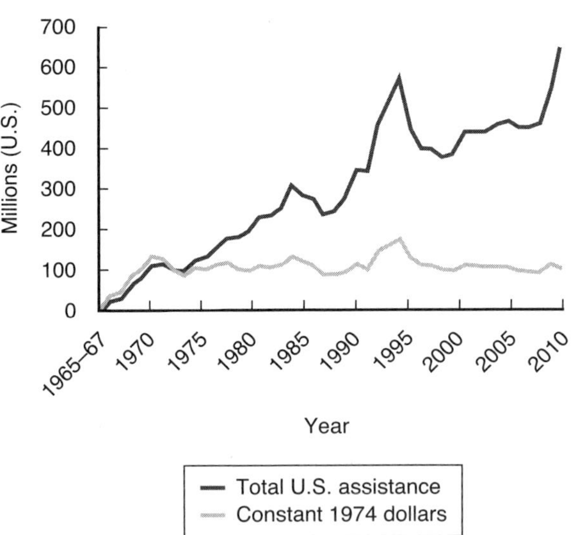

Figure 14–2 Trends in U.S. Population Assistance

Reproduced from Population Action International. Trends in U.S. population assistance. 2011. http://populationaction.org/articles/trends-in-us-population-assistance/. Accessed January 8, 2014.

higher in the mid-1970s than it is now (and world population has grown by 3 billion).[37]

Other issues impeded progress for population and reproductive health programs. Many believe that the continuation of ICPD disputes "translated into a severe de-funding of international population programs."[5(p114)] Another detractor was that international support for the HIV/AIDS epidemic, which began in the 1980s, "took place at the expense of ongoing efforts in family planning and reproductive health."[5(p115)] At the same time, conservative American politicians winnowed down U.S. population assistance in form and substance.[38,39]

Policy issues are cyclic, however. Once again, population and reproductive health, along with their linkages to other global problems, have begun to attract worldwide attention. In the *Population and Reproductive Health Policies* chapter, John May discusses "signs that the population factor could return to the forefront of global policy concerns."[26] Similarly, in her book *The Means of Reproduction*, Michelle Goldberg comments, "For years, talk of too rapid population growth has been politically incorrect on the feminist left and on the socially conservative right, but the issue is poised to reappear"[40(p230)] (Box 14–4).

Addressing global population and reproductive health is certainly feasible. In 2012, it was estimated it would cost $8.1 billion to "fully meet the existing need for modern contraceptive methods for all women in the developing world"—only $4.1 billion more than the $4 billion expenditure for contraceptive care that year.[41(p1)] It is important to note that "no contradiction needs to

---

Box 14–4   Feminists and the Reemergence of Population

For global women's rights activists, the reemergence of the population question might seem like a defeat, a return to a discourse in which women's welfare is seen as merely instrumental. But it can also be an opportunity, if it's used to force the world to pay attention to reproductive justice. An unhappy lesson of the last few decades is that men in power will rarely work to advance women's rights for their own sake, but they will do so in the service of some other grand objective, be it demographic or economic. One can decry this reality and try to change it, while also taking advantage of it.

Reproduced from Goldberg M. *The Means of Reproduction: Sex, Power, and Reproduction.* New York: Penguin Press; 2009.

exist between respect for reproductive rights and strong advocacy for smaller families and for mass adoption of effective contraceptive methods."[42(p1810)]

Clearly, more policy actions and financial support are necessary. Moreover, as John May writes in his book *World Population Policies*, "To a large extent, population policies or their absence will determine the demographic evolution of human societies."[5(p286)] It is no exaggeration to say that the world's future depends on how global population and reproductive health are addressed in our time.

CONCLUSION

World population will continue to increase for the foreseeable future. At the same time, great disparities exist among and within countries in terms of reproductive health and fertility outcomes. There is synergy between effective policy interventions for improving reproductive health and slowing population growth. To harness that synergy, international cooperation, leadership, policy action, and research are needed. Global attention to population and reproductive health is now on the upswing.

DISCUSSION QUESTIONS

1. Why is it important for the United States to exercise a leadership role in global reproductive health and population? What do you think it would take for this to happen?
2. Explain the synergy between the population perspective and the reproductive health perspective. Why are the ICPD disputes "dated and unproductive"?
3. Why has world attention recently returned to population and reproductive health after years of neglect?
4. In 2012, it was estimated that an additional $4.1 billion would cover the cost of providing family planning services to all women in developing countries who wanted to use modern contraception. How does this amount compare to other global expenditures?
5. Why does Michelle Goldberg think that the current focus on world population growth offers an opportunity for feminist reproductive health advocates?
6. What will be the ramifications if world attention turns away from population and reproductive health?

REFERENCES

1. U.S. Census Bureau, International Programs. World population. http://www.census.gov/population/international/data/worldpop/table_population.php.
2. Ehrlich PR, Erlich AH. Symposium on population law: the population explosion: why we should care and what we should do about it. *Environ Law.* 1997;27:1187–1373.
3. Hinrichsen D, Robey B. Population and the environment: the global challenge. *Pop Rep.* Fall 2000;XXVIII(3, Series M):15. http://www.k4health.org/toolkits/info-publications/population-and-environment-global-challenge. Accessed September 28, 2013.
4. Haub C, Kaneda T. *2013 World Population Data Sheet.* Washington, DC: Population Reference Bureau; 2013. http://www.prb.org/Publications/Datasheets/2013/family-planning-worldwide-2013.aspx.
5. May JF. *World Population Policies: Their Origin, Evolution, and Impact.* New York/London: Springer; 2012.
6. Haub C. *World Population Trends, 2012.* Washington, DC: Population Reference Bureau; 2012.
7. United Nations Maternal Mortality Estimation Inter-agency Group. Maternal mortality estimates. http://www.maternalmortalitydata.org/. Accessed December 20, 2013.
8. United Nations Population Fund (UNFPA). Safe motherhood: stepping up efforts to save mothers' lives. http://www.unfpa.org/public/home/mothers. Accessed December 20, 2013.
9. World Health Organization (WHO). Trends in Maternal Mortality 1990–2010: WHO, UNICEF, and The World Bank estimates. Geneva, Switzerland: WHO; 2012. http://www.unfpa.org/public/home/publications/pid/10728. Accessed March 22, 2014.
10. World Health Organization (WHO). *Sexually Transmitted Infections: The Importance of a Renewed Commitment to STI Prevention and Control in Achieving Global Sexual and Reproductive Health.* Geneva: Switzerland: WHO/RHR/13.02. http://www.who.int/reproductivehealth/publications/rtis/rhr13_02/en/index.html. Accessed December 20, 2013.
11. Singh S, Sedgh G, Hussain R. Unintended pregnancy: worldwide levels, trends, and outcomes. *Stud Family Plan.* 2010;41(4):245.
12. World Health Organization (WHO). U*nsafe Abortion: Global and Regional Estimates of the incidence of Unsafe Abortion and Associated Mortality in 2008.* Geneva, Switzerland: WHO; 2011.
13. Marston C, Cleland J. Relationships between contraception and abortion: a review of the evidence. *Intl Family Plan Perspect.* 2003;29(1):6–13.
14. Sedgh G, Singh S, Shah IH, et al. Induced abortion: incidence and trends worldwide from 1995–2008. *Lancet.* January 19, 2012. doi: 10.1016/S0140-6736(11)61786-8.

15. Myers JE, Self MW. Global perspective of legal abortion: trends analysis and accessibility. *Best Pract Res Clin Obstet Gynaecol.* 2010;24:457–466.

16. Denton EH. Benefits of family planning, In: McFarlane DR, ed. *Global Population and Reproductive Health.* Burlington, MA: Jones & Bartlett Learning; 2015.

17. Sieverding M. Gender and reproductive health. *Intl Encyclopedia Soc Behav Sci.* 2001;5969–5972.

18. Pillsbury B, Maynard-Tucker G, Nyguen F. *Women's Empowerment and Reproductive Health: Links Throughout the Life Cycle.* New York/Los Angeles: United Nations Population Fund and Pacific Institute for Women's Health; 2000.

19. Greene G, Joshi S, Robles O. *UNFPA State of the World Population 2012: By Choice, Not by Chance: Family Planning, Human Rights, and Development.* New York: United Nations Population Fund; 2012.

20. Hodgson D. Abortion, family planning, and population policy: prospects for the common-ground approach. *Pop Develop Rev.* 2009;35(3):479–518. http://www.jstor.org/stable/25593662. Accessed June 8, 2013.

21. Gore A. *Our Choice: A Plan to Solve the Climate Crisis.* Emmanuas, PA: Rodale; 2009.

22. Becker JE, McFarlane DR. Sustainability, population, and environmental degradation. In: McFarlane DR, ed. *Global Population and Reproductive Health.* Burlington, MA: Jones & Bartlett Learning; 2015.

23. Hardee K. Climate change, population, and reproductive health. In: McFarlane DR, ed. *Global Population and Reproductive Health.* Burlington, MA: Jones & Bartlett Learning; 2015.

24. White RE. Food security, population, and reproductive health. In: McFarlane DR, ed. *Global Population and Reproductive Health.* Burlington, MA: Jones & Bartlett Learning; 2015.

25. McFarlane DR. U.S. domestic population and family planning policies: 1970–1983: management lessons learned and pitfalls to avoid. *J Health Admin Educ.* 1987;5(1):63–82.

26. Donaldson PJ. *Nature Against Us: The United States and the World Population Crisis, 1965–1980.* Chapel Hill, NC: University of North Carolina Press; 1990.

27. May JF. Population and reproductive health policies In: McFarlane DR, ed. *Global Population and Reproductive Health.* Burlington, MA: Jones & Bartlett Learning; 2015.

28. Hoff DE. *The State and the Stork: The Population Debate and Policy Making in US History.* Chicago: University of Chicago Press; 2012.

29. McFarlane DR, Grossman R. Contraceptive history and practice. In: McFarlane DR, ed. *Global Population and Reproductive Health.* Burlington, MA: Jones & Bartlett Learning; 2015.

30. Kantner JF, Kantner A. *The Struggle for Consensus on Population and Development.* New York: Palgrave Macmillan; 2006: Appendix B, Levels and Trends in Development Assistance and International Population Funding.

31. Ashford L. *Resource Flows for International Population Assistance and UNFPA.* Washington, DC: Center for Global Development Working Group; November 2010.

32. Lester JP, Stewart J Jr. *Public Policy: An Evolutionary Approach.* Belmont, CA: Wadsworth Thomson Learning; 2000.

33. Tsui AO. Commentary. In: Harkavy O. *Curbing Population Growth: An Insider's Perspective on the Population Movement.* New York: Plenum Press; 1995:237–249.

34. Kulczycki A. Abortion and reproductive health. In: McFarlane DR, ed. *Global Population and Reproductive Health.* Burlington, MA: Jones & Bartlett Learning; 2015.

35. World Health Organization (WHO). *Preventing Unsafe Abortion.* Geneva, Switzerland: UNDP/UNFPA/WHO/World Bank Special Programme of Research, Development and Research Training in Human Reproduction (HRP); 2013. http://www.who.int/reproductivehealth/topics/unsafe_abortion/magnitude/en/index.html. Accessed December 25, 2013.

36. Nass SJ, Strauss JF. *New Frontiers in Contraceptive Research: A Blue Print for Action.* Washington DC: Institute of Medicine of the National Academies, National Academy Press; 2004.

37. Population Action International. Trends in U.S. population assistance 2011. http://populationaction.org/articles/trends-in-us-population-assistance/. Accessed December 24, 2013.

38. McFarlane DR. Reproductive health policies in President Bush's second term: old battles and new fronts in the United States and internationally. *J Public Health Policy.* 2006;27:4.

39. Sinding SW. Overview and perspective. In: Robinson WC, Ross JA, eds. *The Global Family Planning Revolution: Three Decades of Policies and Programs.* Washington, DC: World Bank; 2007: 1–12.

40. Goldberg M. *The Means of Reproduction: Sex, Power, and Reproduction.* New York: Penguin Press; 2004.

41. Singh S, Darroch JE. *Adding it Up: Costs and Benefits of Contraceptive Services—Estimates for 2012.* New York: Guttmacher Institute and United Nations Population Fund; 2012. Cited in Greene G, Joshi S, Robles O. *UNFPA State of the World Population 2012: By Choice, Not by Chance: Family Planning, Human Rights, and Development.* New York: United Nations Population Fund; 2012.

42. Cleland J, Bernstein S, Ezeh A, et al. Family planning: the unfinished agenda. *Lancet.* 2006;368:1810–1827.

Index

Note: Page numbers followed by *b*, *f*, or *t* indicate materials in boxes, figures, or tables respectively.

A

abortion, 21, 97–98, 171, 208–209, 381
  to contraception, relationship of, 209
  data, 173–174
  global and regional, 175–177, 175*t*
  health-related controversies, 184–185
  in history, 172–173
  induced, 11–12
  laws and consequences, 188–190
  postabortion care, 187–188
  prenatal diagnosis and sex-selective, 185–187
  rate, 98, 99*f*
  ratio, 98
  safe and unsafe, 180–183
  spontaneous abortions, 174
  structural and individual-level determinants of, 177–179
  techniques and changing medical practice, 183–184
  United States
    levels and differentials in, 179–180
    service in, 190–192
acacia, tips of, 145
ACS. *See* American Communities Survey
adaptation, 289, 296–298
administrative data systems, costs of, 59
adult lifetime risk of maternal death, 103
Advisory Group on Energy and Climate Change (AGECC), 273
Africa FGM, prevalence of, 235*f*
African family planning poster, 360
age at menarche, 88
age at sexual initiation, 88
Age of Enlightenment in Europe, 111
age-specific death rates (ASDRs), 70
  of Japan, Thailand, the United Kingdom, and India, 71*f*
age-specific fertility rates (ASFRs), 75, 76

age-specific probability of death, 72
age structure, 64, 65, 67
AGECC. *See* Advisory Group on Energy and Climate Change
Agenda 21, 276
aging, 291
  of society, 134
aging populations, 133
  impact of, 42
agricultural investments, 334
agricultural productivity, 331
  long-term stability of, 321
Agricultural Revolution, 27, 28, 32
agroecology, 327
AMA. *See* American Medical Association
Amazon rainforest, agricultural conversion of, 269*f*
American and French Revolutions, 111, 243
American Birth Control League, 154
American Communities Survey (ACS), 57, 78
American contraceptive policies (1940–1960), 155–156
American foreign policy and abortion, 192
American future, population growth and, 260*b*
American healthcare reform and abortion, 191–192
American Medical Association (AMA), 155
ancient cultures, 146–147
ancient Egyptians, 145
ancient Greeks, 146
ancient Hebrews, 145
antenatal care, 103–104
ART. *See* Assisted Reproductive Technology
ASDRs. *See* age-specific death rates
ASFRs. *See* age-specific fertility rates

L

M

N